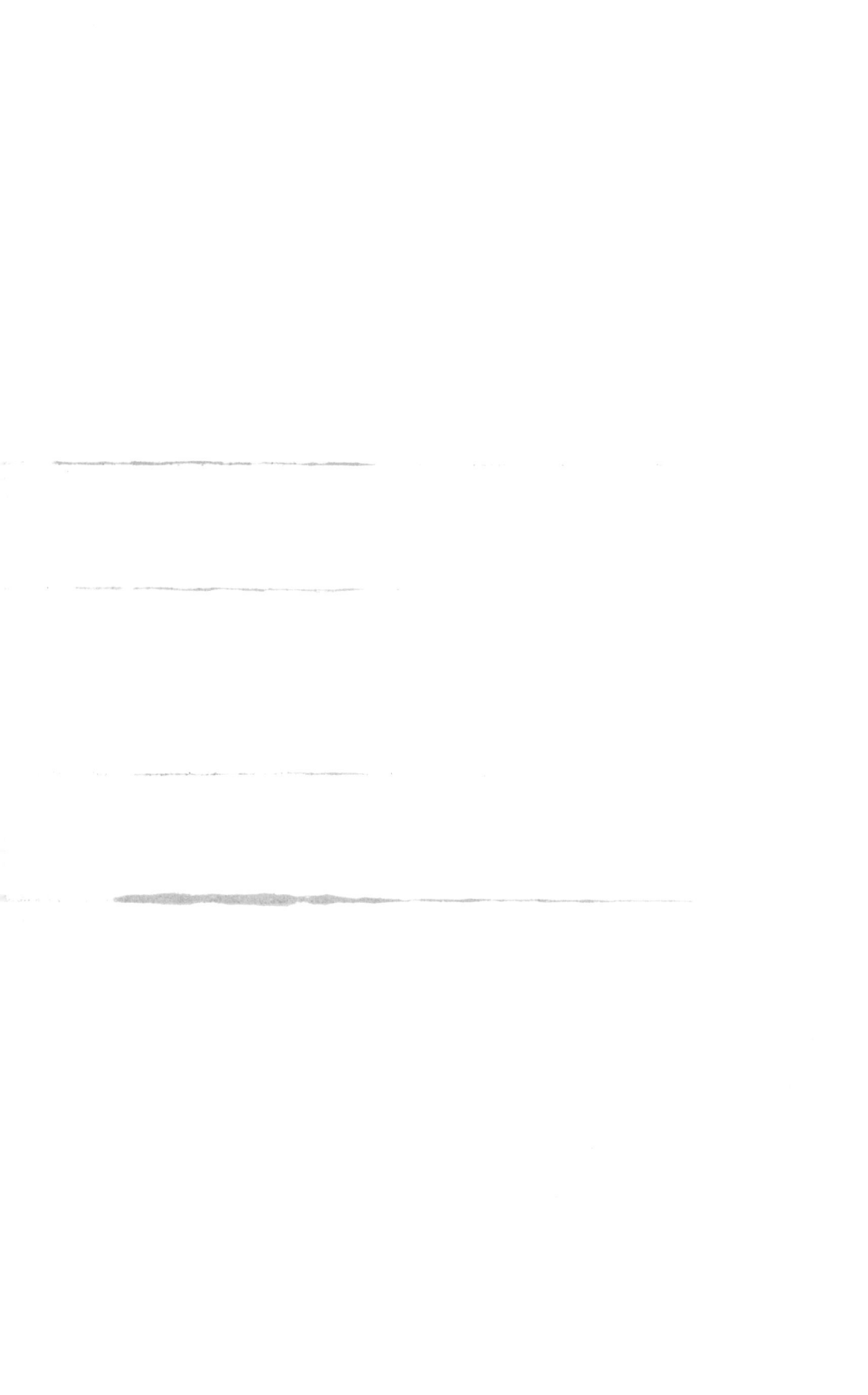

IET CONTROL, ROBOTICS AND SENSORS SERIES 121

# Integrated Fault Diagnosis and Control Design of Linear Complex Systems

# Other volumes in this series:

# Integrated Fault Diagnosis and Control Design of Linear Complex Systems

Mohammadreza Davoodi, Nader Meskin
and Khashayar Khorasani

The Institution of Engineering and Technology

Published by The Institution of Engineering and Technology, London, United Kingdom

The Institution of Engineering and Technology is registered as a Charity in England & Wales (no. 211014) and Scotland (no. SC038698).

© The Institution of Engineering and Technology 2018

First published 2018

The Institution of Engineering and Technology
Michael Faraday House
Six Hills Way, Stevenage
Herts, SG1 2AY, United Kingdom

www.theiet.org

**British Library Cataloguing in Publication Data**
A catalogue record for this product is available from the British Library

**ISBN 978-1-78561-705-8 (hardback)**
**ISBN 978-1-78561-706-5 (PDF)**

Typeset in India by MPS Limited
Printed in the UK by CPI Group (UK) Ltd, Croydon

*To Mania, my wife
and
to Reza and Parvaneh, my parents
for their unwavering love
Mohammadreza Davoodi*

# Contents

# Preface

Due to increasing demands for stricter system performance and capabilities in many engineering and industrial systems, more feedback control loops are being integrated into automatic control systems. There is an ever-increasing demand on reliability of safety critical systems such as robots operating in hazardous environments, aircraft and spacecraft, aero engines, chemical processes, manufacturing systems, power networks, electric machines, wind energy conversion systems, and industrial electronic equipment, among others.

In many cases, the loss of performance or stability due to potential process abnormalities and component faults may cause serious damage. To avoid these problems, innovative methods are being developed to diagnose the occurrence of any potential abnormality and fault as early as possible for implementing fault-tolerant controllers to minimize the performance degradation and to avoid catastrophic situations.

Notwithstanding the above, there are scenarios where it is possible and more importantly, desirable to consider an integrated design of feedback controllers and fault diagnosis filters. This simultaneous design unifies both the control and diagnosis modules into a single framework. Consequently, it is conceivable and expected that an integrated fault detection and control (IFDC) design strategy should lead to a far less overall complexity as compared to an approach where the two modules are designed separately.

The main objectives of this book are to provide the reader with state-of-the-art design methodologies for simultaneous fault diagnosis and control (SFDC) of complex linear systems. In essence, we design and develop novel SFDC schemes for the following class of systems, namely,

- linear time-invariant (LTI) systems,
- linear Markovian jump systems,
- networked control systems, and
- teams of linear multiagent systems.

By using the proposed methodologies, the above systems can simultaneously achieve the desired diagnosis and control objectives in presence of uncertainties and external disturbances.

The approaches and frameworks that are proposed and presented are developed based on different types of robust state observers and filters that lead to lower computational complexity and less conservativeness as compared to separate designs of fault detection and isolation (FDI) and fault-tolerant control modules.

This book is organized as follows. In Chapter 1, an overview of model-based FDI literature, in general, and IFDC, in particular, are provided. In Chapter 2, the focus is on the problem of dynamic observer-based IFDC design for continuous-time LTI systems subject to faults and external disturbances. The formulation and solution of the IFDC problem are presented in Sections 2.2 and 2.3, respectively.

In Chapter 3, the integrated fault detection, isolation, and tracking (IFDIT) control problem is investigated for linear systems subject to both bounded energy and bounded peak external disturbances. The proposed methodology is capable of simultaneously accomplishing detection and isolation of faults as well as tracking control of closed-loop system in presence of both bounded energy and bounded peak disturbances. The IFDIT problem is formulated in Section 3.2 and a systematic linear matrix inequality (LMI) approach for solving the proposed IFDIT problem is proposed in Section 3.3.

In Chapter 4, the focus is on the problem of integrated fault detection, isolation, and control (IFDIC) design of continuous-time Markovian jump linear systems with uncertain transition probabilities. Using this methodology, not only the considered uncertain Markovian jump system can be effectively controlled but also the occurrence of faults in the system can be detected and isolated. We describe the system and the IFDIC module governing equations in Section 4.2.1. The IFDIC problem is then formulated for the Markovian jump linear systems with uncertain transition probabilities in Section 4.2.2. An LMI-based solution to the developed IFDIC problem is provided in Section 4.3.

To reduce the usage of computational and communication resources without degrading the overall desired closed-loop system performance results, the problem of event-triggered integrated fault detection, isolation, and control (E-IFDIC) design is defined in Chapter 5. The governing equations for the considered discrete-time LTI system under investigation as well as the statement of the problem for E-IFDIC design are presented in Section 5.2. An LMI-based solution to the proposed problem is developed and implemented in Section 5.3.

In Chapter 6, the problem of event-triggered active fault-tolerant control (E-AFTC) of discrete-time linear systems is addressed. In Section 6.2, the event-triggered fault/state observer and the fault-tolerant controller are described, and the definition of the E-AFTC problem is presented. In Section 6.3, sufficient LMI conditions are derived to simultaneously obtain the observer and the controller parameters as well as the sufficient event-triggered conditions.

In Chapter 7, the problem of integrated fault detection and consensus control (IFDCC) design of multiagent systems is developed. Using the proposed methodology, all the agents reach either a state consensus or a model reference consensus, while simultaneously, the agents collaborate with one another to detect the occurrence of faults in the team. Section 7.2 starts by presenting the statement of the distributed IFDCC problem for a team of multiagent systems. Next, model transformation and decomposition techniques are employed in Sections 7.3.1 and 7.3.2, respectively, to reduce the computational complexity of the design as well as to specify the IFDCC modules having a distributed architecture. An LMI formulation and design solution to the IFDCC problem is developed in Section 7.4.

In the final chapter, i.e., Chapter 8, the outline of future directions of study in which further investigation is needed to be conducted is provided.

The funding for much of the research described in this book was provided by the NPRP grant no. 5-045-2-017 from the Qatar National Research Fund (a member of the Qatar Foundation).

*Mohammadreza Davoodi*
Doha, Qatar

*Nader Meskin*
Doha, Qatar

*Khashayar Khorasani*
Montreal, Canada

# Acknowledgments

I would like to offer special thanks to Dr. R. Amirifar, my MS adviser, who is, although no longer with us, continues to inspire by his dedications to me and his other students. I would like to express my special thanks of gratitude to my academic advisors Prof. H. A. Talebi and Prof. H. R. Momeni for all their motivation, guidance, and support throughout my Ph.D. program at Tarbiat Modares University. I am deeply indebted to Prof. K. Khorasani who provided me the chance of doing some complementary researches during my Ph.D. leave at Concordia University and latter related researches when I was a postdoctoral researcher at Qatar University and finally for giving me the golden opportunity to cowrite this book. I would like to include a special note of thanks to my postdoctoral supervisor Dr. N. Meskin of Qatar University; his valuable advise helped me a lot in conducting my research from which I gathered so many professional experiences at Qatar University. Sincere thanks to my father whose teachings will remain with me forever and to my mother who gave me the desire to learn more through her pure love. Finally, my mind knows no bounds in expressing my cordial gratitude to my wife, Mania. Her keen interest and encouragement was a great help throughout this work. I humbly extend my thanks to all my friends, colleagues, and all other concerned persons who cooperated with me in this regard.

*Mohammadreza Davoodi*
Qatar University

N. Meskin would like to express his appreciations to Qatar National Research Fund (a member of the Qatar Foundation) for funding much of the research that is described in this book under the NPRP grant no. 5-045-2-017.

*Nader Meskin*

K. Khorasani would like to express his appreciations for funding of much of the research that is described in this book by the Natural Sciences and Engineering Research Council of Canada (NSERC) under the Discovery Grants Program and by the NPRP grant no. 5-045-2-017 from the Qatar National Research Fund (a member of the Qatar Foundation).

*Khashayar Khorasani*

# List of acronyms

| | |
|---|---|
| AUVs | autonomous underwater vehicles |
| BMI | bilinear matrix inequality |
| ERG | embedded residual generator |
| E-IFDIC | event-triggered integrated fault detection, isolation, and control |
| E-AFTC | event-triggered active fault-tolerant control |
| FDI | fault detection and isolation |
| FTC | fault tolerant control |
| IFDC | integrated fault detection and control |
| IFDCC | integrated fault detection and consensus control |
| IFDIT | integrated fault detection, isolation, and tracking |
| IFDIC | integrated fault detection, isolation, and control |
| LMI | linear matrix inequality |
| LTI | linear time-invariant |
| PWM | pulse width modulation |
| ROV | remotely operated vehicle |
| SDRE | state-dependent Riccati equation |
| SFDC | simultaneous fault diagnosis and control |
| SoS | system of systems |
| UAVs | unmanned aerial vehicles |
| UGVs | unmanned ground vehicles |
| UUVs | unmanned underwater vehicles |

# Long author biographies

Mohammadreza Davoodi is a Post-Doctoral Fellow with School of Electrical and Computer Engineering, University of Georgia, Athens, USA. His research interests include fault diagnosis, simultaneous fault detection and control (SFDC), robust control, convex optimization, multiagent systems, autonomous network of unmanned vehicles, and hybrid systems.

Nader Meskin is an Associate Professor with Qatar University, Doha, and a member of the Department of Electrical Engineering. His research interests include fault detection and isolation (FDI), multiagent systems, active control for clinical pharmacology, and linear parameter varying systems.

Khashayar Khorasani is a Professor and Research Chair with the Department of Electrical and Computer Engineering and the Concordia Institute of Aerospace Design and Innovation, Montreal, Canada. His research interests include nonlinear and adaptive control; intelligent and autonomous cooperative control of networked unmanned systems; fault diagnosis, isolation, and recovery (FDIR); diagnosis, prognosis, and health monitoring (DPHM); cyber-physical systems security and protection; resilient control against cyberattacks; and computational intelligence and machine learning.

*Chapter 1*

# Introduction

Associated with the increasing demands for higher system performance and enhanced product quality on one hand and improved cost efficiency on the other hand, complexity and automation levels of engineering processes are continuously growing. These trends and challenges call for development of stringent system safety and reliability requirements. In order to achieve this goal, systems should have the capability to detect and isolate the occurrence of possible faults and at the same time reconfigure their control algorithms to compensate for these anomalies.

However, in most modern control systems that are equipped with fault detection capabilities, it is highly desirable to unify and integrate the control and fault detection modules into a single unit. Motivation for this is drawn from the fact that majority of control and detection schemes implement a type of state observer or filter. Hence, it is plausible that an integrated fault detection and control (IFDC) design would lead to a far less overall complexity as compared to a design where the two modules are designed separately. The main objective of this book is to explore the advantages, capabilities, and benefits of developing simultaneous fault diagnosis and control (SFDC) methodologies for complex linear systems.

## 1.1  Statement of the problem

In this book, the problem of SFDC design for complex linear systems is addressed. The SFDC problem is among the most critical and central challenges in engineering systems. The systems that are considered consist of

- linear time-invariant (LTI) systems,
- linear Markovian jump systems,
- networked control systems (NCSs), and
- teams of linear multiagent systems (MASs).

The hallmark of LTI systems is that their dynamics can be modeled by using differential equations. Most physical systems fall into this category. LTI systems are the easiest class of systems to work with and have a number of properties that make them ideal to study. Therefore, the study and control of many industrial and engineering systems is accomplished by using their linearized models. In this book, we first focus on the IFDC design problem for continuous-time LTI systems. Since the conservatism in the IFDC design problem depends on the type of filters that are employed in the structure of the IFDC block, one of the main challenges in IFDC design algorithms is to determine an appropriate filter or state observer that should be employed in the structure of the IFDC module.

Consequently, the development of a novel IFDC design problem based on dynamic observers for LTI systems will be investigated to eliminate the disadvantages of static observer structures in designing the IFDC module. Moreover, the developed dynamic observer-based IFDC design is extended to the problem of integrated fault detection, isolation, and tracking control (IFDIT) for LTI systems. The proposed methodology is capable of simultaneously accomplishing the detection and isolation of faults as well as tracking control of the closed-loop system in presence of both bounded energy and bounded peak disturbances.

The IFDC design problem is subsequently considered for a more advanced class of systems, namely, the Markovian jump systems that belong to an important class of switching systems and have recently received a great deal of attention. Markovian jump systems are popular in modeling many practical systems that are subject to random failures and structural changes, such as electric power systems, communication systems, aircraft flight control, control of nuclear power plants, and manufacturing systems. Moreover, one of the main challenges in the design of IFDC schemes is to propose strategies that remain robust to various types of uncertainties and external disturbances. Hence, an integrated fault detection, isolation, and control (IFDIC) design strategy is addressed and developed here for uncertain Markovian jump systems in presence of both bound energy and bounded peak disturbances.

Moreover, in recent years, NCS have received extensive research interest and attention due to technological advances in computational modules, communication channels, and smart sensors and actuators [4]. The role of NCS is not only to reduce the wiring and installation costs but also to offer greater flexibility for evolving existing installation and support diagnosis and maintenance actions. Due to these distinctive benefits, applications of NCS cover a wide range of fields, such as automotive, mobile robotics, advanced aircraft, etc. [5].

In addition, wide usage and employment of communication networks introduce new challenging problems that make the analysis and synthesis of NCS more complex and demanding. Specifically, in the NCS framework, data is transmitted in discrete packets over a communication network of limited bandwidth. Once a network is overloaded, network-induced delay and packet dropout rates begin to increase rapidly, which will significantly degrade the overall system performance [6,7]. To reduce the usage of computational and communication resources without degrading the overall desired closed-loop system performance, we specify and design the problem of event-triggered integrated fault detection, isolation, and control (E-IFDIC) of NCS.

Given that the proposed E-IFDIC scheme is a passive fault-tolerant control (FTC) methodology, the fault detection information is not explicitly used in the controller design. However, the generated diagnostic information can be used subsequently in the framework of design of an active FTC system. Consequently, by using an integrated design of event-triggered fault/state estimator with a fault-tolerant controller, the problem of event-triggered active fault-tolerant control (E-AFTC) in discrete-time NCS will be addressed.

Finally, there has been a growing interest on development of a network of MASs for solving problems that are difficult or impossible to address by using an individual agent or a monolithic system. These networks may be potentially made up of a large number of dynamical systems (agents), such as unmanned aerial vehicles (UAVs), unmanned ground vehicles (UGVs), or unmanned underwater vehicles (UUVs). Any of these systems can commonly consist of a number of sensors, actuators, and decision makers, and therefore, the network of these systems is a network of large number of sensors and actuators, or as is commonly known in the literature, a system of systems (SoS) [8].

Subsequently, we address the problem of integrated fault detection and consensus control (IFDCC) of linear continuous-time MASs so that all the agents reach either a state consensus or a model reference consensus while simultaneously collaborating with one another to detect the occurrence of faults in the team. We also employ a decomposition technique in the IFDCC design procedure that allows the system to be partitioned into a set of lower order subsystems for reducing the computational complexity of the design as well as for specifying the IFDCC modules having a distributed architecture.

## 1.2   Literature review

### 1.2.1   Fault diagnosis

In order to avoid the cost incurred and the impact due to failures, it is desirable to have an advanced failure detection system that detects anomalies early enough to minimize the damage and that can identify as many failures as possible, quickly and efficiently, prior to catastrophes. The safe operation of any complex system rests upon the reliability of the control and fault detection systems and the speed of detection and identification of component, subsystem, or system failures.

In the proposed research program, we intend to study the issue of recovery from failures. Here, fault diagnosis and control recovery become closely interconnected. This results in a complex problem requiring a systematic and a formal approach. Systematic generation of computer code for diagnosis and recovery is less prone to human error than manual code generation. In addition to enhancing the reliability and accuracy of the diagnostic system, the use of systematic approaches can potentially result in faster design process. It may also reduce the cost of future revisions and maintenance of the diagnosis code. Therefore, the proposed research is of considerable practical importance.

Common approaches to monitoring and fault detection and isolation (FDI) are (i) temporal redundancy, (ii) hardware redundancy (HR), and (iii) analytical redundancy (AR) or software redundancy. Temporal redundancy uses limit and trend checking on the measurable inputs and outputs based on some prior normal behavior information about the system. This approach is quite common in many areas today.

Knowledge-based techniques could be thought of as techniques that belong to this class. Another approach is HR combined with a majority vote ruling logic. This approach, although commonly used in certain domains as military aviation or space applications, is not always a feasible alternative and not always practical due to its prohibitive costs, space requirements, and other limitations.

Finally, the AR or model-based approaches use the system's inherent redundancies with the system's model and measurements to generate redundant information about the system. The system's behavior is then monitored by processing this information in a logical and deterministic manner or a statistical decision process. It is our belief that in a truly large-scale complex system all of the above approaches may have to be employed for control and health monitoring of the system. As such, it is the goal of this book to perform research work on these individual areas and at the same time develop a framework for bringing in techniques from each of the above domains to arrive at a unified and integrated approach to advanced control and monitoring of complex systems.

In complex systems, fault diagnosis is typically accomplished using hierarchical approaches. At the lower levels of this hierarchy, continuous-time models are used in diagnosing both minor and major failures. At the higher levels, discrete-event models (e.g., finite state automata and Petri nets) are employed for the detection of drastic failures and coordination among different subsystems. One of the discrete-event models that have been studied for fault diagnosis is the finite-state automata.

In an automata-based approach, it is assumed that the plant under supervision can be modeled as a finite state automaton. The diagnosis system is also a finite state automaton that takes the output sequence of the plant as input and generates at its output an estimate of the condition (failure status) of the plant. In this framework, a failure is considered "diagnosable" if it can be detected and isolated in a finite time. As with any technique based on discrete-event models, computational complexity of the design is perhaps the most difficult obstacle that has to be overcome in the application of the methodology to real-world problems.

At the subsystem levels of the hierarchy, the model-based or AR-based approach to FDI would require a mathematical model of the process or subprocess under consideration. Based on this, knowledge quantities called residuals are generated. The residuals should be small or close to zero when there are no failures in the system. On the other hand, they should become nonzero and grow large if there are malfunctions in the system. This will accomplish the failure detection. The next important task is the design of a fault isolation module that would isolate the faulty subsystems.

There are essentially two distinct approaches that fall under the umbrella of model-based approaches to FDI. First are those techniques that use physically based model of the plant or the process and use more traditional techniques from the estimation and control domain to tackle the problem. The other approach would be to use artificial neural networks, fuzzy logic and genetic and evolutionary algorithms for model development and use those models along with other intelligent networks for FDI purposes. Finally, there is a possibility to use a combination of the two for both model development as well as the FDI logic.

The advantages in using AR-based scheme for FDI are that (i) it does not require costly HR and (ii) it can detect and diagnose transient as well as steady-state faults. This is important since many diagnostics schemes work based on the limit and trend checking of various measurements within the system. These schemes generally require a priori knowledge of such information and by and large only work when the system is in steady state. As such, they are not applicable to dynamical systems that operate over a wide operating regime and often undergo through transients, or systems that encounter intermittent failures.

Knowledge or expert system-based techniques are representative of this latter class of diagnostic schemes and therefore are not very good in detecting small incipient failures in systems that operate over a wide operating range and go through dynamic changes very often. On the other hand, knowledge- and rule-based systems may be desirable in static systems composed of many slowly varying subsystems.

In addition, in our advanced control and health condition monitoring architecture, they may be more desirable for higher level coordination tasks where uncertainties and time horizon both increase and symbolic reasoning would be more prevalent, and where the AR-based techniques are easily transferable from one class of problems to another, and they have the potential for providing a systematic means for detecting and isolating a number of different failures.

There are also some challenges with regard to model-based schemes in which robustness of the FDI algorithm is a major issue. That is, if the model is inaccurate, there is potential for false alarms or missed alarms. These issues are addressed through research in robust FDI.

Among the common model-based techniques that can be considered is the qualitative causal method [9]. In this method, the cause and effect reasoning about the model behavior is analyzed and investigated. Researchers have used fault-trees [9] to backwardly seek the possible original causes for observed anomalous system behaviors. The main disadvantage of this method is that one has to generate a large number of hypotheses in order to avoid poor resolution, which could result in ambiguous decisions. Furthermore, the large number of potential hypotheses would naturally lead to a high burden on the online computational resources.

Another very common and an important type of model-based technique is known as residual-based methods, where the FDI process can be divided into three distinct stages. The first stage is to generate a signal (also called residual) that is sensitive to the occurrence of a given fault category. Considering and treating faults as inputs to the system, a specific residual must generate sufficient information so that one is able to discriminate and recognize the occurrence of a fault and separate this specific fault from other faults. The second stage is to use residuals to make appropriate decisions according to a preassigned logic corresponding to the location of the fault occurrence. Finally, the recovery task will estimate the severity of the fault and subsequently reconfigure the control system to condition the system into a manageable and desirable operational mode.

Parity space approach [10,11] is among the residual-based techniques considered in the literature. Reference [12] has provided the main idea of parity space approach.

This approach basically checks the consistency of the mathematical governing equations of the system through two types of redundancies: (i) direct redundancy, which makes use of relationships among redundant sensor outputs, or (ii) temporal redundancy, which counts on dynamic relationships between outputs and inputs. The parity space approach leads to a special type of an observer called dead-beat observer. Reference [10] applied the parity space approach to a UAV dynamic system in fully autonomous guidance and navigation systems in a decentralized system architecture.

Other than the parity space approach, many researchers simply use Luenberger observers or Kalman filters to monitor the system performance. In the absence of uncertainties and disturbances, the errors between the state estimates and the actual values of the states diminish to zero, while the presence of faults will increase the errors on the contrary. For example, in [13,14], Clark used this simple observer-based method and checked the increment of the errors by threshold logic. In order to avoid the false alarm caused by uncertainties and disturbances, the threshold must be larger than zero, although this will decrease the sensibility to faults.

Kalman filter approach to fault detection, isolation, and recovery (FDIR) for aircraft control system has also been used widely [15–18]. An approach based on the ratio of two quadratic forms of which matrices are theoretic and selected covariance matrices of Kalman filter innovation sequence for sensor fault detection is presented in [15], where the optimal arguments of the quadratic forms are found to quickly detect the faults in sensors, and the longitudinal dynamics of an aircraft control system is considered. In [16], an algorithm for checking the generalized innovation sequence characterizing the volume of the correlation ellipsoid and application to an aircraft dynamics is investigated. Moreover, a surface fault detection algorithm based on the extended Kalman filter is presented in [17].

Besides, the method based on the use of constrained Kalman filters, which are able to detect and isolate such faults by exploiting functional relationships that exist among various subsets of available actuator input and sensor output date is developed, and the feasibility and efficacy of the approach is demonstrated through simulation in the context of a nonlinear jet engine control system [18].

Furthermore, enhanced capability of isolating faults can be realized by utilizing a bank of observers driven by the actual output vector that is making decisions based on multiple hypothesis testing. In this methodology, each observer is designed for a different fault hypothesis. The hypotheses are then examined with respect to likelihood functions [12,19]. In [20], the fault detection and diagnosis for an aircraft control system is carried out using the multiple model approach, and an eigenstructure assignment technique is used for reconfigurable feedback control law design.

Beard [21] and Jones [22] have done further work on the observer-based techniques. The Beard–Jones detection filter is a type of Luenberger observer. However, the appealing feature of their approach is that by selecting an appropriate observer gain matrix that makes residuals caused by the actuator faults belong to the output modes, the residual caused by a given fault becomes independent with other faults. Note, however, that they only proved a sufficient condition for detecting faults by designing their special observer as a detection filter. Also, the sufficient condition is quite limited as the fault modes have to satisfy a strong mutual detectability condition.

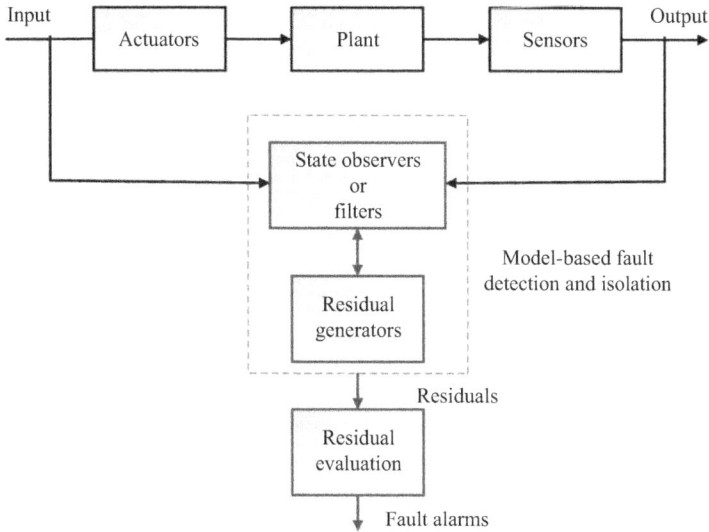

*Figure 1.1 The schematic of a model-based fault detection and isolation (FDI) approach*

Massoumnia [23] restated Beard–Jones approach in a geometric sense and successfully extended this idea into a more general case, in which residuals resulted from actuator faults could belong to more than the output modes. More importantly, sufficient and necessary conditions of solvability of the residual generation for an LTI system were obtained.

Another residual-based method is the parameter estimation method. Frank [12] and Isermann [24] have presented the main idea behind this method. It makes use of the fact that faults of a dynamic system are reflected in the physical parameters of the system. It detects the faults through estimating the parameters of the mathematical model of the system. If the estimates of the parameters deviate from the nominal values, it may be declared that a fault has occurred.

Model-based fault diagnosis (i.e., fault detection, isolation, and identification) schemes have attracted considerable interest over the recent years (see, e.g., [25–31] and the references therein). Among model-based approaches, the most common FDI methodology is to use state observers or filters to construct residual signals that are then evaluated against predefined thresholds. When the so-called residual evaluation function exceeds a given threshold, an alarm is generated [30], as shown in Figure 1.1. However, noise and disturbances may impact in significant ways the residual signals, leading to declaration of false alarms. Consequently, fault diagnosis filters have to be made sensitive to faults and simultaneously robust to noise and disturbances.

In [27], the fault detection filter design is formulated as an $H_\infty$-filtering problem, where the error between the reference residual and the actual residual in presence of parametric uncertainty is minimized. In [28], the problems of $H_-$ index and multi-objective $H_-/H_\infty$ fault detection observer design via linear matrix inequality (LMI) conditions are considered. Different performance indices are considered for optimal selection of postfilters as well as optimization of fault detection filters in [32].

In [33], the problem of robust detection and accommodation of incipient component faults using nonlinear detection observers for a class of nonlinear distributed processes is addressed. In [34], the problem of fault detection and identification for a class of uncertain nonlinear discrete-time systems with both multiplicative actuator and additive system faults is considered. For more comprehensive details on model-based fault diagnosis techniques, refer to the book [30], the survey paper [35], and the references therein.

## 1.2.2  Integrated fault detection and control

In the FDI research domain, diagnostic systems are often designed separately from the control strategies. A fundamental problem in decoupled design of controller and fault detection (FD) modules is the neglecting of rather significant interactions that occur between these two modules in closed-loop system operation. Indeed, a fault diagnosis module designed for an uncontrolled plant may not perform satisfactorily with the controlled plant.

Furthermore, a fault diagnosis module designed for a controlled plant may not perform satisfactorily due to inherent limitations inadvertently imposed on the achievable diagnosis performance during the control module design [36]. In such situations, faults may be masked or covered up by the control action, and the early detection of the process faults becomes clearly more challenging [37]. On the other hand, if the quality of the available model is poor, design of the control and FD systems has to be undertaken simultaneously in order to improve the overall system functionality.

Based on the above observations, it can be argued that a suitable trade-off between the control and FD objectives must be made. For this reason, in the past two decades in the theoretical and application domains, the problem of tackling IFDC design has received significant interest [36,38–45]. The goal of the simultaneous design is to unify control and diagnosis objectives into a single module that can then lead to lower design, computation, and implementation complexity as to when compared with separate design approaches. It also provides a desirable and formal mechanism since design of each module should take into account the other module's considerations.

According to the classification proposed in [46], the existing IFDC techniques can be divided into the following two categories, namely,

1.  Without embedded residual generator (ERG) [47] and
2.  With the ERG [37].

One of the main objectives in the IFDC design with or without ERG is to generate an appropriate signal (or a bank of signals), $z_f$, to perform the FD task. In certain literature, $z_f$ represents the residual signal, that is $r$, which is sensitive to the fault. Specifically, $z_f = r = 0$, when there is no fault in the system and under faulty conditions $r \neq 0$. In other techniques, $z_f$ is defined as the difference between the fault $f$,

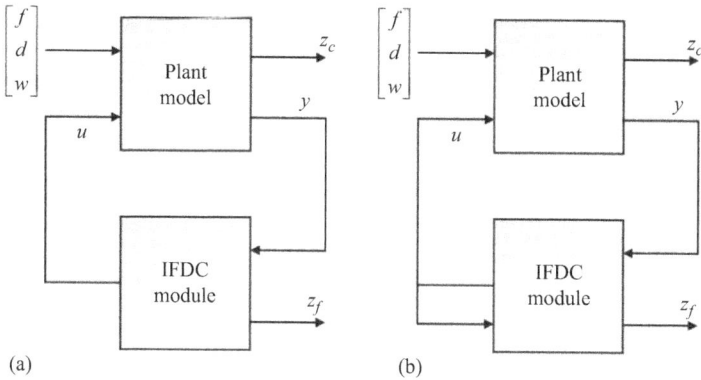

*Figure 1.2* *The IFDC block diagram: (a) Without ERG and (b) With ERG;*
*f denotes the fault signal, d denotes the external disturbances,*
*w denotes the reference signal, u denotes the control input, y denotes*
*the measured output, $z_c$ denotes the control output, and $z_f$ denotes the*
*diagnosis output*

and the residual signals as $z_f = f - r$, which gives a weighted estimate of the fault. However, in the IFDC with the ERG, the controller output is fed back and utilized for designing the FD. The IFDC block diagrams with and without ERG are depicted in Figure 1.2.

The distinguishing and distinctive features between the IFDC frameworks with and without the ERG from structural, interactions, robustness, and simplicity perspectives can be described as follows:

- **Controller/detector configuration:** The controllers/detectors in integrated schemes without an ERG are configured as feed forward from the measurement $y$ (with or without the reference signal $w$) to $u$ and $r$. In comparison, the ones with an ERG include a feedback loop from $u$ to $r$. This structural difference plays a critical role in the following other comparisons.
- **Interactions between the control and diagnosis:** In the system configurations without ERG, changes in the controller parameters and performance will cause changes in the diagnosis performance. Moreover, the diagnostic signal $r$ is structurally coupled with the reference signal $w$ which may deteriorate the FD performance. In contrast, the diagnostic signal provided by an ERG is independent of the changes in the control laws and decoupled from $w$.
- **Robustness against model uncertainties:** Since the residual signal is independent of the control laws, the diagnosis performance of the control configurations with an ERG will be sensitive to modeling uncertainties. One way to address this problem is to apply advanced observer-based FDI schemes to improve the robustness of the residual generator against modeling uncertainties, which may, on the other hand, lead to complex and involved design processes. In comparison,

control configurations without the ERG will offer higher robustness both in the control and diagnosis performance characteristics.

- **Design requirements:** Typically, additional design degrees of freedom is incorporated into the IFDC scheme without the ERG, which increases the computational complexity of the methodology.

In [42,43], the problem of IFDC based on dynamic observers for linear continuous-time systems is considered. The work in [44] considers an $H_\infty$ formulation of the IFDC problem by employing a dynamic observer detector and state feedback controller for switched linear systems. In [48], the problem of the IFDC for linear switched discrete- and continuous-time systems under a mixed $H_\infty/H_-$ framework is considered. In [41], a two-step procedure is developed to solve the IFDC problem for linear discrete-time systems in presence of polytopic uncertainties.

In [49], an $H_\infty$ formulation of the robust IFDC design for a class of uncertain linear systems with state time delay is considered. In [50], a switching law and an output feedback-based controller/detector are designed to accomplish simultaneous control and detection objectives for discrete-time switched systems. In [51], the problem of IFDC design for discrete-time switched systems with actuator faults is studied. The problem of IFDC design for switched systems with two quantized signals is studied in [52]. In essence, two dynamic quantizers are, respectively, employed before the measured output signal is transmitted to the fault detector, and before the control input is sent to the system.

Based on a trade-off analysis between fault detectability and closed-loop performance, the IFDC design problem for a chemical plant in presence of stochastic parametric faults is investigated in [53]. In [54], the design of IFDC problem for a class of nonlinear stochastic switched systems under asynchronous switching and subject to time-varying state delay and parameter uncertainties is addressed. In [55], the problem of robust simultaneous finite-time control and fault detection for linear switched systems with state time-varying delay and parameter uncertainties is studied. In [56], the IFDC design problem for a switched linear parameter-varying system with inexact scheduling parameters is studied and simulation results for an aero-engine model is provided to show the effectiveness and capabilities of the proposed methodology.

In [57], the IFDC design problem for networked-based discrete-time switched systems with time-varying delays under an arbitrary switching signal is addressed. In [4], a mixed $H_2/H_\infty$ IFDC approach is proposed to achieve the desired detection and control objectives for a class of discrete-time networked systems with multiple stochastic delays and data missing. In [58], an event-triggered IFDC design problem for uncertain discrete-time stochastic systems with limited communication is proposed. A survey of integrated design of controllers and fault diagnosers is provided in [46].

## 1.2.3 Multiagent systems

In recent years, the number of applications in which interactions between humans and machines are becoming ever more challenging has increased. This has motivated many researchers to solve and address these types of complicated engineering problems and applications. Toward this end, one can transform the problem into a distributed network of smaller and simpler autonomous subsystems that can operate without a human involvement. The collective behavior of animal groups in nature has shown that distributed decisions made by each individual for its own position, direction, and speed of motion can make the entire group to behave as a single entity, which has its own rules of motion and decision-making.

Consideration of problems in the above teams is currently one of the strategic areas of research. The motivation for this focus may be traced to applications where direct human intervention is not feasible due to either environmental hazards, extraordinary complexity of the tasks, or other restrictions. On the other hand, observations made based on natural behavior of animals operating as a team have inspired scientists in various disciplines to investigate the possibilities of networking a group of systems to accomplish a given set of tasks without requiring an explicit supervisor.

Some examples of such natural behaviors can be found in migration of birds, motion of fish searching for food, and team work of other animals that have a group living style. Examples of such a collective behavior can be seen in birds formations, flocks of birds, schools of fish, and mammal herds. In all the above examples, the animals work together in a team with an intergroup cooperation and with no supervision from outside the team to perform complex tasks. Furthermore, advances in wireless communication networks have made it feasible to connect a number of systems distributed over a large geographic area.

Inspired by these natural collective behaviors of animals in their groups, scientists and engineers are encouraged to network group of systems to let them exchange their information. Each agent then uses its locally available information and a cooperative control strategy to respond so that the overall team performs the required tasks without a need of using external supervisors.

The above advances and observations have led the scientists to the area of design of networked unmanned systems (NUMS) or MASs. The advantages of NUMS are enormous, and applications in various fields of research are being developed.

Some of the advantages for deploying autonomous network of unmanned systems are the enhanced group robustness to individual faults, increased and improved instrument sensing and resolution, reduced cost of operation, and adaptive reconfigurability capabilities that have been discussed in [59]. A team of agents can cooperate with one another to accomplish a complicated task that is impossible or highly challenging to be performed by a single agent.

Based on the above discussion and due to the multidisciplinary nature of the problem, design of a network of sensors, actuators, and decision-makers is currently one of the important trends of research in various disciplines, such as communications, control theory, and mechanics. In order to fully take advantage of these large-scale networks, several prerequisites have to be satisfied. Some of these prerequisites are in development of reliable communication, optimal power consumption management, security, optimal cooperation, and team collaboration that are discussed in [60].

Team cooperation and coordination for accomplishing predefined goals and requirements are some of the main prerequisites for these networked agents that are intended to be deployed in challenging missions. Although, a large body of work has been devoted to design requirements for NUMS, unsolved problems still exist. Some of these challenges are (i) lack of complete information by and for all the agents in the network, (ii) cohesion requirement and connectivity of the team in presence of uncertainties and partial information, (iii) tackling inaccurate information due to the scale of the network, (iv) presence of adversarial and environmental uncertainties, (v) robustness issues, and (vi) fault diagnosis and recovery, to name a few. Other issues are to address dynamic nature of the team and to use more complicated agent dynamical models rather than point-mass models. Hence, it is imperative that one designs reliable and high-performance networks that can ensure and accommodate these requirements.

Cooperative control of NUMS or MASs covers a wide range of applications such as autonomous underwater vehicles (AUVs) [61–63], UAVs [64–66], mobile robots [67–69], and satellite clusters [70–72]. Some recent review papers and books on recent progress in the study of these systems can be found in [73–78] and [79–81]. However, each of these areas has its own specific challenges although some common underlying characteristics can be identified. Many practical and theoretical challenges are present in cooperative control of MASs. Instead of a single system, we have a SoS which needs to communicate with one another subject to limitations and constraints on the communication channels bandwidths.

It is a difficult task to determine which agents should communicate with whom at each time and what to communicate. Moreover, there is a compromise between the individual's goals and the team goal. In MASs, there are a number of research problems that have resulted in development of many useful tools and theories. Among different problems in MASs research area, consensus problem is one of the most favorable, which is to provide in a distributed manner, and with minimal computation and communication requirements determine the average of a shared quantity in a network of agents or computational units. Here, agents are commonly coupled since they are performing operations without directly influencing each other. Each agent in the team makes its own decisions by using only limited data that is obtained by its own measurements or communication with the neighboring agents.

Cooperative control of NUMS or autonomous MASs has been extensively investigated in the past few years [82], and the work in this field can be categorized into two general areas [83]. The first category is consensus-based formation of agents including mobile robots [84], UAVs [85], satellites [86], AUVs [87], and automated highway systems [88]. The second category is nonconsensus-based cooperative control algorithms such as task assignment [89], payload transport [90], role assignment [91], air traffic control [92], and search and timing. The consensus problem involves convergence of outputs or states of all agents to a common value [93]. It implies that each agent has access to other agents state, known as the neighboring agents by using either a communication network or sensing devices [94,95].

Depending on the amount of data exchange between agents and the available data of the other agents, centralized or distributed cooperative control strategies can be used [96]. The centralized strategy relies on the assumption that each team member has the access to the data of all the other team members, and in the distributed strategy, it is assumed that it only has access to data of some neighboring team members. Usually, it is preferred to use distributed algorithm to achieve consensus.

Despite dedication of a large body of work to study multiagent networks and cooperative control, there are still many unsolved problems in this area—among them are uncertainties and failure in the agents, limited communication, and actuation capability. Most of introduced consensus algorithms are focused on systems with single-integrator or double integrator kinematics, while practical applications generally consist of LTI and nonlinear dynamical systems.

Moreover, control algorithms are generally designed based on an implicit assumption of unlimited computational resources, non-delayed sensing and actuation, unlimited bandwidth, and perfect communication environment. However, in MASs, computational and communication resources are limited and often shared among multiple agents. Therefore, due to presence of the above inherent constraints, in practical applications, it is necessary to investigate cooperative control of MASs subject to practical factors such as communication constraints and actuator saturations.

## 1.2.4 Cooperative control and consensus control concepts and notions

In this section, we first introduce basic definitions for MASs and then provide an overview and application of cooperative control and consensus achievement problems.

**Definition 1.1.** *An agent is a dynamical system with a state vector that evolves through time based on its past value and a control input vector. Here, the state of an agent is not dependent on any other agent, but the control input is a function of the agent and other agents' state vectors.*

Since the state of an agent is decoupled from other agent's state and interactions among agents are through their control inputs, basically without a common control strategy and information exchange among different agents, agents represent completely independent systems.

**Definition 1.2.** *A MAS is a set of agents that are exchanging information and collaborating with each other based on a common control strategy to achieve a goal as a single entity that cannot be achieved by each agent alone.*

The connection between MASs and algebraic graph theory is their necessity to exchange information that can be best modeled by an information flow graph $\mathscr{G}$. Let us label each agent with a number and let each node in the graph $\mathscr{G}$ represent an agent and each edge from node $v_i$ to node $v_j$ shows information flow from agent $i$ to agent $j$. Note that the exchanged information can be the entire state vector of agents (agent state) or a function of that (agent output).

**Definition 1.3.** *A MAS is called homogenous if the dynamics and the exchanged information of all the agents are the same; otherwise, it is called heterogeneous MASs.*

**Definition 1.4.** *A MAS is said to follow a distributed control strategy having a topology $\mathcal{G}$ if the control input of each agent is a function of its own state (or output) and states (or outputs) of other agents that are in the in-neighbor set of the agent in the graph.*

It is worth noting that the MAS and notations that are used here are mainly common in the control systems community [97,98] and are different from those that are used by the computer science community [99]. In view of the above, one of the main concerns of cooperative control strategies is to solve the consensus problem.

**Definition 1.5.** *The consensus problem in MASs is to determine a distributed control strategy that ensures all agents agree on a common value for a variable of interest. This value is usually called the consensus value and can represent a state or an output of the agent.*

## 1.2.5  Applications

Over the past decade, the consensus problem has been studied extensively in the literature due to its applications in numerous areas such as cooperative control of UAVs [100], formation of mobile robots [101], UUVs distributed control [102], and sensor networks [103], among others. Various aspects of consensus achievement of MASs have been investigated in the literature including consensus problem of first-order or second-order integrators with fixed and switching graph topologies [104], communication delays [105], graph connectivity preservation [106], and reference signals [107].

In each cooperative mission, the agreed quantity may be the agents' positions, velocities, heading angles, altitudes, etc. For instance, consider a formation control problem in a team of unmanned air vehicles. In this scenario, a consensus protocol is required in order to agree upon an altitude in which the formation is to be accomplished.

Wireless NUMS provide significant capabilities and hence has received extensive attention in the past several years, and numerous applications in various fields of research are being considered and developed. Some applications that necessitate development of these systems are in space explorations; satellite deployment for distributed deep space observations; automated factories; maneuvers of a group of UAVs; utilization of network of UGVs, e.g., mobile robotics, and UUVs for search and rescue; and teams of robots deployed in a hazardous environment where human involvement is dangerous. In [60], more applications are mentioned such as home and building automation, intelligent transportation systems, health monitoring and assisting, and commercial applications. There are also military applications in intelligence, surveillance, and reconnaissance (ISR) missions in presence of environmental disturbances, vehicle failures, and in battlefields subject to unanticipated uncertainties and adversarial actions [108].

The consensus problem has been considered in more practical problems some of which are stated below:

- **Formation control in multivehicle systems:** Maintaining relative positions among agents in a desired formation is one of the challenging problems in the area of MASs. A formation of agents has a centroid, size, and attitude each of which can be obtained via an agreement problem [109,110].
- **Rendezvous in multivehicle systems:** In some cooperative tasks, agents should be in the visibility range of each other. In the rendezvous of a team of agents, the objective is to reach a common position. Therefore, a consensus problem can be solved to agree upon a given position [111–113].
- **Flocking in a team of multiagents:** Inspired by aggregation of biological organisms in groups such as flocks of birds, schools of fish, herds of animals, and colonies of bacteria, coordinated behavior of autonomous agents have found many engineering applications in ISR, and monitoring missions. These applications are motivated by the increasing chance of finding foods and avoiding predators in living collective behaviors. Achieving agreement on the average velocity and average heading angle is an important issue in flocking of MASs [88,114–117].
- **Attitude alignment:** In some applications, it is necessary for the agents to reach consensus regarding their attitudes through local interactions. For instance, in a formation flying of spacecraft, it may be necessary for the agents to achieve a common attitude or an identical relative attitude. Therefore, consensus strategies can be considered for these missions [95,118].
- **Filtering in sensor networks:** Sensor networks have numerous applications in surveillance and monitoring missions. The objective is to detect a target through a network of sensors; therefore, distributed filtering should be employed to track the average of all sensor measurements [103,119,120].

Below, we provide detailed description and overview of cooperative control and consensus control problems and their associated research challenges.

## 1.2.6   Cooperative control

Cooperation in a network of unmanned systems, known as formation, network agreement, collective behavior, flocking, consensus, or swarming in different contexts, has received extensive attention in the past few years. Several approaches to this problem have been investigated within different frameworks and by considering different architectures [121–131]. Moreover, the problem of cooperation in a network has been considered at different levels. At the high-level, one can refer to task assignment, timing and scheduling, navigation and path planning, reconnaissance and map building [132–134], to name a few. In the mid-level, cooperative rendezvous, formation keeping, application of consensus algorithms, collective motion, and formal methods based on flocking/swarming ideas can be mentioned [59,113,126,127,131,135].

Since the present work is mostly aimed at cooperation in the mid-level, most of the literature reviewed in this section are on the low-level cooperation, i.e., formation keeping, application of consensus algorithms, and formal methods based on flocking/swarming ideas. The formation control problems can be distinguished from the consensus seeking and flocking/swarming-based approaches based on the degree of autonomy as well as the degree of distribution of the proposed algorithm. Generally speaking, these two problems are characterized (these will be described in more detail in the next two subsections) as follows:

1.  **Formation control:** Based on the definition presented in the survey paper [136], a formation control law couples the dynamics of each member of a group of vehicles through a common control law. Basically, the formation control should have two properties: (i) at least one member of the group must track a predefined state relative to another member and (ii) the corresponding control law should be dependent on the state of this member [136]. Some of the characteristics of this type of problem are as follows:
    i.   The formation control is usually solved based on a centralized approach to the problem [137], although decentralized solutions have also been suggested, e.g., [138]. However, the assumption of modular architecture is common in the formation problem [59,139,140].
    ii.  Usually, conventional control methods are used, e.g., adaptive and nonlinear control [137].
    iii. Autonomy is up to the level where a tracking path is given, a leader guides the group, or a coordination vector is provided [137].
    iv.  For different structures, e.g., virtual structure [137], and multiinput–multioutput (MIMO) [141], the problem of formation has been addressed for complicated dynamics such as aircraft, UAVs, and spacecraft.
    v.   Various work has covered different aspects of the control problem in this area, such as formation keeping, tracking, formation stability (e.g., input–output stability), and fuel optimality [59,142], as well as high-level tasks, e.g., initialization and reconfiguration [143,144].
    vi.  The present challenges are on experimental and practical issues such as design of high-resolution space instruments, fuel optimality, and reducing the computational complexity and time. Also, the estimation problem in formation keeping [145–147] and design of decentralized and distributed algorithms are some of the current research trends.
2.  **Flocking/consensus/agreement:** The problem is to have network agreement on a scalar state or a vector of states or on a function of states, while other behaviors (e.g., formation) are guaranteed. Some of the characteristics of these types of problems are as follows:
    i.   They are based on a distributed approach, and a large number of agents can be addressed [88].
    ii.  Information is considered as one part of modeling, where there are influences of communication topology on the stability and other dynamical properties [148].
    iii. The mathematical tool for the problem formulation is graph theory for a majority of the work except for few references, e.g., [123,149–154].

    iv.  Very simple agents' dynamical model (first or second order integrator models) is generally assumed except in few works, e.g., [148,149,151,152,155].

    v.  Usually, there is an analysis of the response and not a synthesis, except in few work, e.g., [149–152].

    vi.  Different aspects of the problem that have already been considered are as follows:

        a.  Basic properties such as convergence [124], finding equilibrium state [95], and in [156] controllability definition are provided.

        b.  Behaviors such as formation keeping, collision avoidance, obstacle avoidance, as well as generating feasible planar trajectories to get the maximum coverage of the region, and obtaining the minimal information structure needed for stability of a swarm are addressed in [100,124,153,157].

        c.  In [158], the main purpose is to develop a consensus algorithm weighting such that the fastest convergence speed is achieved.

    vii.  The main challenges and problems in recent years have been

        a.  To define the basic properties, e.g., group stability, controllability, and observability conditions.

        b.  To extend the existing methods to conditions where more complicated agents' dynamical models are considered.

        c.  To add conditions such as uncertainty in the mission plan (leader command) for the followers, model uncertainties, disturbances, communication/sensor noise, appearance of fault in the leader and the followers.

        d.  To put constraints on the input signal or other practical constraints.

        e.  To assume the time-varying neighboring set, switching structure, and dynamic network topology.

        f.  To formulate the estimation problem in the network framework.

        g.  To assume stochastic frameworks with probabilistic information links.

        h.  To consider delayed information exchange between the leader and the followers or among the followers.

In the following subsections, we will present a detailed review on different issues that arise in the cooperative control of MASs.

## 1.2.7 Formation control

Inspired by biological behavior such as flocking of geese, formation control of multivehicle systems has become an interesting problem in the area of MASs. This is expressed as maintaining relative positions/distances among agents under desired geometric patterns. Formation control can be accomplished in a group of UAVs [159,160], UGVs [161,162], and UUVs [163,164], with a cooperative task that has become one of the interesting research areas in recent years with extensive applications in surveillance, discovering, atmospheric studies, rescue missions, fire monitoring, and operations in hazardous environments, etc.

Based on its definition, in formation control, the main focus is that the team achieves a predefined and given geometry and shape. This shape (formation) should be preserved during the mission and so the team of agents should act as a rigid body. Based on this property, a predefined trajectory is usually provided for the team motion, e.g., a leader command, or a trajectory for the virtual structure mass center. The team should track this trajectory, while its members keep their relative positions and preserve the required shape, i.e., the stability of the formation should be maintained. Other requirements can be added to these objectives as well. The main prerequisite to keep in formation is guidance and control of vehicles, while they perform tasks such as initialization, contraction, and expansion.

As discussed in [136], different architectures can be considered for the formation of a team of agents, namely, (i) considering the formation as a single MIMO system [141], (ii) leader-follower [165], (iii) virtual structure [59] (or virtual leader [166], in which the entire formation is considered as a virtual structure), (iv) cyclic with nonhierarchical control architecture [167], and (v) behavioral [168].

In the *MIMO architecture*, the entire system dynamics is considered as one MIMO model. Hence, in this architecture, any of the conventional control strategies, e.g., optimal, nonlinear, or robust control strategies can be applied to the system.

The *leader–follower architecture* has been used very often in the literature [95, 108,165,169–172]. In this approach, a hierarchical control architecture is considered with one or more of the agents as the leader(s), and other agents as the followers. The followers should track the position and orientation of the leader(s). This structure can also be constructed in a tree form, in which an agent is the leader of some other agents who are the leaders of some other ones and so on. The advantages of this approach are that it has an easy and understandable behavior, the formation is preserved even if the leader is perturbed, and that group behavior can be inspected by defining the behavior of the leader. However, lack of an explicit feedback to the formation, from the followers to the leader, is a disadvantage of this structure. Also, the failure of the leader implies the failure of the formation as mentioned in [59,121].

In the *virtual structure approach*, the entire formation is treated as a unit. In this approach, three steps are considered for control design: (i) first one defines the desired dynamics of the virtual structure, (ii) then one transforms the states of the virtual structure into the states of individual agents, and (iii) finally one designs the control laws for each agent. In [59,121], the advantages of this method are considered as its simplicity in defining the coordinated behavior of the group, keeping the formation during different maneuvers, and existence of feedback from the agents to the virtual structure. The weakness of this structure is in its limitation in applications to time-varying or frequently reconfigurable formations.

In [59,121,140], the idea of adding a feedback from the vehicles to the coordination unit is presented. Authors in [121] proposed the virtual structure as a solution to the problem of multiple-spacecraft formation. In their approach, a feedback from a vehicle to the virtual structure is considered for keeping the vehicles in formation and for improving the robustness of the formation in the case of disturbances or when the virtual structure moves very fast and some of the slow members may be lost from the group.

In the *cyclic architecture*, the agents are connected to each other in a cyclic form rather than a hierarchical architecture [167]. The disadvantage of this method is that the stability analysis of the proposed controller is not straight forward due to the dependency of individual agent's controller on others' in a cyclic form.

In the *behavioral approach*, several commands are combined to reach different and probably competing goals or several behaviors, e.g., collision avoidance, obstacle avoidance, and formation keeping for the agents. The control law for each agent is a weighted average of the control for each behavior. Since competing behaviors are averaged, occasionally strange and unpredicted behaviors may occur. Despite the advantages of simple derivation of control strategies, explicit feedback to the formation in this approach, and capability of decentralized implementation, there are some weaknesses as well. For some scenarios, group behavior cannot be explicitly defined and mathematical analysis, e.g., stability, is difficult to accomplish as mentioned in [59].

Although the above structures were originally introduced for formation keeping, some of them can be used for flocking and consensus-based algorithms as well. Due to the supervised nature of structures such as leader–follower and virtual structure, they are not commonly used for flocking and consensus seeking, which are autonomous in their very nature (there is no predefined path, trajectory for the leader, or virtual coordinates to be tracked). However, these structures can also be redefined for utilization in flocking/swarming or consensus seeking problems, see, e.g., [95,170].

Strategies for formation control of MASs vary from one mission to another. Given the possibility of various *communication topologies* associated with MASs, two classes of approaches can be considered in formation control of MASs, namely, *acyclic* and *cyclic*. These are described in more detail below.

- **Acyclic approaches:** Most studies on MASs formation control are devoted to acyclic approaches. This is due to the ease of implementation and stability analysis of acyclic approaches. Since no loops exist in the associated communication graph, the stability analysis of these approaches is reduced to a *follower* which tracks another agent as its *leader*. In this scenario, the decisions are made by the leader agent, and no feedbacks are sent from the followers to the leader. This framework decreases the degree of autonomy of these approaches. Moreover, the leader represents as a single point of failure [173–176]. In some applications, to keep a formation precisely as a rigid body, the agents track a *virtual leader* as a reference signal. In other words, the agents are aware of the desired trajectory of the virtual leader; therefore, they can track relative positions/distances with respect to the trajectory of the virtual leader. In this situation, the formation can be kept rigid during the maneuvers with a high degree of precision. However, a virtual leader cannot make decisions autonomously, which represents as the main disadvantage of the virtual leader-based approaches [160,177–179].
- **Cyclic approaches (consensus-based approaches):** In contrast to the leader–follower and virtual leader-based approaches, in the cyclic approach, each agent is the leader of a group of agents, while it follows the other agents as its leaders. Hence, there is no specific leader in the multiagent team, and this issue increases

the degree of autonomy in these approaches. Under the associated communication graph, there is a directed path between any two agents, and hence the graph is strongly connected. Achieving consensus on some quantities of interest, such as formation centroid, formation radius, etc., through the cyclic communication graph is one of the main objectives of cyclic approaches. However, the stability analysis of the overall MAS in presence of cyclic graphs represents as the main challenge of this approach [180–184]. For instance, in [185], this approach was employed for formation control of single integrator MASs around a centroid obtained from a consensus problem. The results were extended to unicycle kinematic agents with constant forward speeds in [182,186]. According to these results, since the agents were considered with identical constant forward speeds, they rotated after achieving a regular polygon formation. Hence, the direction of rotation was determined based on the initial relative conditions of the agents. In [187], a cyclic formation control approach for first-order kinematics with unit speeds under bidirected time-varying topologies was proposed, and in [112,188], cyclic approaches for first-order kinematics rendezvous were introduced. Moreover, cyclic formation control of double-integrator mobile agents was proposed in [184,189], where the agents reached a formation around an a priori unknown centroid obtained from a consensus problem. In the above results, the formation radius was an a priori unknown information as well.

In certain missions, the agents may have *multiobjective behaviors*. In this case, no definite formation is considered, and this structure is useful for multiobjective missions such as seeking a target, obstacle avoidance, etc. In other words, while keeping a desired acyclic or cyclic formation, collision or obstacle avoidance are important issues that can also be considered. In this situation, the agents' control commands can be obtained based on the weighted control strategies that are considered for different competing behaviors [88,161,162].

One of the main challenges that arises in the development of a formation keeping strategy in a team of unmanned systems is lack of complete information and presence of uncertainties, faults and unpredictable events in the team. This necessitates the design and application of adaptive methods for formation control in some applications [108,130,137,165,190]. In [108], leader commands are unknown for the follower vehicle in a leader–follower architecture, and therefore an adaptive controller is used for formation keeping. Missing leader commands may occur during a mission when the leader spots an approaching threat and quickly reacts to avoid it. In this case, there is no adequate time for the leader to send its new commands to the follower vehicles, and hence the leader commands need to be assumed to be unknown in the control design.

Also, in [165], the authors considered uncertainties in the vehicle dynamics, in which vortex forces are considered as unknown functions. In a formation mode, each vehicle experiences an upwash field generated by the other vehicles, and hence the aircraft motion is affected by the vortex of the adjacent vehicles. These effects are usually unknown and depend on the area, gross mass, span, and dynamic pressure as well as velocity and position of the vehicles.

The ideas in [108,165] are integrated in [130] by using the framework that is introduced in [108] for design of an adaptive controller in order to compensate for time-varying unknown leader commands and vortex forces. The idea in [108] was extended to the case where the vortex forces are presented in the dynamical model of the system and both the vortex forces of the follower and leader commands are treated as time-varying unknown parameters. The control objective is to design the follower control input such that the relative distances between the followers and the leader are maintained close to their desired values in presence of these uncertainties.

For the case that vortex forces are considered in the velocity dynamics, adaptive updating laws are introduced for two cases of time-varying and constant forces. In this case, the stability and tracking of relative distances are guaranteed. If in addition, the forces are applied in the heading angle dynamics, stability and tracking of relative distances are guaranteed for the case of constant vortex forces. The proposed algorithm is applied to formation control of UAVs.

Some of the works performed in coordination in a group of vehicles assume graph theory as the mathematical framework for modeling a distributed network of agents [100,122,148,191]. In some approaches, the analysis is also done in this framework, using graph properties. The main difference between the previously reviewed work on formation and the above references is that in the latter, the final goal is to achieve a stabilized formation in an autonomous manner. No path or trajectory is provided for the group motion, and the formation is stabilized based on the intergroup exchange of information and the desired shape provided a priori. In some other work, the graph theory tool is used only for modeling, and other tools are used for mathematical analysis.

In [100], a local distributed bounded control input is designed for formation stabilization in a group of multiple autonomous vehicles. Point-mass dynamical models are assumed for agents, and the information flow graph can be either directed or undirected. The concept proposed is to use potential functions that are constructed by using desired properties, such as collision-free and stable formation. Similarly, results for a global stabilization and tracking are presented in [122] for a group of agents with linear dynamics. The problem is first solved for two agents and then extended to the general case using "dynamic node augmentation." The collision avoidance is also guaranteed and shown. In this approach, the problems of stabilization and tracking are decoupled into three subproblems, namely, control of shape dynamics, rotational dynamics, and translational dynamics.

In [191], the same problem is considered where different maneuvers such as split, rejoin, and reconfiguration are assumed for the group. In [148], graph theory is used to model the communication network and to find the relation between the topology and the stability of the formation. Based on the graph Laplacian matrix properties, a Nyquist criterion is obtained to show the formation stabilization for groups of agents with linear dynamics. The formation stability is divided into two parts: stabilization of the information flow and stabilization of the individual vehicles. The leader–follower architecture can be addressed in this framework.

In [123], the formation and alignment goals are translated into an error framework. A decentralized robust controller is then designed for the error dynamics based on an overlapping design methodology. The assumed structure is leader–follower, and a constant velocity command and formation structure is provided for the entire team. In [137], an adaptive approach is proposed to achieve formation of three satellites, which in turn are used as a free-flying interferometer. The goal is that the formation follows a desired attitude trajectory. In [192], formation flight control of UAVs is addressed.

There are other important issues that have been discussed in the literature on the formation of MASs. In the work of Smith *et al.* [145], a parallel estimation structure based on the error covariance is suggested for control of a formation; however, the solution is not necessarily an optimal one. In their approach, each vehicle with a linear dynamical model estimates the state vector of the entire group based on the noisy output it receives. Then in design of the controller, each vehicle uses its own estimation information. The same authors have investigated results on adding communication to remove disagreement dynamics in [146,193]. For this purpose, they have assumed that some of the nodes (at least $N - 1$ receivers) send their estimation results to the rest of the group. In [147,194], a similar idea is used for applications in spacecraft formation. In [195], convex optimization is used to develop a framework for distributed estimation, or equivalently for data fusion. For solving the corresponding optimization problem, sub-gradient method and dual decomposition are utilized.

## 1.2.8   Flocking/swarming-based approaches

As presented in [196], the definition of flocking in a group of agents is

> A group of mobile agents is said to (asymptotically) flock, when all agents attain the same velocity vectors, and furthermore distances between the agents are asymptotically stabilized to constant values.

In [124], a dynamic graph theoretic framework is presented for formalizing the problem of flocking in presence of some obstacles that are assumed to be in convex and compact sets, and their boundaries are closed differentiable Jordan curves. An energy function is constructed for the team of agents in which different tasks of flocking are considered. Dissipation of this energy functions through the protocols that are inspired by the Reynolds rules [114], namely, (i) alignment, (ii) flock centering, and (iii) collision avoidance, results in achieving all the predefined goals of the team.

The main contribution of [124] is to derive and analyze an advanced form of Reynolds rules, specifically the last two rules. The authors have considered the point-mass dynamics for the agents, and a flocking protocol is defined for the interactions among the agents that result in reduction of the constructed potential function. This protocol results in alignment, i.e., convergence of each agent to the weighted average position of its neighbors, and obstacle avoidance in the network of agents. Two tasks of split/rejoin and squeezing were presented in the provided simulation results. Similarly, in [88], a particle-based approach to flocking is considered for two scenarios, i.e., in presence of multiple obstacles and in a free space. The suggested algorithms address the three rules of Reynolds as mentioned above.

In [135,197,198], a stable flocking motion law is introduced for a group of mobile agents with a connected graph. This law guarantees a collision free and cohesive motion and an alignment in the headings of the agents. The authors assumed a two-component control law in which the first part is produced by using a potential function that regulates the relative distances among the agents as well as avoiding the collision, and the second part that regulates the velocities. The potential function can define the final shape of the formation. In the case of a switching topology [135,198], control laws may be switching, and so Filippov and nonsmooth system frameworks are used for stability analysis. In this case, neighboring is based on the distance between the agents that renders it dynamic. In [199,200], the flocking problem is solved by decomposing the entire team dynamics into the dynamics of the group formation and dynamics of the motion of the center-of-mass. Each of these dynamics are analyzed and stabilized separately such that both formation keeping and velocity regulation to a constant value are guaranteed.

### 1.2.9 Consensus algorithms

Investigation of effects of information structure on a control decision was initiated by the early work on the theory of teams and was first introduced in [201] and later followed by [202–206]. These works can be considered as the first results in which the control in a team is discussed where only part of the information is available to the team members. However, since these early works were performed, and especially in recent years, a large body of research has been conducted in which the effects of information on the control design problems are discussed, see, for example, [95, 124,125,127,129,148,207]. In most of these works, each team member has access to limited information from other agents, or to information of its neighbors. The final state of the team is decided by the team members.

Consensus algorithms are one of the tools that are used for analysis of distributed systems where the network information structure has a vital effect on the control design where only part of the information is available to each member. As mentioned in [127], consensus problems deal with the agreement of a group of agents upon specific "quantities of interest." In this configuration, the agents try to decide and agree among themselves upon what the final state should be.

The state where all the quantities of interest are the same is called the consensus state. In other works, one maybe specifically interested in problems in which the quantities of interest are related to the motion of the agents, e.g., their velocities or positions. However, as indicated in [208], several applications for consensus algorithms exist, such as applications in decentralized computation, clock synchronization, sensor networks, and in coordinated control of multiagent networks.

In [127], linear and nonlinear consensus protocols are applied to directed and undirected networks with fixed and switching topologies. A disagreement function was introduced as a Lyapunov function to provide a tool for convergence analysis of an agreement protocol in a switching network topology. The authors have shown that the maximum time-delay that can be tolerated by a network of integrators applying a linear consensus protocol is inversely proportional to the largest eigenvalue of the

Laplacian of the information flow graph or the maximum degree of the nodes of the network. Similar results are obtained in [125] where convergence analysis is developed by using Nyquist plots for linear protocols. For nonlinear protocols, the notion of action graphs and disagreement cost are introduced, and the problem was solved in a distributed manner.

In [95], the coordination problem is discussed for a team of agents using nearest neighbor rule for both leaderless and leader–follower configurations. The main focus of this work is on heading angle alignment in undirected graphs where the agents have simple integrator dynamics, and the agents have the same speed but have different headings. In the leader–follower case, the leader can affect the followers whenever it is in their neighboring set. However, there is no feedback from the followers to the leader.

It is shown that the connectivity of the graph on average (connection of union of graphs) is sufficient for convergence of the agents heading angles. The neighboring set assignment is switching, and therefore the team structure is dynamic. In [209], asynchronous protocols for consensus seeking are introduced. Some updating rules for the control input of agents with discrete-time dynamical equations are suggested so that the consensus state would take a desirable predefined value.

In [149], passivity is used as a tool to achieve network agreement (or consensus) for a class of agents with dynamics which can satisfy the passivity conditions. The group's main goal is to reach at a predefined common velocity (or any other interpretation of the derivative of a state), while the relative positions (the difference between a common state in the group) converge to a desired compact set. Based on this method, a Lyapunov function can be constructed for stability analysis in a distributed communication network with bidirectional links. The designed controller is a filter which has a nonlinear function of the relative states as its input and is designed based on passivity properties. The relation between the topology and the stability of the formation is provided. In [210], a wider class of systems, i.e., nonlinear dissipative systems are considered, and synchronization in a strongly connected network of agents with this dynamical property is discussed.

In [156,211,212], an interpretation of controllability is defined and shown for a first-order integrator model. The main purpose of these work is to obtain the effects of external decisions on the agreement dynamics, in particular the conditions where some of the nodes do not follow the agreement protocols (decisions). In other words, there are some nodes that follow the agreement protocol, while others have external inputs. The authors try to answer the question whether these "anchored" nodes are able to guide the rest of the group to the desired point.

Similar to the ideas presented in [156] for a fixed network topology, the controllability conditions for a network with fixed and switching topologies are discussed in [213]. The authors have considered a leader–follower structure with one-way links from the leader to the followers. They have shown that controllability of the team is highly dependent on how the followers are connected to the leader.

The goal in [153] is to find the minimum communication that is required for guaranteeing the stability of a swarm of vehicles. The approach is to first define a centralized cost function for the group and then divide it into individual costs. A vector of parameters is introduced for quantization of the information which should be exchanged among the agents in order to achieve a stable formation, and the optimization is accomplished with respect to these parameters. Similar approach is pursued in [154].

In [158], a fixed and a given network structure is assumed, and the question that is addressed is how to find the weights of the interconnection links such that the convergence to consensus value is achieved at the fastest rate. To solve the problem, a set of criteria is introduced to be minimized. The resulting optimization problem is nonconvex which is then converted into a convex one. In [214], an estimate of the convergence rate of the consensus seeking is obtained. The communication links are assumed to be time-varying. In [215], a lower bound on convergence rate of some of the consensus algorithms is provided.

Toward the above end, two approaches based on the properties of stochastic matrices and the concept of random walks are used. In [216], it is shown that connectivity of a network with a fixed number of links can be significantly increased by selecting the interagent information flow links properly. This in turn can result in an ultrafast consensus seeking procedure. The idea is best applicable to small-world networks where any two nodes can be connected using a few links though the total size of the network can be large.

In some work, the communication delay is considered in representing the agents network with point-mass model. For an example, one can refer to [127] in which directed and undirected networks with fixed and switching topologies are considered. It is assumed that the delayed information from other agents are compared with the delayed value of the agent's own dynamics at each time step. On the other hand, in [217], the delayed information of the neighbors are compared with the current value of the agents' state. In this work, uniformly delayed communication links are analyzed for consensus algorithms.

In [218], the agreement protocol is analyzed when there are nonuniform time-delays in the links amongst the agents. Linear protocols are used for fixed networks with and without communication time-delays and communication channels that have filtering effects. Similarly, nonlinear protocols are applied to dynamic networks to achieve consensus. In all these cases, the effects of time-delay are analyzed for the agreement protocol only, and the analysis is performed in the frequency domain. In [219], the authors have considered time-varying time-delays in communication links and presented conditions for consensus achievement in a network of nonlinear, locally passive systems. Considering time-delays in other coordination problems such as formation control and target tracking is still an open area of research.

The problem of team cooperation, and specifically consensus seeking with switching topology, has received a wide attention in recent years and has been discussed in the literature from different perspectives [220–230]. The work in [220] can be considered as one of the pioneers in which algorithms for distributed computation in a network with a time-varying structure are analyzed. Specifically, in [221], for a discrete-time model of processors and a given number of tasks, convergence of a consensus algorithm in a time-varying structure is discussed given that some restrictions are imposed on the frequency of availability of the interagent communication links.

One of the underlying assumptions in many of the related works on switching networks is that the graph describing the information exchange structure is a balanced graph. The authors in [230] considered balanced information graphs and showed the stability under switching time-delayed communication links. The analysis is performed by introducing a Lyapunov functional and then by showing the feasibility of a set of LMIs.

In [129,224], switching control laws are designed for a network of agents with undirected and connected underlying graphs, whereas in [228], consensus in a directed, jointly connected, and balanced network is considered. Necessary conditions for achieving consensus in a network are discussed in [226]. The concept of preleader–follower is introduced as a new approach to achieve consensus in a network of discrete-time systems. The basic properties of stochastic matrices are used to guarantee consensus achievement in a network with switching topology and time-delayed communication links.

In [223], higher order consensus algorithms are discussed. The author's approach to handle the switching network structure with a spanning tree is to find an appropriate dwell time with its own provided definition. It is shown that the final consensus depends on the information exchange structure as well as the controller weights. In [98], analysis is performed for a time-varying network of agents with discrete-time models. In this work, milder assumptions on connectivity of the agents over time are imposed when compared to [95], and necessary and sufficient results for consensus achieving are presented. The works in [135,197,198] are extensions of the approach in [95] for second-order dynamics for fixed and dynamic topologies in undirected and connected graphs.

In [225], the authors used a similar approach to the one used in [135,197,198] to analyze consensus achievement in a team with fixed and switching topologies. They divided the control law into several parts and used nonsmooth analysis framework to address the problem. In [127], consensus achievement for a connected graph subject to certain switching in the network structure is addressed. The underlying assumption there is that the graph under consideration is balanced.

In [207,228], consensus in directed, jointly connected, and balanced network is considered. The authors in [207] have considered information consensus in multi-agent networks with dynamically changing interaction topologies in presence of limited information. They have shown that consensus can be achieved asymptotically under these conditions if the union of the directed interaction graphs has a spanning tree frequently enough for agents with discrete and continuous-time dynamics. This condition is weaker than the assumption of connectedness that is made in [127], [95] and implies that one half of the information exchange links that are required in [95] can be removed without affecting the convergence result. However, the finally achieved equilibrium points will depend on the property of the directed graph, e.g., its connectedness. This work is an extension of [95] to the digraphs case with more flexible weight selections in information update schemes. Some simulation examples of this work are presented in [231].

One of the recent research topics in consensus seeking is analysis of behavior of consensus algorithms in presence of measurement noise or even design of consensus algorithms that can compensate for lack of measurements or inaccuracies in them. In [232], performance of first and second order consensus algorithms is discussed in presence of measurement noise. A relationship between the measurement error and the consensus error is derived. In [233], the measurement noise is considered for a leader–follower structure. Having used the stochastic analysis and by assuming time-varying weights, the authors could guarantee a mean square consensus achievement in presence of measurement noise. Similar idea of using time-varying weights is used in [234] to guarantee consensus on a Gaussian random variable. The authors in [235] have linked the consensus problem into a multiinventory system control problem, where bounded disturbances affect the first order agents' dynamics.

## 1.3 Features and objectives of the book

The main features and objectives of this book are as follows:

- **IFDC design based on dynamic observers**
  The IFDC design problem based on dynamic observers for continuous-time LTI systems is investigated. In essence, the structure of the dynamic observer is employed to eliminate the disadvantages of static observer structures in designing the IFDC. It is shown that the proposed method has major advantages in comparison with previous results in which it leads to strict LMI conditions for designing the dynamic observer parameters and the controller gains.

- **IFDIT control problem**
  The developed above dynamic observer-based IFDC design problem is extended to the problem of IFDIT control problem in LTI systems. The proposed methodology is capable of simultaneously accomplishing the detection and isolation of faults as well as tracking control of the closed-loop system in presence of both bounded energy and bounded peak disturbances. Consequently, the computational complexity as far as the number and dimension of the required observers are significantly reduced when compared to the existing techniques in the literature. Moreover, LMI feasibility conditions for solving the IFDIT problem are developed such that no products of Lyapunov matrices with the system matrices are present that reduces the conservatism in obtaining a solution to our multiobjective diagnosis and control problem.

- **Integrated design of fault detection, isolation, and control for Markovian jump systems**
  The problem of IFDIC design for uncertain Markovian jump systems subject to both bounded energy and bounded peak disturbances is investigated. Using this methodology, not only the considered uncertain Markovian jump system can

be effectively controlled but also the occurrence of faults in the system can be detected and isolated. Moreover, the developed scheme can also handle isolation of occurrence of simultaneous faults.

- **Event-triggered integrated fault diagnosis and control**
  To reduce the usage of computational and communication resources without degrading the overall desired closed-loop system performance, a general problem known as E-IFDIC is defined. By utilizing a filter to represent, characterize, and specify the E-IFDIC module, a multiobjective formulation of the problem is developed based on $H_\infty$, $H_-$, $l_1$ and generalized $H_2$ performance criteria and sufficient conditions for solvability of the problem are obtained in terms of LMI feasibility conditions. Consequently, it is shown that certain existing problems in the fields of time and event-trigged control and fault diagnosis can be considered as special cases of the proposed methodology.

- **Event-triggered integrated fault estimation and accommodation design**
  By using an integrated design of event-triggered fault/state estimator with a fault-tolerant controller, the problem of E-AFTC of discrete-time linear systems is investigated. A robust $H_\infty$ formulation of the problem is provided that guarantees stability of the resulting closed-loop system, attenuates the effects of external disturbances, and compensates for the effects of faults. Sufficient LMI conditions are derived to simultaneously obtain the observer and controller parameters and the event-triggered conditions.

- **Integrated fault detection and consensus control design for a network of MASs**
  The problem of IFDCC of linear continuous-time MASs is investigated. A mixed $H_\infty/H_-$ formulation of the IFDCC problem is presented, and distributed detection filters are designed using only relative output information among the agents. With the developed methodology, all agents reach either a state consensus or a model reference consensus while simultaneously collaborate with one another to detect the occurrence of faults in the team. A decomposition technique is employed in the IFDCC design procedure that allows the system to be partitioned into a set of lower order subsystems for reducing the computational complexity of the design as well as for specifying the IFDCC modules with a distributed architecture.

To summarize, the objective of this book is to provide the first step toward the design of novel IFDC systems for complex systems within the domain of LTI, Markovian jump, multiagent and NCSs. Multiple and various issues in the current conventional IFDC systems such as robustness with respect to external disturbances, compensations for the overprovisioning of the computational and communication resources, and different IFDC architectures from centralized to distributed, from deterministic to stochastic, and from time-triggered to event-triggered cases are considered and developed.

## 1.4 Outline of the book

This book is organized as follows. In Chapter 2, the focus is on the problem of IFDC design for continuous-time LTI systems subject to faults and external disturbances. In Section 2.2, the IFDC problem based on the $H_2$ and $H_\infty$ performance measures is first described and formulated. The solution of the proposed IFDC problem is obtained in Section 2.3. To demonstrate the validity of the proposed approach, two case studies corresponding to a four-tank process and an AUV are provided in Section 2.4.

In Chapter 3, the IFDIT control problem for linear systems subject to both bounded energy and bounded peak external disturbances is investigated. A brief literature review is first provided in Section 3.1. In Section 3.2, the IFDIT problem for the considered LTI systems under investigation is formulated. In Section 3.3, a systematic LMI approach for solving the proposed IFDIT problem and for generating a set of residual and control signals is presented. In Section 3.4, the proposed approach is applied to the problem of integrated FDI, and tracking control of the Subzero II AUV.

In Chapter 4, the focus is on the problem of fault detection, isolation, and control design (IFDIC) for continuous-time Markovian jump linear systems with uncertain transition probabilities. After a brief literature review in Section 4.1, the system and the IFDIC module governing equations are described in Section 4.2.1. The IFDIC problem is then formulated for the Markovian jump linear systems with uncertain transition probabilities in Section 4.2.2. In essence, the IFDIC design problem is transformed into a mixed $H_\infty/H_-/L_1$ optimization problem. An LMI-based solution to the developed IFDIC problem is provided in Section 4.3. In Section 4.4, the proposed approach is applied to the problem of IFDIC design for the GE F-404 aircraft engine system.

In Chapter 5, the problems of event-triggered multiobjective synthesis of feedback controllers and fault diagnosis filters are investigated through a unified framework. After a brief overview of the relevant literature in Section 5.1, in Section 5.2, the statement of the problem for E-IFDIC design and the governing equations for the considered discrete-time LTI system under investigation. In Section 5.3, an LMI-based solution to the proposed E-IFDIC problem is developed and implemented. In Section 5.4, different aspects and certain capabilities of the methodology in event-triggered multiobjective synthesis of feedback controllers and fault diagnosis filters are demonstrated by providing two industrial case studies under different scenarios.

In Chapter 6, the problem of E-AFTC of discrete-time linear systems is addressed. In Section 6.2, the event-triggered fault/state observer and the fault-tolerant controller are described, and the definition of E-AFTC problem is presented. The main results are given in Section 6.3, where the LMI-based solution to the E-AFTC problem is obtained. To show the validity of the proposed approach, the event-triggered AFTC design for the Subzero III ROV is considered in Section 6.4.

In Chapter 7, a team of multiagents is considered, and an IFDCC scheme is developed so that all the agents reach either a state consensus or a model reference consensus while simultaneously collaborating with one another to detect the occurrence of faults in the team. In Section 7.2, the statement of the distributed IFDCC problem for a team of MASs is presented. Next, a model transformation methodology is employed in Section 7.3.1 and a model decomposition technique is employed in Section 7.3.2, to reduce the computational complexity of the design as well as to specify the IFDCC modules with a distributed architecture. An LMI formulation and design solution to the IFDCC problem is obtained in Section 7.4. Moreover, the residual evaluation criterion and the proposed distributed fault isolation scheme are studied in Sections 7.4.1 and 7.4.2, respectively. The effectiveness and capabilities of the proposed results are demonstrated and provided in Section 7.5.

Finally, in Chapter 8, closing remarks and perspectives as well as future directions of research are outlined and provided.

## 1.5   Notations and preliminary lemmas

Let $\mathbb{R}^n$ denote the $n$ dimensional Euclidean space, $\mathbb{R}^{n \times m}$ denote the set of all $n \times m$ real matrices, and $\mathbb{Z}^+$ is the set of positive integers. Moreover, let $(\Omega, \mathfrak{F}, \mathrm{P})$ denote a complete probability space where $\Omega$ is the sample space, $\mathfrak{F}$ is the algebra of events, and P is the probability measure defined on $\mathfrak{F}$. $\mathbb{E}\{.\}$ denotes the expectation operator with respect to the probability measure P. The term $\mathbb{L}_2$ ($l_2$) stands for the space of square integrable (square-summable) vector functions (sequences). The term $\mathbb{L}_\infty$ ($l_\infty$) denotes the space of bounded vector functions (sequences). The norm $|| \cdot ||$ of a real vector function and a matrix are defined according to their Euclidean counterparts.

For a vector $x \in \mathbb{R}^n$, its norm is denoted by $||x|| = \sqrt{x^\mathsf{T} x}$. For $x(t) \in \mathbb{L}_2, t \geq 0$ ($x(k) \in l_2$, $k \geq 0$), its norm is given by $||x||_2 = \sqrt{\int_0^{+\infty} ||x(t)||^2 dt}$ $\left( ||x||_2 = \sqrt{\sum_{k=0}^\infty ||x(k)||^2} \right)$. For $x(t) \in \mathbb{L}_\infty$, $t \geq 0$ ($x(k) \in l_\infty$, $k \geq 0$), it is implied that $||x||_\infty = \sup_{t \geq 0} ||x(t)||$ ($||x||_\infty = \sup_{k \in \mathbb{Z}^+} ||x(k)||$). For the stochastic signal $x(t)$, $||x||_\infty^\mathbb{E} = \sup_{t \geq 0} \sqrt{\mathbb{E}\{|x(t)|^2\}}$ and $||x||_2^\mathbb{E} = \mathbb{E}\left\{ \left( \int_0^{+\infty} |x(t)|^2 dt \right)^{1/2} \right\}$.

The operator $T_{yx} : x \to y$ denotes the transfer matrix (or the mapping) from $x$ to $y$. For the operator $T_{yx}$, the $H_\infty$, $L_1$ norms and $H_-$ index are defined as $||T_{yx}||_\infty = \sup((||y||_2)/(||x||_2))$, $||T_{yx}||_1 = \sup((||y||_\infty)/(||x||_\infty))$, and $||T_{yx}||_- = \inf((||y||_2)/(||x||_2))$, respectively. $H_-^+$ index is defined as the minimum nonzero singular value of a transfer matrix, namely, $||T||_-^+ = \inf_{\omega \in \mathbb{R}} \underline{\sigma}^+(T(j\omega))$, where $\underline{\sigma}^+$ denotes the minimum nonzero singular values of $T(s)$.

For a matrix $A$, $A^\mathsf{T}$ denotes its transpose, $\mathbf{1}_n$ denotes the $n \times 1$ column vector whose elements are all one. $I$ and $0$ denote the identity and zero matrices with appropriate dimensions, respectively, where $I_n$ and $0_n$ denote the $n \times n$ identity and null matrices, respectively. $S_m$ denotes the set of all symmetric $m \times m$ matrices. For a given set $N$, $|N|$ denotes the cardinality of $N$. The Hermitian part of a square matrix $A$ is denoted by $\mathrm{Herm}(A) = A + A^\mathsf{T}$, and $*$ denotes the symmetric entries of a matrix.

The communication topology among $N$ agents is represented by an undirected graph $\mathfrak{G} = (\mathcal{V}, \mathcal{E})$, consisting of the node set $\mathcal{V} = \{1, 2, \ldots, N\}$ and the edge set $\mathcal{E} \subseteq \mathcal{V} \times \mathcal{V}$. The set of nearest neighbors of the $i$th node is designated by $\mathcal{N}_i = \{j \in \mathcal{V} : (i, j) \in \mathcal{E}\}$. The Laplacian matrix $L = [L_{ij}]_{N \times N}$ associated with the graph $\mathfrak{G}$ is defined by $L_{ii} = |\mathcal{N}_i|$, $L_{ij} = 0$ when $j \notin \mathcal{N}_i$ and $L_{ij} = -1$ when $j \in \mathcal{N}_i$. The Kronecker product of two matrices, namely, $A = [a_{ij}] \in \mathbb{R}^{m \times n}$ and $B = [b_{ij}] \in \mathbb{R}^{p \times q}$ is a $mp \times nq$ matrix, which is denoted by $A \otimes B$ and is defined as follows [236]:

$$A \otimes B = \begin{bmatrix} a_{11}b_{11} & a_{11}b_{12} & \cdots & a_{11}b_{1q} & \cdots & \cdots & a_{1n}b_{11} & a_{1n}b_{12} & \cdots & a_{1n}b_{1q} \\ a_{11}b_{21} & a_{11}b_{22} & \cdots & a_{11}b_{2q} & \cdots & \cdots & a_{1n}b_{21} & a_{1n}b_{22} & \cdots & a_{1n}b_{2q} \\ \vdots & \vdots & \ddots & \vdots & & & \vdots & \vdots & \ddots & \vdots \\ a_{11}b_{p1} & a_{11}b_{p2} & \cdots & a_{11}b_{pq} & \cdots & \cdots & a_{1n}b_{p1} & a_{1n}b_{p2} & \cdots & a_{1n}b_{pq} \\ a_{21}b_{11} & a_{21}b_{12} & \cdots & a_{21}b_{1q} & \cdots & \cdots & a_{2n}b_{11} & a_{2n}b_{12} & \cdots & a_{2n}b_{1q} \\ a_{21}b_{21} & a_{21}b_{22} & \cdots & a_{21}b_{2q} & \cdots & \cdots & a_{2n}b_{21} & a_{2n}b_{22} & \cdots & a_{2n}b_{2q} \\ \vdots & \vdots & \ddots & \vdots & \ddots & & \vdots & \vdots & \ddots & \vdots \\ a_{21}b_{p1} & a_{21}b_{p2} & \cdots & a_{21}b_{pq} & \cdots & \cdots & a_{2n}b_{p1} & a_{2n}b_{p2} & \cdots & a_{2n}b_{pq} \\ \vdots & \vdots & \ddots & \vdots & \ddots & & \vdots & \vdots & \ddots & \vdots \\ \vdots & \vdots & \ddots & \vdots & & \ddots & \vdots & \vdots & \ddots & \vdots \\ a_{m1}b_{11} & a_{m1}b_{12} & \cdots & a_{m1}b_{1q} & \cdots & \cdots & a_{mn}b_{11} & a_{mn}b_{12} & \cdots & a_{mn}b_{1q} \\ a_{m1}b_{21} & a_{m1}b_{22} & \cdots & a_{m1}b_{2q} & \cdots & \cdots & a_{mn}b_{21} & a_{mn}b_{22} & \cdots & a_{mn}b_{2q} \\ \vdots & \vdots & \ddots & \vdots & & & \vdots & \vdots & \ddots & \vdots \\ a_{m1}b_{p1} & a_{m1}b_{p2} & \cdots & a_{m1}b_{pq} & \cdots & \cdots & a_{mn}b_{p1} & a_{mn}b_{p2} & \cdots & a_{mn}b_{pq} \end{bmatrix}$$

$$(1.1)$$

For simplicity, the following notation may be used to define the Kronecker product of two matrices:

$$A \otimes B = \begin{bmatrix} a_{11}B & a_{12}B & \cdots & a_{1n}B \\ a_{21}B & a_{22}B & \cdots & a_{2n}B \\ \vdots & \vdots & \ddots & \vdots \\ a_{m1}B & a_{m2}B & \cdots & a_{mn}B \end{bmatrix} \qquad (1.2)$$

The Kronecker product has some interesting properties that are used in this book, including [237]

$$A \otimes (B + C) = A \otimes B + A \otimes C, \qquad (1.3)$$

$$(A + B) \otimes C = A \otimes C + B \otimes C, \qquad (1.4)$$

$$(\alpha A) \otimes B = A \otimes (\alpha B) = \alpha A \otimes B, \qquad (1.5)$$

$$(A \otimes B) \otimes C = A \otimes (B \otimes C), \qquad (1.6)$$

$$(A \otimes B)^{\mathrm{T}} = A^{\mathrm{T}} \otimes B^{\mathrm{T}}. \qquad (1.7)$$

The following lemmas are used in the subsequent chapters.

**Lemma 1.1.** *(Projection Lemma [238]) Given a symmetric matrix $Z \in S_m$ and matrices U and V of column dimension m, there exists a matrix X satisfying:*

$$U^{\mathrm{T}}XV + V^{\mathrm{T}}X^{\mathrm{T}}U + Z < 0, \tag{1.8}$$

*if and only if the following projection inequalities with respect to X are satisfied:*

$$N_U^{\mathrm{T}}ZN_U < 0, \tag{1.9a}$$

$$N_V^{\mathrm{T}}ZN_V < 0, \tag{1.9b}$$

*where $N_U$ and $N_V$ are arbitrary matrices whose columns form a basis of the null spaces of U and V, respectively.*    □

The next result is known as the matrix inversion formula [239].

**Lemma 1.2.** *Suppose $A_{11}$, $A_{22}$, and $\begin{bmatrix} A_{11} & A_{12} \\ A_{21} & A_{22} \end{bmatrix}$ are nonsingular matrices. Define $\Delta := A_{11} - A_{12}A_{22}^{-1}A_{21}$. Then,*

$$\begin{bmatrix} A_{11} & A_{12} \\ A_{21} & A_{22} \end{bmatrix}^{-1} = \begin{bmatrix} \Delta^{-1} & -\Delta^{-1}A_{12}A_{22}^{-1} \\ -A_{22}^{-1}A_{21}\Delta^{-1} & A_{22}^{-1} + A_{22}^{-1}A_{21}\Delta^{-1}A_{12}A_{22}^{-1} \end{bmatrix}. \tag{1.10}$$

□

**Lemma 1.3.** *[240] Let F, X, and Y denote real matrices of appropriate dimension with $F^{\mathrm{T}}F \leq I$. Then for any scalar $\theta > 0$, it follows that*

$$XFY + Y^{\mathrm{T}}F^{\mathrm{T}}X^{\mathrm{T}} < \theta XX^{\mathrm{T}} + \theta^{-1}Y^{\mathrm{T}}Y. \tag{1.11}$$

□

**Lemma 1.4.** *[241] Let $L_C = [L_{C_{ij}}] \in S_N$ denote a symmetric matrix with $L_{C_{ii}} = ((N-1)/N)$ and $L_{C_{ij}} = (-1/N)$ for $i \neq j$, where N is a positive integer number. There exists an orthogonal matrix $U \in \mathbb{R}^{N \times N}$ such that $U^{\mathrm{T}}L_C U = \mathrm{diag}(I_{N-1}, 0)$ and the last column of U is $\mathbf{1}_N/\sqrt{N}$. Furthermore, let $L \in \mathbb{R}^{N \times N}$ denote the Laplacian of any undirected graph, then $U^{\mathrm{T}}LU = \mathrm{diag}(L_1, 0)$, where $L_1 \in \mathbb{R}^{(N-1) \times (N-1)}$ is positive definite if and only if the graph is connected.*    □

**Lemma 1.5.** *(Schur complement) For any symmetric matrix, M, of the form*

$$M = \begin{bmatrix} A & B \\ B^{\mathrm{T}} & C \end{bmatrix}, \tag{1.12}$$

*if A is invertible, the following properties hold*

(a)  $M > 0$ iff $A > 0$ and $C - B^{\mathrm{T}}A^{-1}B > 0$.
(b)  If $A > 0$, then $M \geq 0$ iff $C - B^{\mathrm{T}}A^{-1}B \geq 0$.

□

**Lemma 1.6.** *Suppose the following system is asymptotically stable*

$$\begin{cases} \dot{x}(t) = Ax(t) + Bw(t), \\ z(t) = Cx(t) + Dw(t), \end{cases} \tag{1.13}$$

*and let $T(s) = C(sI - A)^{-1}B + D$ denote its transfer matrix. If $D = 0$, the following statements are equivalent:*

(a) *There exists a prescribed positive constant $\gamma$ such that*

$$||T(s)||_2 < \gamma, \tag{1.14}$$

(b) *There exist matrices $P = P^{\mathrm{T}}$ and $Z$ such that*

$$\begin{bmatrix} A^{\mathrm{T}}P + PA & PB \\ * & -\gamma I \end{bmatrix} < 0,$$

$$\begin{bmatrix} P & C^{\mathrm{T}} \\ * & Z \end{bmatrix} > 0, \ \mathrm{trace}(Z) < \gamma. \tag{1.15}$$

□

**Lemma 1.7.** *(Bounded real lemma) For the system* (1.13), *the $H_\infty$ performance, with $\gamma > 0$ is equivalent to the existence of $X > 0$ satisfying*

$$\begin{bmatrix} A^{\mathrm{T}}X + XA + C^{\mathrm{T}}C & XB + C^{\mathrm{T}}D \\ * & D^{\mathrm{T}}D - \gamma^2 I \end{bmatrix} < 0, \tag{1.16}$$

*or*

$$\begin{bmatrix} AX + XA^{\mathrm{T}} & B & XC^{\mathrm{T}} \\ * & -\gamma I & D^{\mathrm{T}} \\ * & * & -\gamma I \end{bmatrix} < 0. \tag{1.17}$$

□

**Lemma 1.8.** *($H_-$ index [242]) Let $\beta > 0$ be a constant scalar. Let us denote $T(s) = C(sI - A)^{-1}B + D$, and*

$$H = \begin{bmatrix} A + BR^{-1}D^{\mathrm{T}}C & BR^{-1}B^{\mathrm{T}} \\ -C^{\mathrm{T}}(I + DR^{-1}D^{\mathrm{T}})C & -(A + BR^{-1}D^{\mathrm{T}}C)^{\mathrm{T}} \end{bmatrix}, \tag{1.18}$$

$$R = \beta^2 I - D^{\mathrm{T}}D.$$

*Then the following properties are equivalent:*

(a) *$||T(j\omega)||_-^{[0,\infty)} > \beta$, i.e., $\underline{\sigma}(T(j\omega)) > \beta, \ \forall \omega \in [0, \infty)$.*

(b) *$\underline{\sigma}(D) > \beta$ and there exists a symmetric matrix $X$ such that*

$$X(A + BR^{-1}D^{\mathrm{T}}C) + (A + BR^{-1}D^{\mathrm{T}}C)^{\mathrm{T}}X +$$

$$XBR^{-1}B^{\mathrm{T}}X + C^{\mathrm{T}}(I + DR^{-1}D^{\mathrm{T}})C = 0, \tag{1.19}$$

*and $A + BR^{-1}D^{\mathrm{T}}C + BR^{-1}B^{\mathrm{T}}X$ has no eigenvalues on the imaginary axis.*

(c)  $\underline{\sigma}(D) > \beta$ and there exists a symmetric matrix $X$ such that

$$X(A + BR^{-1}D^{\mathrm{T}}C) + (A + BR^{-1}D^{\mathrm{T}}C)^{\mathrm{T}}X + \\ XBR^{-1}B^{\mathrm{T}}X + C^{\mathrm{T}}(I + DR^{-1}D^{\mathrm{T}})C > 0. \tag{1.20}$$

□

By using the Schur complement and Lemma 1.8, the following LMI conditions for the $H_-$ measure are readily obtained.

**Lemma 1.9.** *($H_-$ index [242]) Let $\beta > 0$ be a constant scalar and denote $T(s) = C(sI - A)^{-1}B + D$. Then $||T(j\omega)||_-^{[0,\infty)} > \beta$, if and only if there exists a symmetric matrix $X$ such that*

$$\begin{bmatrix} A^{\mathrm{T}}X + XA + C^{\mathrm{T}}C & XB + C^{\mathrm{T}}D \\ * & D^{\mathrm{T}}D - \beta^2 I \end{bmatrix} > 0. \tag{1.21}$$

□

From (1.21), it is obvious that

$$D^{\mathrm{T}}D - \beta^2 I > 0.$$

Therefore, a positive $\beta$ implies that the matrix $D$ has full column rank. However, in practical applications, the matrix $D$ in $T(s) = C(sI - A)^{-1}B + D$ could have any value. For strictly proper systems, $D = 0$, and $H_-$ index over $[0, \infty)$ is always zero. Consequently, it is necessary to evaluate the worst case performance even within a specific frequency range, especially the steady state (DC) and low frequency part $[0, \omega_l]$ [242]. Furthermore, it is clear that the $H_-$ measure within a finite frequency range could be much higher than that for $[0, \infty)$. The minimum $H_-$ index in the low frequency range can be obtained by invoking the following lemma.

**Lemma 1.10.** *(Finite frequency $H_-$ index [243]) Let $\beta > 0$ be a constant scalar, a positive scalar $\omega_l$ be given and let us denote $T(s) = C(sI - A)^{-1}B + D$. Then, the following properties are equivalent:*

(a)  $||T(j\omega)||_-^{[-\omega_l,\omega_l]} > \beta$, *i.e.,* $\underline{\sigma}(T(j\omega)) > \beta, \forall|\omega| \le \omega_l$.
(b)  $\underline{\sigma}(T(j\omega)) > \beta, \forall|\omega| \le \omega_l$ *if there exist Hermitian matrices $P$ and $Q > 0$ satisfying*

$$\begin{bmatrix} A & B \\ I & 0 \end{bmatrix}^{\mathrm{T}} \Phi \begin{bmatrix} A & B \\ I & 0 \end{bmatrix} + \begin{bmatrix} C & D \\ 0 & I \end{bmatrix}^{\mathrm{T}} \Xi \begin{bmatrix} C & D \\ 0 & I \end{bmatrix} < 0, \tag{1.22}$$

*where* $\Phi = \begin{bmatrix} -Q & P \\ P & \omega_l^2 Q \end{bmatrix}$ *and* $\Xi = \begin{bmatrix} -I & 0 \\ 0 & \beta^2 I \end{bmatrix}$.

□

*Chapter 2*
# Integrated fault detection and control design based on dynamic observer

In this chapter, we briefly review the advantages of employing dynamic observers in the structure of an integrated fault detection and control (IFDC) design problem. A mixed $H_2/H_\infty$ formulation of the IFDC problem using dynamic observer for linear continuous-time systems is presented. Effectively, a single unit designated as detector/controller is designed where the detector is a dynamic observer and the controller is a state feedback controller based on the dynamic observer.

Consequently, the detector/controller module produces two signals, i.e., the detection and control signals. It is shown that the dynamic observer can be used effectively to tackle the drawbacks of the existing IFDC design methods. Sufficient conditions for solvability of the problem are obtained in terms of the linear matrix inequality (LMI) feasibility conditions. Applications of our methodology to two case studies are presented to demonstrate and illustrate the effectiveness of the proposed approach. The work presented in this chapter has partially appeared in [43,244].

The remainder of this chapter is organized as follows. In Section 2.2, the IFDC problem for a continuous-time linear time-invariant (LTI) system subject to faults and external disturbances is formulated based on the $H_2$ and $H_\infty$ performance indices. The LMI-based solution to the IFDC problem is presented in Section 2.3. To demonstrate the validity of the proposed approach, two case studies corresponding to a four-tank process and an autonomous underwater vehicle (AUV) are given in Section 2.4.

## 2.1 Introduction

Most of the existing IFDC observers have been simply confined to traditional static observers (classic Kalman–Luenberger observers) [245]. In order to distinguish between static observers, the term "dynamic observers" is used, which is an extension of static observers in its configuration and puts dynamics in the observer gain [246]. In [247], a dynamic observer design method is proposed as a dual of the control design for the state estimation. A similar work is the Lipschitz unknown input observer (UIO) [248], where two dynamic compensators are introduced to tackle Lipschitz nonlinearities. In [246], a zero assignment approach for $H_2/H_\infty$ dynamic filter design with application to fault detection is proposed.

In this chapter, we propose a mixed $H_2/H_\infty$ formulation of the IFDC problem by using a dynamic observer detector and a state feedback controller. In fact, a single module designated as the detector/controller, where the detector is a dynamic observer and the controller is a state feedback is designed which produces two signals, namely, the detection and control signals, which are used to detect faults and satisfy certain control objectives, respectively.

It should be pointed that the conservatism in the IFDC design problem depends on the type of the filter that is used in the structure of the IFDC block. The IFDC problem by using a dynamic output feedback structure has been studied in [40], where conditions are proposed in terms of bilinear matrix inequality which are heavily dependent on the initial conditions of the iteration and are not globally convergent. Furthermore, as stated earlier, most of the existing fault detection observers have been simply confined to traditional static observers. By using equality constraints [249], the LMI conditions can be obtained for solving such observer-based IFDC problems. It should be pointed that by applying equality constraints, certain degrees of conservativeness in the design problem are introduced.

In this chapter, the structure of the dynamic observer is employed to eliminate the disadvantages of the above observer structures in designing the IFDC. Consequently, our proposed method has major advantages with respect to previous results that we obtain strict LMI conditions for designing the dynamic observer parameters and the controller gains. Two case studies corresponding to a four-tank process and an AUV are provided to demonstrate the capabilities and effectiveness of the proposed results.

## 2.2   The problem statement and definitions

In this section, we first describe a continuous-time LTI system subject to faults and external disturbances. The IFDC problem is then formulated based on the $H_2$ and $H_\infty$ performance indices.

### 2.2.1   System model

Consider the following LTI system:

$$G: \begin{cases} \dot{x}(t) = Ax(t) + B_1u(t) + B_2w(t) + B_3d(t) + B_4f(t) \\ y(t) = Cx(t) + D_1u(t) + D_2w(t) + D_3d(t) + D_4f(t) \;, \\ z(t) = Ex(t) + F_1u(t) + F_2d(t) + F_3f(t) \end{cases} \tag{2.1}$$

where $x(t) \in \mathbb{R}^n$ is the state, $u(t) \in \mathbb{R}^{n_u}$ is the control input, $y(t) \in \mathbb{R}^{n_y}$ is the measured output, and $z(t) \in \mathbb{R}^{n_z}$ denotes the regulated output. The unknown input $w(t) \in \mathbb{R}^{n_w}$ is assumed to be a fixed spectral density process/measurement noise, $d(t) \in \mathbb{R}^{n_d}$ is assumed to be a finite energy disturbance modeling errors due to exogenous signals, linearization process, or parameter uncertainties. Moreover, the unknown input $f(t) \in \mathbb{R}^{n_f}$ denotes a possible additive fault.

Typical additive faults that are studied in practice are classified as offset in sensors and actuators or drift in the sensors. The former can be described by a constant, while the later by a ramp function. Moreover, other types of actuator or sensor faults can be modeled based on the additive fault $f(t)$. For example, loss of effectiveness (LOE) and outage that are two common types of actuator faults can be modeled by considering $f(t) = -\Gamma u(t)$, where $\Gamma = \text{diag}(\Gamma_1, \ldots, \Gamma_k, \ldots, \Gamma_{n_u})$, and $0 < \Gamma_k \leq 1$, $k = 1, \ldots, n_u$. Note that $\Gamma_k$ is the LOE factor of the $k$th actuator. If $0 < \Gamma_k < 1$, the fault is designated as the LOE and if $\Gamma_k = 1$, then the fault is called an outage fault. The matrices $A, B_i$s, $C, D_i$s, $E$, and $F_i$s are assumed to be known constants of appropriate dimensions. Note that the fault matrices $B_4$, $D_4$, and $F_3$ are specified according to the nature of the faults that are considered and to be detected in the system corresponding to the components, actuators, or sensors. For simplicity, the modeling uncertainties are assumed to be recast as disturbances.

**Remark 2.1.** *In practice, malfunctions in the process or in sensors and actuators can cause changes in the model parameters. These circumstances are called multiplicative faults, and they are generally modeled in terms of the system parameter changes as follows:*

$$\dot{x}(t) = (A + \Delta A(t))x(t) + (B_1 + \Delta B_1(t))u(t) + B_2 w(t) + B_3 d(t),$$
$$y(t) = (C + \Delta C(t))x(t) + (D_1 + \Delta D_1(t))u(t) + D_2 w(t) + D_3 d(t),$$

(2.2)

*where $\Delta A(t), \Delta B_1(t), \Delta C(t), \Delta D_1(t)$ denote time-varying multiplicative faults in the plant, actuators, and sensors, respectively. Let us represent*

$$\Delta A(t) = \sum_{i=1}^{l_A} \bar{A}_i \theta_{\bar{A}_i}(t), \quad \Delta B_1(t) = \sum_{i=1}^{l_B} \bar{B}_i \theta_{\bar{B}_i}(t),$$

$$\Delta C(t) = \sum_{i=1}^{l_C} \bar{C}_i \theta_{\bar{C}_i}(t), \quad \Delta D_1(t) = \sum_{i=1}^{l_D} \bar{D}_i \theta_{\bar{D}_i}(t),$$

*where $\bar{A}_i, \ i = 1, \ldots, l_A, \ \bar{B}_i, \ i = 1, \ldots, l_B, \ \bar{C}_i, \ i = 1, \ldots, l_C, \ \bar{D}_i, \ i = 1, \ldots, l_D$ are known and of appropriate dimensions, and $\theta_{\bar{A}_i}(t), \theta_{\bar{B}_i}(t), \theta_{\bar{C}_i}(t), \theta_{\bar{D}_i}(t)$ are unknown time-varying functions. By introducing the following variables and matrices, namely,*

$$B_4 = [\bar{A}_1 \ldots \bar{A}_{l_A}, \bar{B}_1 \ldots \bar{B}_{l_B}], \quad D_4 = [\bar{C}_1 \ldots \bar{C}_{l_C}, \bar{D}_1 \ldots \bar{D}_{l_D}], q_M(x(t), u(t))$$

$$= G_F x(t) + H_F u(t), f_{1M}(x(t), u(t), t) = \Delta_{1F}(t) q_M(x(t), u(t)), f_{2M}(x(t), u(t), t)$$

$$= \Delta_{2F}(t) q_M(x(t), u(t)), G_F = [I_{n \times n} \ldots I_{n \times n}, 0 \ldots 0]^{\mathrm{T}},$$

$$H_F = [0 \ldots 0, I_{n_u \times n_u} \ldots I_{n_u \times n_u}]^{\mathrm{T}},$$

$$\Delta_{1F}(t) = \text{diag}(\theta_{\bar{A}_1}(t) I_{n \times n}, \ldots, \theta_{\bar{A}_{l_A}}(t) I_{n \times n}, \theta_{\bar{B}_1}(t) I_{n_u \times n_u}, \ldots, \theta_{\bar{B}_{l_B}}(t) I_{n_u \times n_u}),$$

$$\Delta_{2F}(t) = \text{diag}(\theta_{\bar{C}_1}(t) I_{n \times n}, \ldots, \theta_{\bar{C}_{l_C}}(t) I_{n \times n}, \theta_{\bar{D}_1}(t) I_{n_u \times n_u}, \ldots, \theta_{\bar{D}_{l_D}}(t) I_{n_u \times n_u}),$$

*it is possible to rewrite the system (2.2) as follows:*

$$\dot{x}(t) = Ax(t) + B_1u(t) + B_2w(t) + B_3d(t) + B_4f_{1M}(x(t), u(t), t),$$
$$y(t) = Cx(t) + D_1u(t) + D_2w(t) + D_3d(t) + D_4f_{2M}(x(t), u(t), t).$$

*Consequently, it now follows that the multiplicative faults are modeled as additive faults. Also for this reason, the major focus of our study in this chapter is on detection of additive faults. However, it should be noted that $f_{1M}(x(t), u(t), t)$ and $f_{2M}(x(t), u(t), t)$ are functions of the state and input variables of the system. For simplicity, they are represented as only functions of t, i.e., $f_{1M}(t)$ and $f_{2M}(t)$. This is a very common practice and notation in the literature [30].*

The following model is proposed for the detector (dynamic observer)/controller (observer-based state feedback) throughout this chapter:

$$F: \begin{cases} \dot{\hat{x}}(t) = A\hat{x}(t) + B_1u(t) + n(t) \\ \hat{y}(t) = C\hat{x}(t) + D_1u(t) \\ u(t) = -K\hat{x}(t) \end{cases}, \tag{2.3}$$

where $n(t) \in \mathbb{R}^n$ denotes the correction signal, the dynamics of which is given by

$$\begin{cases} \dot{x}_d(t) = A_dx_d(t) + B_dr(t) \\ n(t) = C_dx_d(t) + D_dr(t) \\ r(t) = y(t) - \hat{y}(t) \end{cases}, \tag{2.4}$$

where $\hat{x}(t) \in \mathbb{R}^n$ denotes the estimation of $x(t)$, $\hat{y}(t) \in \mathbb{R}^{n_y}$ denotes the observer output, $x_d(t) \in \mathbb{R}^n$ denotes an auxiliary vector, $K \in \mathbb{R}^{n_u \times n}$ denotes the controller gain, $r(t) \in \mathbb{R}^{n_y}$ denotes the residual signal and the constant matrices $A_d$, $B_d$, $C_d$, $D_d$ denote the observer parameters to be designed subsequently.

By substituting the detector/controller (2.3) into the system equations (2.1), results in the following closed-loop system equations:

$$\begin{cases} \dot{\xi}(t) = \overline{A}\xi(t) + \overline{B}_ww(t) + \overline{B}_dd(t) + \overline{B}_ff(t) \\ r(t) = \overline{C}_1\xi(t) + D_2w(t) + D_3d(t) + D_4f(t) \\ z(t) = \overline{C}_2\xi(t) + F_2d(t) + F_3f(t) \end{cases}, \tag{2.5}$$

where

$$\overline{A} = \begin{bmatrix} A - B_1K & D_dC & C_d \\ 0 & A - D_dC & -C_d \\ 0 & B_dC & A_d \end{bmatrix}, \quad \xi = \begin{bmatrix} \hat{x}^T(t) & e^T(t) & x_d^T(t) \end{bmatrix}^T,$$

$$\overline{B}_d = \begin{bmatrix} D_dD_3 \\ B_3 - D_dD_3 \\ B_dD_3 \end{bmatrix}, \quad \overline{B}_f = \begin{bmatrix} D_dD_4 \\ B_4 - D_dD_4 \\ B_dD_4 \end{bmatrix}, \quad \overline{B}_w = \begin{bmatrix} D_dD_2 \\ B_2 - D_dD_2 \\ B_dD_2 \end{bmatrix}, \tag{2.6}$$

$$\overline{C}_1 = \begin{bmatrix} 0 & C & 0 \end{bmatrix}, \quad \overline{C}_2 = \begin{bmatrix} E - F_1K & E & 0 \end{bmatrix}, \quad e(t) = x(t) - \hat{x}(t).$$

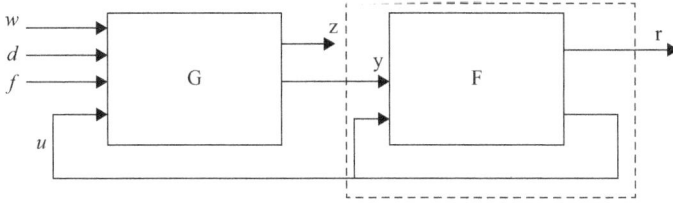

*Figure 2.1   The block diagram of the IFDC problem*

The block diagram of the IFDC problem is depicted in Figure 2.1.

In the next subsection, the IFDC design problem to be addressed in this chapter will be transformed into a mixed $H_2/H_\infty$ optimization problem.

## 2.2.2   Problem formulation

The IFDC problem to be addressed in this chapter can be stated as follows.

**The IFDC problem**:
Given the system (2.1), design a detector/controller (2.3) such that the closed-loop system (2.5) is stable, the effects of disturbance and noise on regulated output $z(t)$ and the residual output $r(t)$ are minimized, and the effects of fault on $z(t)$ are minimized, while the effects of the fault on the residual output $r(t)$ are maximized.

More specifically, we are to find a filter such that the closed-loop system is stable and the following conditions are satisfied:

$$
\begin{array}{lll}
\text{(i)} & ||G_{zd}(s)||_\infty < \gamma_1, & \text{(iv)} \ ||G_{rd}(s)||_\infty < \gamma_4, \\
\text{(ii)} & ||G_{zw}(s)||_2 < \gamma_2, & \text{(v)} \ ||G_{rw}(s)||_2 < \gamma_5, \\
\text{(iii)} & ||G_{zf}(s)||_\infty < \gamma_3, & \text{(vi)} \ ||G_{rf}(s)||_- < \gamma_6,
\end{array}
\tag{2.7}
$$

where

$$
\begin{aligned}
G_{zd}(s) &= \overline{C}_2(sI - \overline{A})^{-1}\overline{B}_d + F_2 \\
G_{zf}(s) &= \overline{C}_2(sI - \overline{A})^{-1}\overline{B}_f + F_3 \\
G_{zw}(s) &= \overline{C}_2(sI - \overline{A})^{-1}\overline{B}_w \\
G_{rd}(s) &= \overline{C}_1(sI - \overline{A})^{-1}\overline{B}_d + D_3 \\
G_{rw}(s) &= \overline{C}_1(sI - \overline{A})^{-1}\overline{B}_w \\
G_{rf}(s) &= \overline{C}_1(sI - \overline{A})^{-1}\overline{B}_f + D_4
\end{aligned}
\tag{2.8}
$$

Note that, the matrix $D_2$ is excluded from the transfer matrix $G_{rw}$ for the same reasons that are stated in [250].

For simplicity, the condition (vi) is replaced by a standard $H_\infty$ model matching problem as follows [251]:

$$
||W_f - G_{rf}||_\infty < \gamma_6.
\tag{2.9}
$$

The above condition implies that the residual signal $r(t)$ robustly tracks a filtered version of the fault signals, $W_f f$, with $W_f$ appropriately to be chosen. Assume that one selects $W_f$ having the following form:

$$W_f = \left[ \begin{array}{c|c} A_F & B_F \\ \hline C_F & D_F \end{array} \right], \tag{2.10}$$

where $A_F$ is a Hurwitz matrix. Then,

$$W_f - G_{rf}(s) = \tilde{C}(sI - \tilde{A})^{-1}\tilde{B} + \tilde{D}, \tag{2.11}$$

where

$$\left[ \begin{array}{cc} \tilde{A} & \tilde{B} \\ \tilde{C} & \tilde{D} \end{array} \right] = \left[ \begin{array}{c|cccc|c} A_F & & 0 & & & B_F \\ \hline & A - B_1 K & D_d C & C_d & & D_d D_4 \\ 0 & 0 & A - D_d C & -C_d & B_4 - D_d D_4 \\ & 0 & B_d C & A_d & B_d D_4 \\ \hline C_F & 0 & -C & 0 & D_F - D_4 \end{array} \right]. \tag{2.12}$$

## 2.3　Integrated fault detection and control problem

There are six performance indices (i)–(vi) that must be satisfied simultaneously for solving the IFDC problem. At first, each performance index will be transformed into an LMI feasibility condition in Theorems 2.1–2.6. Then, in Corollary 2.1, a feasible solution to the IFDC problem is obtained by considering Theorems 2.1–2.6 simultaneously.

First, the design objective (i) is transformed into LMI feasibility constraints in the following theorem.

**Theorem 2.1.** *The closed-loop system* (2.5) *is stable and the condition*

$$\|G_{zd}(s)\|_\infty < \gamma_1, \tag{2.13}$$

*holds if there exist symmetric positive-definite matrices* $Q_{11}$, $P_{11}$, $X$, *matrices* $A_k$, $B_k$, $C_k$, $D_k$, $M$, *and a prescribed positive constant* $\gamma_1$, *such that the following inequalities are satisfied:*

$$\left[ \begin{array}{cc} Q_{11} & I \\ I & P_{11} \end{array} \right] > 0, \tag{2.14}$$

$$\left[ \begin{array}{ccccc} E_{11} & E_{12} & E_{13} & D_k D_3 & XE^\mathrm{T} - M^\mathrm{T} F_1^\mathrm{T} \\ * & E_{22} & E_{23} & B_3 - D_k D_3 & Q_{11} E^\mathrm{T} \\ * & * & E_{33} & P_{11} B_3 + B_k D_3 & E^\mathrm{T} \\ * & * & * & -\gamma_1^2 I & F_2^\mathrm{T} \\ * & * & * & * & -I \end{array} \right] < 0, \tag{2.15}$$

*where*

$$E_{11} = AX - B_1M + XA^T - M^TB_1^T,$$

$$E_{12} = -C_k, \ E_{13} = D_kC,$$

$$E_{22} = AQ_{11} + C_k + C_k^T + Q_{11}A^T, \tag{2.16}$$

$$E_{23} = A + A_k^T - D_kC,$$

$$E_{33} = P_{11}A + B_kC + C^TB_k^T + A^TP_{11},$$

*The control gain K and the dynamic observer parameters $A_d$, $B_d$, $C_d$, $D_d$ are given by*

$$A_d = P_{12}^{-1}\left(A_k - P_{11}(A - D_dC)Q_{11} - P_{12}B_dCQ_{11} + P_{11}C_dQ_{12}^T\right)(Q_{12}^T)^{-1},$$

$$B_d = P_{12}^{-1}(B_k + P_{11}D_d), \ K = MX^{-1}, \tag{2.17}$$

$$C_d = -(C_k + D_dCQ_{11})(Q_{12}^T)^{-1}, \ D_d = D_k,$$

*where $P_{12}$ and $Q_{12}$ are invertible matrices satisfying the following condition:*

$$P_{12}Q_{12}^T = I - P_{11}Q_{11}. \tag{2.18}$$

*Proof.* First, note that by applying the bounded-real lemma, the condition (2.13) is satisfied if and only if the following inequality holds:

$$\begin{bmatrix} \overline{A}^TP + P\overline{A} & P\overline{B}_d & \overline{C}_2^T \\ * & -\gamma_1^2I & F_2^T \\ * & * & -I \end{bmatrix} < 0, \tag{2.19}$$

where it is assumed that $P$ has the following structure:

$$P = \begin{bmatrix} P_{1(n\times n)} & 0 \\ 0 & P_{2(2n\times 2n)} \end{bmatrix}, \ P_2 = \begin{bmatrix} P_{11} & P_{12} \\ P_{12}^T & P_{22} \end{bmatrix}. \tag{2.20}$$

Using the structure as defined for $P$ in (2.20), the condition (2.19) can be rewritten as

$$\Omega = \begin{bmatrix} \Omega_{11} & \Omega_{12} & \Omega_{13} & \Omega_{14} & \Omega_{15} \\ * & \Omega_{22} & \Omega_{23} & \Omega_{24} & \Omega_{25} \\ * & * & \Omega_{33} & \Omega_{34} & \Omega_{35} \\ * & * & * & \Omega_{44} & \Omega_{45} \\ * & * & * & * & \Omega_{55} \end{bmatrix} < 0, \tag{2.21}$$

where

$$\Omega_{11} = P_1(A - B_1 K) + (A - B_1 K)^{\mathrm{T}} P_1, \ \Omega_{12} = P_1 D_d C, \ \Omega_{13} = P_1 C_d,$$

$$\Omega_{14} = P_1 D_d D_3, \ \Omega_{15} = E - F_1 K, \ \Omega_{25} = E, \ \Omega_{35} = 0, \ \Omega_{55} = -I,$$

$$\begin{bmatrix} \Omega_{22} & \Omega_{23} \\ * & \Omega_{33} \end{bmatrix} = P_2 \begin{bmatrix} A - D_d C & -C_d \\ B_d C & A_d \end{bmatrix} + \begin{bmatrix} A - D_d C & -C_d \\ B_d C & A_d \end{bmatrix}^{\mathrm{T}} P_2, \quad (2.22)$$

$$\begin{bmatrix} \Omega_{24} \\ \Omega_{34} \end{bmatrix} = P_2 \begin{bmatrix} B_3 - D_d D_3 \\ B_d D_3 \end{bmatrix}, \ \Omega_{44} = -\gamma_1^2 I, \Omega_{45} = F_2^{\mathrm{T}}.$$

Let us suppose that $X = P_1^{-1}$ and $Q = P_2^{-1}$, and $Q$ is partitioned according to

$$Q = \begin{bmatrix} Q_{11} & Q_{12} \\ Q_{12}^{\mathrm{T}} & Q_{22} \end{bmatrix}. \quad (2.23)$$

Let us define the matrices $\Pi_1$ and $\Pi_2$ as follows:

$$\Pi_1 = \begin{bmatrix} Q_{11} & I \\ Q_{12}^{\mathrm{T}} & 0 \end{bmatrix}, \ \Pi_2 = P_2 \Pi_1 = \begin{bmatrix} I & P_{11} \\ 0 & P_{12}^{\mathrm{T}} \end{bmatrix}. \quad (2.24)$$

Now, pre- and postmultiplication of (2.21) by $\mathrm{diag}(X, \Pi_1^{\mathrm{T}}, I, I)$ and $\mathrm{diag}(X, \Pi_1, I, I)$, respectively, yields in

$$\begin{bmatrix} \Sigma & -C_k & D_k C & D_k D_3 & XE^{\mathrm{T}} - M^{\mathrm{T}} F_1^{\mathrm{T}} \\ * & \mathrm{Herm}(AQ_{11} + C_k) & A - D_k C + A_k^{\mathrm{T}} & B_3 - D_k D_3 & Q_{11} E^{\mathrm{T}} \\ * & * & \mathrm{Herm}(P_{11}A + B_k C) & P_{11} B_3 + B_k D_3 & E^{\mathrm{T}} \\ * & * & * & -\gamma_1^2 I & F_2^{\mathrm{T}} \\ * & * & * & * & -I \end{bmatrix} < 0,$$

$$(2.25)$$

where $\Sigma = \mathrm{Herm}(AX - B_1 M)$ and

$$A_k = P_{11}(A - D_d C)Q_{11} + P_{12} B_d C Q_{11} - P_{11} C_d Q_{12}^{\mathrm{T}} + P_{12} A_d Q_{12}^{\mathrm{T}},$$

$$B_k = -P_{11} D_d + P_{12} B_d, \ C_k = -D_d C Q_{11} - C_d Q_{12}^{\mathrm{T}}, \quad (2.26)$$

$$D_k = D_d, \ M = KX.$$

Note that $X > 0$ and $P_2 > 0$. Using the definition of $\Pi_1$ and $\Pi_2$ in (2.24), $P_2 > 0$ is equivalent to

$$\Pi_1^{\mathrm{T}} P_2 \Pi_1 = \Pi_2^{\mathrm{T}} \Pi_1 = \begin{bmatrix} Q_{11} & I \\ I & P_{11} \end{bmatrix} > 0. \quad (2.27)$$

This completes the proof of the theorem. □

The LMI constraints for the condition (iv) are given by the following theorem.

**Theorem 2.2.** *The closed-loop system* (2.5) *is stable and guarantees the performance index* (iv) *if there exist symmetric positive-definite matrices* $Q_{11}$, $P_{11}$, $X$, *matrices* $A_k$, $B_k$, $C_k$, $D_k$, $M$, *and a prescribed positive constant* $\gamma_4$, *such that the following inequalities are satisfied:*

$$\begin{bmatrix} Q_{11} & I \\ I & P_{11} \end{bmatrix} > 0, \tag{2.28}$$

$$\begin{bmatrix} E_{11} & E_{12} & E_{13} & D_k D_3 & 0 \\ * & E_{22} & E_{23} & B_3 - D_k D_3 & Q_{11} C^{\mathrm{T}} \\ * & * & E_{33} & P_{11} B_3 + B_k D_3 & C^{\mathrm{T}} \\ * & * & * & -\gamma_4^2 I & D_3^{\mathrm{T}} \\ * & * & * & * & -I \end{bmatrix} < 0, \tag{2.29}$$

*where* $E_{11}$, $E_{12}$, $E_{13}$, $E_{22}$, $E_{23}$, $E_{33}$, *are defined in* (2.16).

   *The control gain* $K$ *and the dynamic observer parameters* $A_d$, $B_d$, $C_d$, $D_d$ *are obtained from* (2.17).

*Proof.* The proof of this theorem is similar to that of Theorem 2.1, so it is omitted for the sake of brevity.   □

   The following theorem gives the LMI constraints for the condition (iii).

**Theorem 2.3.** *The closed-loop system* (2.5) *is stable and guarantees the performance index* (iii) *if there exist symmetric positive-definite matrices* $Q_{11}$, $P_{11}$, $X$, *matrices* $A_k$, $B_k$, $C_k$, $D_k$, $M$, *and a prescribed positive constant* $\gamma_3$, *such that the following inequalities are satisfied:*

$$\begin{bmatrix} Q_{11} & I \\ I & P_{11} \end{bmatrix} > 0, \tag{2.30}$$

$$\begin{bmatrix} E_{11} & E_{12} & E_{13} & D_k D_4 & XE^{\mathrm{T}} - M^{\mathrm{T}} F_1^{\mathrm{T}} \\ * & E_{22} & E_{23} & B_4 - D_k D_4 & Q_{11} E^{\mathrm{T}} \\ * & * & E_{33} & P_{11} B_4 + B_k D_4 & E^{\mathrm{T}} \\ * & * & * & -\gamma_3^2 I & F_3^{\mathrm{T}} \\ * & * & * & * & -I \end{bmatrix} < 0, \tag{2.31}$$

*where* $E_{11}$, $E_{12}$, $E_{13}$, $E_{22}$, $E_{23}$, $E_{33}$ *are defined in* (2.16).

   *The control gain* $K$ *and the dynamic observer parameters* $A_d$, $B_d$, $C_d$, $D_d$ *are obtained from* (2.17).

*Proof.* The proof of this theorem is similar to that of Theorem 2.1, so it is omitted for the sake of brevity.   □

   The LMI constraints for the condition (vi) are given by the following theorem.

**Theorem 2.4.** *The closed-loop system* (2.5) *is stable and guarantees the performance index (vi) if there exist symmetric positive-definite matrices* $Q_{11}$, $P_{11}$, $X$, $P_F$, *matrices* $A_k$, $B_k$, $C_k$, $D_k$, $M$, *and a prescribed positive constant* $\gamma_6$, *such that the following inequalities are satisfied:*

$$\begin{bmatrix} Q_{11} & I \\ I & P_{11} \end{bmatrix} > 0, \tag{2.32}$$

$$\begin{bmatrix} A_F^T P_F + P_F A_F & 0 & 0 & 0 & P_F B_F & C_F^T \\ * & E_{11} & E_{12} & E_{13} & D_k D_4 & 0 \\ * & * & E_{22} & E_{23} & B_4 - D_k D_4 & -Q_{11} C^T \\ * & * & * & E_{33} & P_{11} B_4 + B_k D_4 & -C^T \\ * & * & * & * & -\gamma_6^2 I & D_F^T - D_4^T \\ * & * & * & * & * & -I \end{bmatrix} < 0, \tag{2.33}$$

*where* $E_{11}$, $E_{12}$, $E_{13}$, $E_{22}$, $E_{23}$, $E_{33}$ *are defined in* (2.16).
    *The filter gains* $A_d$, $B_d$, $C_d$, $D_d$, *and the controller gain* $K$ *are obtained from* (2.17).

*Proof.* The proof follows by employing the same steps as in the proof of Theorem 2.1. Hence, the detailed procedure is omitted here.                    □

The LMI constraints for the condition (ii) are presented in the following theorem.

**Theorem 2.5.** *The closed-loop system* (2.5) *is stable and the condition:*

$$||G_{zw}(s)||_2 < \gamma_2, \tag{2.34}$$

*holds if there exist symmetric positive-definite matrices* $Q_{11}$, $P_{11}$, $X$, *matrices* $A_k$, $B_k$, $C_k$, $D_k$, $M$, $Z$, *and a prescribed positive constant* $\gamma_2$, *such that the following inequalities are satisfied:*

$$\begin{bmatrix} E_{11} & E_{12} & E_{13} & D_k D_2 \\ * & E_{22} & E_{23} & B_2 - D_k D_2 \\ * & * & E_{33} & P_{11} B_2 + B_k D_2 \\ * & * & * & -\gamma_2 I \end{bmatrix} < 0, \tag{2.35}$$

$$\begin{bmatrix} X & 0 & 0 & XE^T - M^T F_1^T \\ * & Q_{11} & I & Q_{11} E^T \\ * & * & P_{11} & E^T \\ * & * & * & Z \end{bmatrix} < 0, \tag{2.36}$$

    $trace(Z) < \gamma_2,$

*where* $E_{11}$, $E_{12}$, $E_{13}$, $E_{22}$, $E_{23}$, $E_{33}$ *are defined in* (2.16).
    *The filter gains* $A_d$, $B_d$, $C_d$, $D_d$, *and the controller gain* $K$ *are obtained from* (2.17).

*Proof.* First, the condition (2.34) is transformed into the following inequalities by using (1.15) in Lemma 1.6 as follows:

$$\begin{bmatrix} \overline{A}^T P + P\overline{A} & P\overline{B}_w \\ * & -\gamma_2 I \end{bmatrix} < 0, \tag{2.37}$$

$$\begin{bmatrix} P & \overline{C}_2^T \\ * & Z \end{bmatrix} > 0, \tag{2.38}$$

$$trace(Z) < \gamma_2.$$

The inequality (2.37) is nonconvex because of the nonlinear terms $P\overline{A}$ and $P\overline{B}_w$. Therefore, the matrix $P$ is partitioned as in (2.20), and $P_2^{-1}$, $\Pi_1$ and $\Pi_2$ are defined the same as in (2.23) and (2.24), respectively. With pre- and post-multiplying the inequality (2.38) by diag$(X, \Pi_1^T, I)$ and diag$(X, \Pi_1, I)$, respectively, the inequality (2.36) is obtained. Similarly, the LMI condition (2.35) is derived from (2.37) by pre- and post-multiplication by diag$(X, \Pi_1^T, I)$ and diag$(X, \Pi_1, I)$, respectively.   □

The LMI constraints for the condition (v) are presented by the following theorem.

**Theorem 2.6.** *The closed-loop system* (2.5) *is stable and guarantees the performance index* (v) *if there exist symmetric positive-definite matrices* $Q_{11}$, $P_{11}$, $X$, *matrices* $A_k$, $B_k$, $C_k$, $D_k$, $M$, $Z$, *and a prescribed positive constant* $\gamma_5$, *such that the following inequalities are satisfied:*

$$\begin{bmatrix} E_{11} & E_{12} & E_{13} & D_k D_2 \\ * & E_{22} & E_{23} & B_2 - D_k D_2 \\ * & * & E_{33} & P_{11}B_2 + B_k D_2 \\ * & * & * & -\gamma_5 I \end{bmatrix} < 0, \tag{2.39}$$

$$\begin{bmatrix} X & 0 & 0 & 0 \\ * & Q_{11} & I & Q_{11}C^T \\ * & * & P_{11} & C^T \\ * & * & * & Z \end{bmatrix} < 0, \tag{2.40}$$

$$trace(Z) < \gamma_5,$$

*where* $E_{11}$, $E_{12}$, $E_{13}$, $E_{22}$, $E_{23}$, $E_{33}$ *are defined in* (2.16).
*The filter gains* $A_d$, $B_d$, $C_d$, $D_d$, *and the controller gain K are obtained from* (2.17).

*Proof.* The proof follows by employing the same steps as in the proof of Theorem 2.5. Hence, the detailed procedure is omitted here.   □

At this point, all the control and fault detection objectives given in (2.7) have been transformed into LMI feasibility constraints. The next corollary unifies the above theorems and provides a procedure for solving the IFDC problem.

**Corollary 2.1.** *Given $\gamma_1$, $\gamma_2$, $\gamma_3$, $\gamma_4$, $\gamma_5$, a feasible solution to the IFDC problem is obtained by solving a sequence of convex optimization problems:*

$$\min_{X,P_{11},Q_{11},P_F,A_k,B_k,C_k,D_k,M,Z} \gamma_6 \tag{2.41}$$

$$s.t.(2.14),(2.15),(2.29),(2.31),(2.33),(2.35),(2.36),(2.39),(2.40)$$

*Proof.* The proof follows in view of the previous theorems (Theorems 2.1–2.6).  □

**Remark 2.2.** *For more perspective in applying dynamic observers in the IFDC design, we compare its structure with the static observer-based IFDC design. Consider, the static observer detector/state feedback controller for the system (2.1) as follows:*

$$\begin{cases} \dot{\hat{x}}(t) = A\hat{x}(t) + B_1 u(t) + L(y(t) - \hat{y}(t)) \\ \hat{y}(t) = C\hat{x}(t) + D_1 u(t) \\ r(t) = y(t) - \hat{y}(t) \\ u(t) = -K\hat{x}(t) \end{cases}, \tag{2.42}$$

*In view of (2.1) and (2.42), the closed-loop system dynamics can be derived as follows:*

$$\begin{cases} \dot{\zeta}(t) = \overline{A}\zeta(t) + \overline{B}_w w(t) + \overline{B}_d d(t) + \overline{B}_f f(t) \\ r(t) = \overline{C}_1 \zeta(t) + D_2 w(t) + D_3 d(t) + D_4 f(t) , \\ z(t) = \overline{C}_2 \zeta(t) + F_2 d(t) + F_3 f(t) \end{cases} \tag{2.43}$$

*where*

$$\overline{A} = \begin{bmatrix} A - B_1 K & B_1 K \\ 0 & A - LC \end{bmatrix}, \quad \zeta = \begin{bmatrix} x^{\mathrm{T}}(t) & e^{\mathrm{T}}(t) \end{bmatrix}^{\mathrm{T}},$$

$$\overline{B}_d = \begin{bmatrix} B_3 \\ B_3 - LD_3 \end{bmatrix}, \quad \overline{B}_f = \begin{bmatrix} B_4 \\ B_4 - LD_4 \end{bmatrix}, \quad \overline{B}_w = \begin{bmatrix} B_2 \\ B_2 - LD_2 \end{bmatrix},$$

$$\overline{C}_1 = \begin{bmatrix} 0 & C \end{bmatrix}, \quad \overline{C}_2 = \begin{bmatrix} E - F_1 K & F_1 K \end{bmatrix}, \quad e(t) = x(t) - \hat{x}(t).$$

*For comparison, consider a static observer-based IFDC for the performance index (i).*

*The closed-loop system (2.43) is stable and satisfy the performance index (i), if there exist positive definite symmetric matrices $P_1$, $P_2$ and matrices $K$ and $N$ such that the following inequality is satisfied:*

$$\begin{bmatrix} \mathrm{Herm}(P_1(A - B_1 K)) & P_1 B_1 K & P_1 B_3 & E^{\mathrm{T}} - K^{\mathrm{T}} F_1^{\mathrm{T}} \\ * & \mathrm{Herm}(P_2 A - NC) & P_2 B_3 - ND_3 & K^{\mathrm{T}} F_1^{\mathrm{T}} \\ * & * & -\gamma_1^2 I & F_2^{\mathrm{T}} \\ * & * & * & -I \end{bmatrix} < 0,$$

$$N = P_2 L, \tag{2.44}$$

*where (2.44) is derived from a procedure similar to the procedure given in the proof of Theorem 2.1. Note that (2.44) is not a strict LMI because of the nonlinear term $P_1 B_1 K$. In this situation, the IFDC design problem can be solved in one step using the equality constraint [252]. It should be pointed out that the equality constraint is applicable only if $F_1 = 0$ in (2.44). However, if $F_1 \neq 0$, then the generically two step procedure [41] can be considered, whereas by using a dynamic observer, one does not face such problems by introducing an extra auxiliary dynamics with the new state variable $x_d$. In fact, the advantage of using this new auxiliary state variable is that one can apply more degrees of freedom by employing $A_d$, $B_d$, $C_d$, and $D_d$ in the closed-loop system dynamics, and therefore, the conditions for designing the controller and the observer parameters can be presented in terms of strict LMIs conditions.*

### 2.3.1   Residual evaluation criterion

Following the construction of the residuals $r_i(t) \in \mathbb{R}, \forall i \in \{1, \ldots, n_y\}$, the final step in accomplishing the IFDC strategy is to determine the threshold functions $J_{\mathrm{th}_i}$ and the evaluation functions $J_{r_i}(t)$. Various evaluation functions can be considered as presented in [30].

The upper and lower threshold values are selected as $J_{\mathrm{th}_i}^u = \sup_{f=0,d,w} r_i(t)$ and $J_{\mathrm{th}_i}^l = \inf_{f=0,d,w} r_i(t)$, respectively. Based on the selected thresholds and the evaluation function taken as $J_{r_i}(t) = r_i(t)$, the occurrence of a fault can then be detected and isolated in our proposed IFDC strategy by using the following decision logic: if $r_i(t) > J_{\mathrm{th}_i}^u$ or $r_i(t) < J_{\mathrm{th}_i}^l \implies f_i \neq 0$.

## 2.4   Case studies

To illustrate the effectiveness of the proposed method, two numerical examples are provided in this section.

### 2.4.1   Case study 1 (four-tank process)

Consider the four-tank process which is depicted in Figure 2.2. A linearized model of the four-tank process is given by [250]

$$\dot{x} = \begin{bmatrix} \frac{-1}{T_1} & 0 & \frac{A_3}{A_1 T_3} & 0 \\ 0 & \frac{-1}{T_2} & 0 & \frac{A_4}{A_2 T_4} \\ 0 & 0 & \frac{1}{T_3} & 0 \\ 0 & 0 & 0 & \frac{-1}{T_4} \end{bmatrix} x + \begin{bmatrix} \frac{\alpha_1 k_1}{A_1} & 0 \\ 0 & \frac{\alpha_2 k_2}{A_3} \\ 0 & \frac{(1-\alpha_2)k_2}{A_3} \\ \frac{(1-\alpha_1)k_1}{A_4} & 0 \end{bmatrix} (u+f) + \begin{bmatrix} 0 & 0 \\ 0 & 0 \\ \frac{-k_{d1}}{A_3} & 0 \\ 0 & \frac{-k_{d2}}{A_4} \end{bmatrix} d,$$

$$y = \begin{bmatrix} k_c & 0 & 0 & 0 \\ 0 & k_c & 0 & 0 \end{bmatrix} x + w,$$

$$z = \begin{bmatrix} k_c & 0 & 0 & 0 \\ 0 & k_c & 0 & 0 \end{bmatrix} x,$$

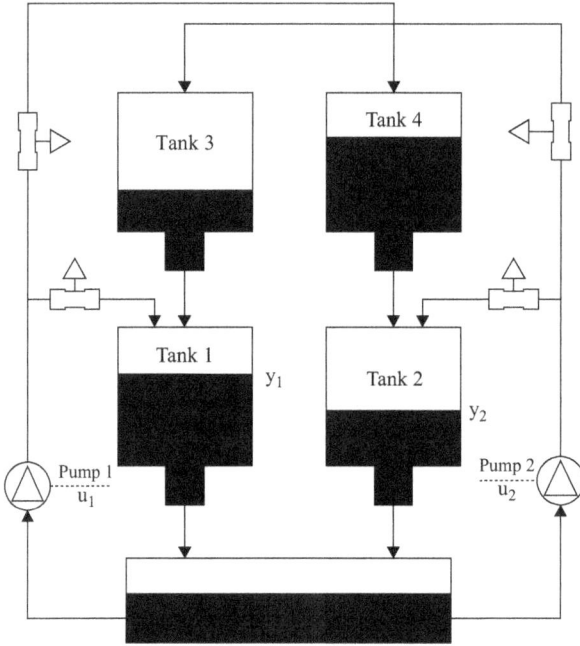

*Figure 2.2   Schematic of the four-tank process*

where $x$ denotes the level of water in the tanks, $u = (u_1, u_2)^T$ denotes the voltage applied to Pumps 1 and 2, $f = (f_1, f_2)^T$ denotes the actuator fault associated with the Pumps 1 and 2, and $d = (d_1, d_2)^T$ denotes the disturbance representing the flow out of the Tanks 3 and 4. $v = (v_1, v_2)^T$ denotes the measurement noise which is assumed to be a zero-mean white noise process with covariance $0.5I_2$ and $T_i = (A_i/a_i)\sqrt{2h_{0i}/g}$.

The following nominal parameter values are selected: $A_1 = A_3 = 28$ [cm$^2$], $A_2 = A_4 = 32$ [cm$^2$], $a_1 = a_3 = 0.071$ [cm$^2$], $a_2 = a_4 = 0.057$ [cm$^2$], $k_c = 0.5$ [V/cm], $g = 981$ [cm/s$^2$], $k_1 = 3.33$ [cm$^3$/V s], $k_2 = 3.35$ [cm$^3$/V s], $k_{d1} = k_{d2} = 1$ [cm$^3$/V s], $\alpha_1 = 0.7$, $\alpha_2 = 0.6$, $h_{01} = 12.4$, $h_{02} = 12.7$, $h_{03} = 1.8$, and $h_{04} = 1.4$. It is desired to detect the actuator fault $f$ in presence of the disturbance $d$ and the measurement noise $w$.

The fault signal $f_1(t)$ is represented as a rectangular pulse with an amplitude of 1, and which is applied during the time interval 50–100 s.

The reference model parameters for the residuals are selected as follows:

$$A_F = \begin{bmatrix} -3 & 0 \\ 0 & -5 \end{bmatrix}, \quad B_F = \begin{bmatrix} 1 & 1 \\ 0.5 & 0.5 \end{bmatrix},$$

$$C_F = \begin{bmatrix} 0.1 & 1 \\ 0.1 & 1 \end{bmatrix}, \quad D_F = \begin{bmatrix} 0 & 0 \\ 0 & 0 \end{bmatrix}.$$

For a given $\gamma_1 = 0.7$, $\gamma_2 = 1.5$, $\gamma_3 = 0.7$, $\gamma_4 = 0.7$, $\gamma_5 = 1.5$, the optimization problem (2.41) was solved and $\gamma_6$ is obtained as 0.87. The regulated output $z(t)$ of

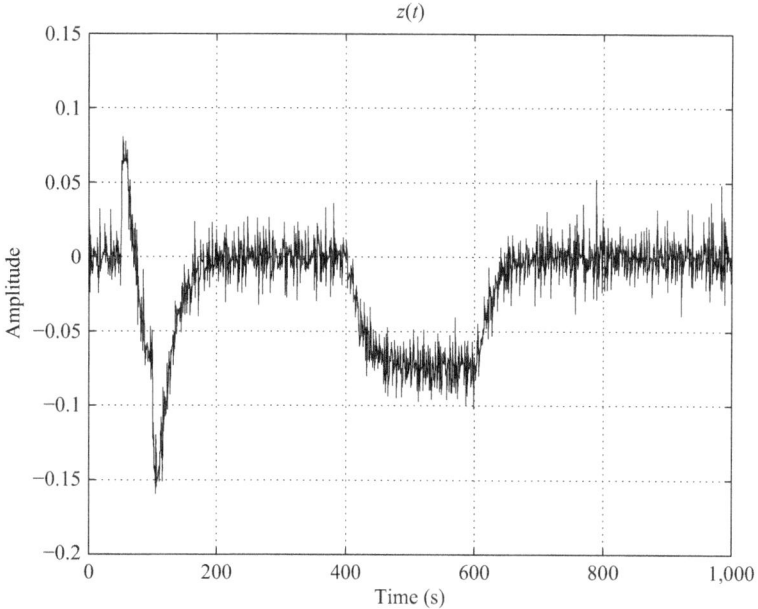

*Figure 2.3    The performance output z(t)*

the closed-loop system is shown in Figure 2.3. From Figure 2.3, it can be concluded that the effects of disturbance $d(t)$, noise $w(t)$, and fault $f(t)$ on the regulated output $z(t)$ have been attenuated. The residual signal is shown in Figure 2.4, where $|r(t)|$ is adopted instead of $r(t)$. From Figure 2.4, it can be seen that the robustness against disturbance and noise, and the fault sensitivity are both enhanced, and the faults are well discriminated from the disturbance and noise. Hence, by using a threshold test, the fault $f(t)$ can be effectively detected.

These results demonstrate the better performance of the proposed technique in comparison with the study that was performed in [250]. To be specific, by comparing Figure 2.4 with Figs. 4–6 in [250], it readily follows that the ratio of fault sensitivity to disturbance attenuation in our proposed methodology is more than the results that are provided in [250]. Consequently, it is expected that the false alarm rates caused by noise, uncertainties, and disturbances on the one hand, and the missing detection rates on the other are respectively decreased and increased by applying our methodology in comparison with the method proposed in [250]. This is the result of sensitivity of the residual signal to the fault signals that was ignored in [250], however which was nevertheless incorporated in our method. Moreover, in comparison with the methodology in [250] that only guarantees the fault detection objective, the IFDC approach in this chapter, in addition to stabilizing the closed-loop system, also simultaneously guarantees that one accomplishes the fault detection and control objectives.

On the other hand, if one employs the observer-based controller structure that is proposed in [252] in our IFDC design, the problem becomes infeasible. As stated

$$|r(t)|$$

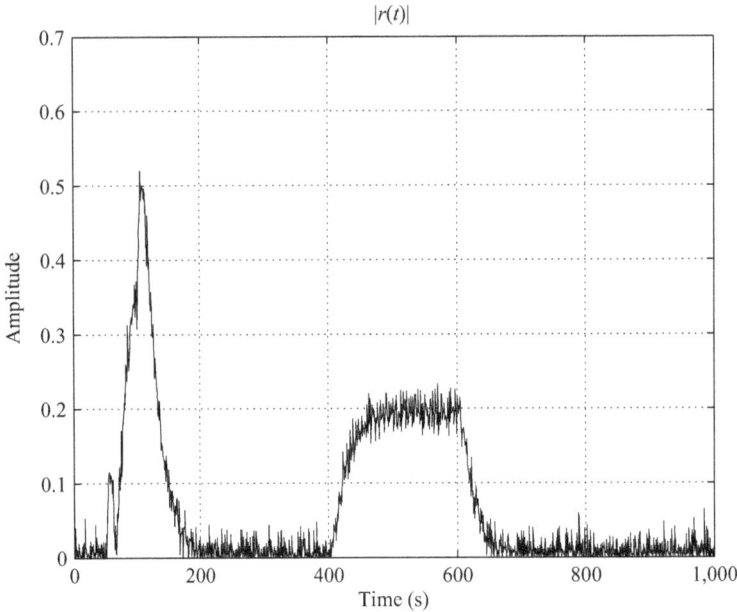

*Figure 2.4    The residual signal $|r(t)|$*

in Remark 2.2, the level of conservativeness that is imposed to the problem by the equality constraint is the reason of this infeasibility.

## 2.4.2    Case study 2 [autonomous underwater vehicle (AUV)]

The oceans occupy approximately 71% of the earth surface with significant unknown and unexplored regions. Therefore, various studies and developments about the oceans such as marine environment, submarine earthquake, ocean life, marine resources research, among others are actively been carried out. Collection of ocean data through survey and observations in actual sea is indispensable for these studies and developments. Due to the fact that oceans have low transparency and one cannot observe the entire deep sea in detail from the surface, survey and observations with ships are not adequate. On the other hand, due to water pressure, one cannot step into the deep sea easily because 1 atm is increasing in every 10 m of diving [253]. Hence, various underwater apparatuses such as manned submersibles and unmanned underwater vehicles (UUVs) have been developed as tools to do survey and observations in the deep sea.

UUVs represent as various types of underwater robots that are operated with minimum or without human operator intervention. In the literature, the phrase is used to describe both a remotely operated vehicle (ROV) and an AUV. ROVs are teleoperated robots that are deployed primarily for underwater installation, inspection, and repair tasks. They have been used extensively in offshore industries due to their advantages over human divers in terms of higher safety, greater depths, longer endurance, and less demand for support equipment. In its operation, the ROV receives instructions from an operator onboard a surface ship (or other mooring platforms) through tethered cable or acoustic link. In order to be useful, the umbilical[1] cable is several hundred meters to several kilometers in length. The combined effect produces a large drag load on the vehicle. ROVs are typically used as underwater work platforms for robotic arms, welding tools, cutters, and related tools. Hence, they are designed to be maneuverable and stable. However, all of these limitations make ROVs unsuitable for survey work where sensors must sweep large areas of ocean floors.

AUVs on the other hand operate without the need of constant monitoring and supervision from a human operator. Consequently, the vehicle does not have the limiting factor in its operation range due to the umbilical cable typically associated with the ROVs. This enables AUVs to be used for certain types of missions such as long-range oceanographic data collection where the use of ROVs are deemed impractical [254].

As AUVs become more common, and in particular as missions become more challenging and take place in less well-known environments, improving the reliability of these vehicles becomes more crucial [255]. Therefore, design of fault diagnosis and fault-tolerant control systems become one of the critical technologies in the field of AUV research.

In [256], the authors have developed a robust model-based fault diagnosis scheme for an UUV using sliding mode-observers. In [257], using the extended Kalman filters, the diagnosis of actuator faults in the AUV Roby 2 is investigated. In [258], the principle component analysis method and Hotelling's $T^2$-test is applied to detect and isolate faults in AUVs. In [259], a nonlinear observer-based fault diagnosis approach is proposed for AUVs.

This observer has a linear component which is available from a linear model of the AUV and a nonlinear component which is a radial basis function neural network to reduce the uncertainties. This network is trained off-line and the training method is a SVM-based method. Wavelet neural networks are also used for detecting and isolating the thruster faults of an AUV in [260].

Among the various fault diagnosis methods that use analytical redundancy, the distinction between model-free and model-based approaches is as follows. Model-free methods are well-suited for large-scale systems, where the development of a model is too complex or expensive. On the other hand, lumped parameter model of an underwater vehicle can be easily described by a small set of well-known equations with highly uncertain parameters. Therefore, it becomes of great significance to design a robust model-based fault diagnosis scheme for AUVs.

To illustrate the effectiveness of our proposed IFDC strategy, it is applied to a linear model of Subzero III AUV [261]. We first describe the kinematic and dynamic equations of motion of an AUV system. We then describe the linearized model of the

---

[1]An umbilical is a cable which connects between the Surface Control Unit and the ROV. For free swimming applications, a tough, flexible, polyurethane sheathed umbilical is used. The umbilical contains power conductors to the ROV as well as control signals and video conductors. Spare conductors are provided for accessories such as sonar, survey sensors, and tools.

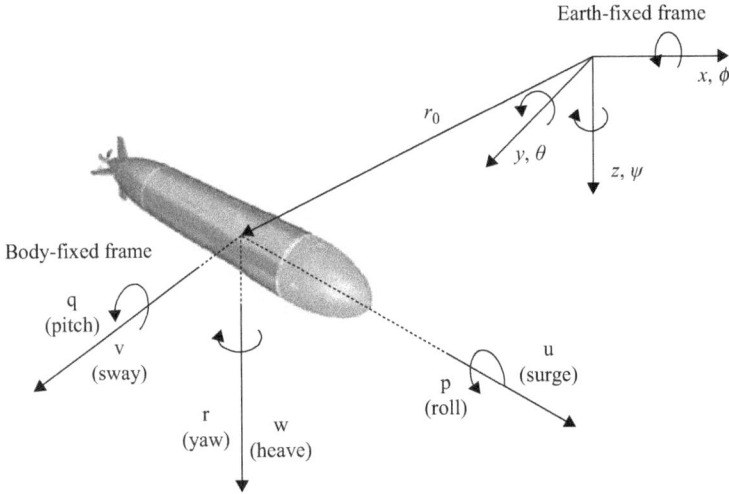

*Figure 2.5   Body-fixed and inertial reference frames*

Subzero III AUV [261]. Finally, we formulate the IFDC problem for the linearized model of the AUV.

### 2.4.2.1   AUV modeling

This section describes the kinematic and dynamic equations of motion of an AUV. The dynamic model of an AUV is developed through Newton–Euler formulation by using conservation laws of linear and angular momentum. The equations of motion of such vehicles are highly nonlinear [262] and coupled due to the hydrodynamic forces which act on the vehicle. The equations of motion of an underwater vehicle in six degrees of freedom with respect to the body-fixed frame (as shown in Figure 2.5) can be written as [262]:

$$M\dot{v} + C(v)v + D(v)v + g(\eta) = \tau, \tag{2.45}$$

where $v = [u, v, w, p, q, r]^T$ denotes the vector of linear and angular velocities in the vehicle coordinate frame and $\eta = [x, y, z, \phi, \theta, \psi]^T$ denotes the vector of absolute positions and orientations (Euler angles) in the inertial frame, $M = M_{RB} + M_A$ and $C(v) = C_{RB}(v) + C_A(v)$ are the system inertia matrix (including added mass) and the Coriolis–centripetal matrix (including added mass), respectively, $D(v)$ denotes the resulting matrix of linear and quadratic drag, $g(\eta)$ denotes the vector of gravitational/ buoyancy forces and moments, and $\tau = [X, Y, Z, K, M, N]^T$ denotes the vector of forces and moments acting on the vehicle in the body-fixed frame. The position of the AUV in the global coordinate frame can be described by [262]:

$$\dot{\eta} = J(\eta)v, \tag{2.46}$$

where $J(\eta)$ denotes the velocity transformation matrix between the body fixed frame and the inertial frame.

### 2.4.2.2 The linearized model

According to [263], and for simplicity, the complete nonlinear model of an AUV can be divided into three subsystems. These noninteracting subsystems are (i) the surge speed control, (ii) the steering (or heading) control, and (iii) the diving (or depth) control.

Note that, the control objective which is derived by the particular choice of the subsystem equations is that the steering system will be responsible for control of the heading errors; the diving system will be responsible for the depth and pitch errors; and the speed system will control the speed commands [263].

For the Subzero III AUV, the state, control input, and output vectors are defined as [261] $x := [u, v, w, p, q, r, \phi, \theta, \psi, z, n, \delta r, \delta s]^T$, $u =: [m_d, \delta r_d, \delta s_d]^T$, and $y =: [u, \psi, z]^T$, respectively; where $n$ in round per second (rps) denotes the rotational speed of the propeller, $\delta r$ and $\delta r_d$ denote the actual and the demanded deflection of the rudder in degrees, $\delta s$ and $\delta s_d$ denote the actual and the demanded deflection of the stern-planes in degrees, $m_d$ denotes the command to adjust the duty cycle ($m_d/2{,}500$) of the pulse width modulation drive to the DC motor which actuates the propeller.

The equilibrium point corresponding to the cruising condition is $u_0 = [522, 0, 0]^T$, $x_0 = [1.30, 0, 0, 0, 0, 0, 0, 0, 0, z_0, 13.67, 0, 0]^T$, and $y_0 = [1.30, 0, z_0]^T$. Due to the slight interactions among the surge motion, heave/pitch motion and the yaw motion for an underwater vehicle, three SISO subsystems (speed, heading, and depth) can be developed for the Subzero III AUV as follows [261].

For the speed subsystem, the state, the control input and the output vectors are selected as $x = [u, n]^T$, $u = m_d$ and $y = u$, respectively. The transfer function from $m_d$ to $u$ is given by

$$T(s) = \frac{0.002864}{(s + 0.439)(s + 4.195)}. \tag{2.47}$$

For the heading subsystem, the state, the control input, and the output vectors are selected as $x = [v, r, \psi, \delta r]^T$, $u = \delta r_d$ and $y = \psi$, respectively, and the transfer function from $\delta r_d$ to $\psi$ is given by

$$T(s) = -\frac{92.94(s + 1.044)}{s(s + 9.121)(s + 7.69)(s + 1.702)}. \tag{2.48}$$

For the depth subsystem, the state, the control input, and the output vectors are selected as $x = [w, q, \theta, z, \delta s]^T$, $u = \delta s_d$, and $y = z$, respectively, and the transfer function from $\delta s_d$ to $z$ is given by

$$T(s) = \frac{-0.3298(s - 5.409)(s + 1.755)}{s(s + 9.063)(s + 11.53)(s + 1.688)(s + 0.08082)}. \tag{2.49}$$

Clearly we have a nonminimum phase system due to the presence of a zero in the open right-half plane.

### 2.4.2.3   Simulation results

It should be noted that the following assumptions are considered for the simulations conducted below. The disturbance $d(t)$ that is injected to the system is band-limited white noise with a power of 0.0005. The fault signal $f(t)$ is simulated as a rectangular pulsed signal with an amplitude of 0.5 that is present from 5 to 5.5 s. Moreover, it is assumed that $w(t) = 0$.

The reference model matrices for the residual analysis are selected as follows:

$$A_F = \begin{bmatrix} -3 & 0 \\ 0 & -5 \end{bmatrix}, \quad B_F = \begin{bmatrix} 1 \\ 0.5 \end{bmatrix},$$

$$C_F = \begin{bmatrix} 0.5 & 1 \end{bmatrix}, \quad D_F = \begin{bmatrix} 0.5 \end{bmatrix}.$$

In the following, we demonstrate the performance of the proposed approach through considering different types of faults in speed, heading, and depth subsystems of the Subzero III AUV (in terms of the location of their occurrence as being in the actuator, the sensor or the component). It is worth noting that by considering the worst case analysis of the residuals corresponding to the healthy operation of the subsystems being subjected to various disturbances, the threshold values $J_{th}^u$ and $J_{th}^l$ are selected.

**Occurrence of a component fault in the speed subsystem**

Consider the AUV model that is governed by (2.1) with the matrices of the Subzero III speed subsystem (2.47), and the fault, disturbance, and the regulated output matrices that are given by $B_4 = B_3 = B_1, B_2 = 0, D_2 = 0, D_3 = 0.5, D_4 = 1,$ $E = C, F_1 = 0.5, F_2 = F_3 = 0.2$. For a given $\gamma_1 = \gamma_3 = \gamma_4 = 0.55, \gamma_2 = \gamma_5 = 0$, the optimization problem (2.41) was solved and $\gamma_6$ is obtained as 0.401.

The residual signal is shown in Figure 2.6(a), where the robustness can be seen against the disturbance and the fault sensitivity is enhanced, and the fault is satisfactorily discriminated from the disturbance. The threshold values of $J_{th}^u$ and $J_{th}^l$ are selected as $J_{th}^u = 0.125$ and $J_{th}^l = -0.075$, respectively. Note that by using our threshold test, the fault $f(t)$ is effectively detected at time $t = 5.01$ s. The regulated output $z(t)$ of the closed-loop system is shown in Figure 2.6(b), and it can be concluded that the effects of the disturbance $d(t)$ and the fault $f(t)$ on the regulated output $z(t)$ have been attenuated.

**Occurrence of a sensor fault in the heading subsystem**

Now, consider the AUV model that is (2.1) with the matrices of the Subzero III heading subsystem (2.48), the fault matrices $B_4 = 0$ and $D_4 = 1$ for the sensor faults and $B_2 = 0, B_3 = 0, D_2 = 0, D_3 = 0.5, E = C, F_1 = F_2 = F_3 = 0.5$. For a given $\gamma_1 = \gamma_3 = \gamma_4 = 0.9, \gamma_2 = \gamma_5 = 0$, the optimization problem (2.41) was solved and $\gamma_6$ is obtained as 0.77. From Figure 2.7(a), it can be seen that the fault $f(t)$ is readily distinguished from the disturbance $d(t)$. This illustrates that the proposed IFDC method satisfies the fault detection requirement. The threshold values of $J_{th}^u$ and $J_{th}^l$ are selected as $J_{th}^u = 0.18$ and $J_{th}^l = -0.09$. Note that by using our threshold test, the fault $f(t)$ is effectively detected at time $t = 5.012$ s.

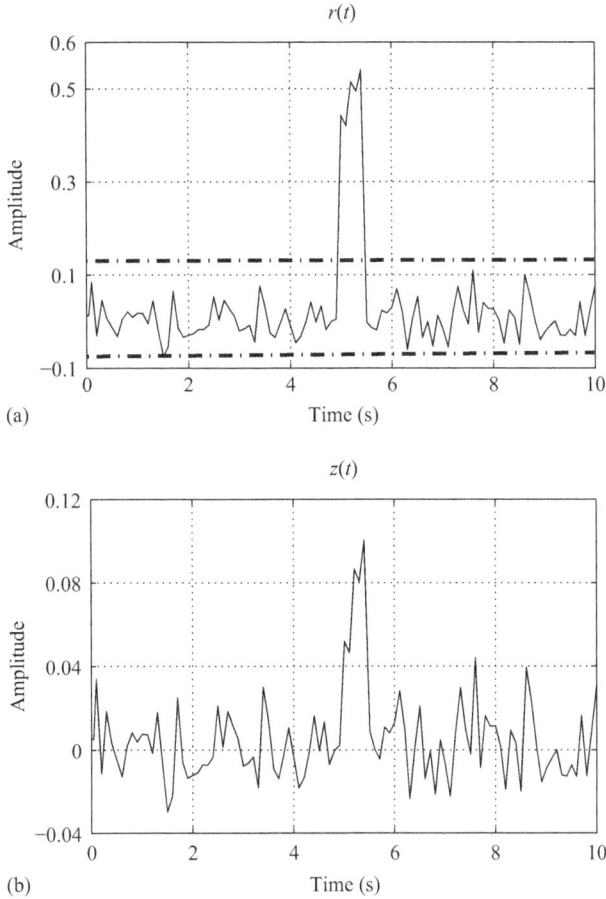

*Figure 2.6* (a) *The residual signal r(t) (the solid line is the residual signal and the dash-dot lines are the residual thresholds); (b) the regulated output z(t)*

From Figure 2.7(b), it can be concluded that the effects of the disturbance $d(t)$ and the fault $f(t)$ on the regulated output $z(t)$ have also been attenuated.

**Occurrence of an actuator fault in the depth subsystem**

Now consider the AUV model (2.1) with the matrices of Subzero III depth subsystem (2.49), and the fault matrices $B_4 = B_1$ and $D_4 = D_1$ for the actuator faults and $B_2 = 0, B_3 = B_1, D_2 = D_3 = D_1 = 0, E = 0.1C, F_1 = F_2 = F_3 = 0$. Note that $B_3 = B_1$ and $D_3 = D_1$ as this is common in industrial applications where disturbances enter the system by corrupting the input signal.

For a given $\gamma_1 = 0.85$, $\gamma_3 = 0.9$, $\gamma_4 = 0.8$, $\gamma_2 = \gamma_5 = 0$, the optimization problem (2.41) was solved and $\gamma_6$ is obtained as 0.328.

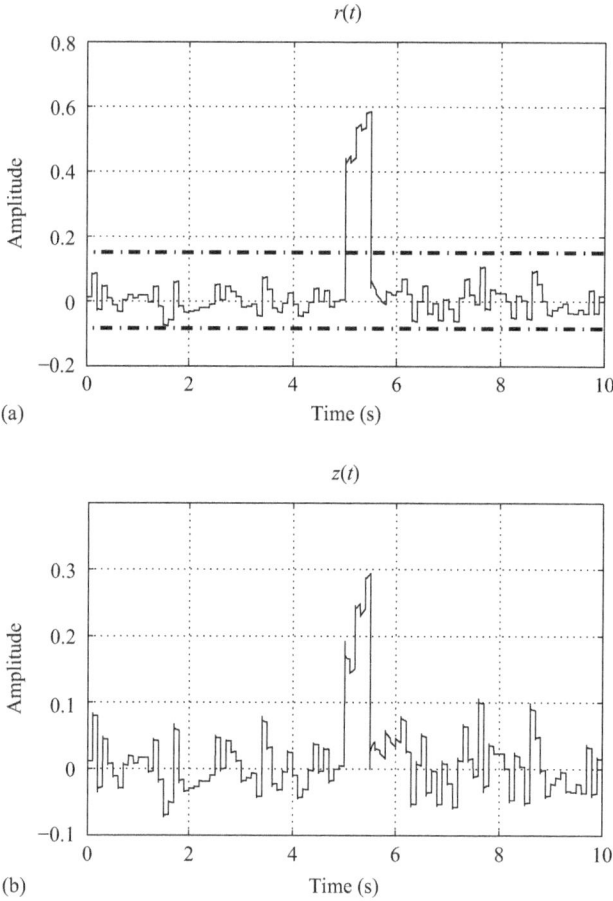

*Figure 2.7* *(a) The residual signal r(t) (the solid line is the residual signal and the dash-dot lines are the residual thresholds); (b) the regulated output z(t)*

The residual signal is shown in Figure 2.8(a) where the robustness against the disturbance and the enhanced fault sensitivity can be seen. The threshold values of $J_{th}^u$ and $J_{th}^l$ are selected as $J_{th}^u = 6e - 5$ and $J_{th}^l = -4e - 5$, respectively. Note that by using our threshold test, the fault $f(t)$ is effectively detected at time $t = 5.484$ s. The regulated output $z(t)$ of the closed-loop system is shown in Figure 2.8(b), where it can be concluded that the effects of disturbance $d(t)$ and fault $f(t)$ have been attenuated.

**Remark 2.3.** *Note that there are no constraints on our methodology in terms of lower bounds on the feasible FDI times. This FDI performance is commonly measured after conducting multiple simulation runs. Therefore, it is possible to qualitatively compare*

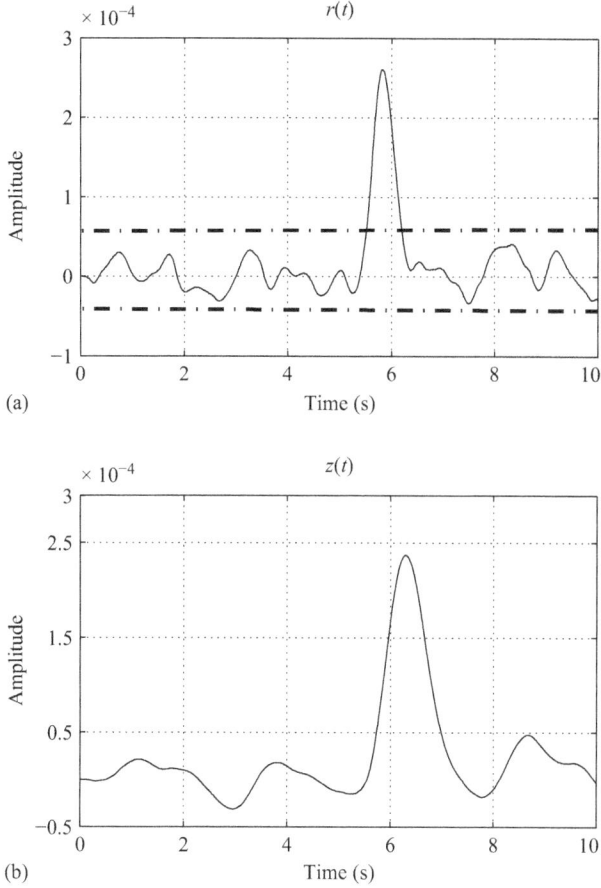

*Figure 2.8* *(a) The residual signal r(t) (the solid line is the residual signal and the dash-dot lines are the residual thresholds); (b) the regulated output z(t)*

our strategy with other available strategies from this perspective. Specifically, let us consider another standard strategy as reported in [30] for the residual evaluation function $J_{r_i}(L)$ and the threshold $J_{th_i}$ as follows:

$$J_{r_i}(L) = ||r_i(t)||_{2,L} = \left( \int_{l_0}^{l_0+L} r_i(v)^{\mathrm{T}} r_i(v) dv \right)^{1/2},$$

$$J_{th_i} = \sup_{f=0,u,d,w} J_{r_i}(L),$$

(2.50)

where $l_0$ denotes the initial evaluation time instant and $L$ denotes the evaluation time window. Note that the length of the time window is finite. Based on the above equation,

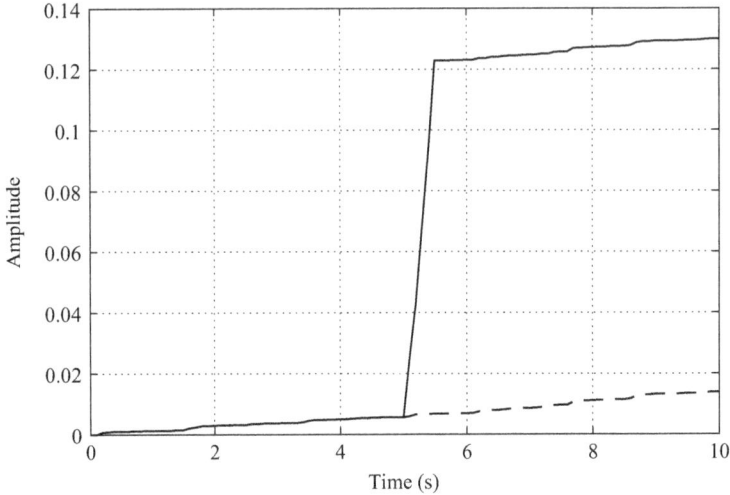

*Figure 2.9    Residual evaluation function for the speed subsystem; the solid line denotes the residual evaluation signal and the dash line denotes the threshold signal*

*the occurrence of faults can be detected by invoking the following logic rule*

$$J_{r_i}(L) > J_{th_i} \implies Faults \implies Alarm,$$

$$J_{r_i}(L) \leq J_{th_i} \implies No\ faults.$$

(2.51)

*Since the norm of the residual signal is used as an evaluation function in the above logic, the number of false alarms that may occur could be less than those that are obtained by using our methodology (where the evaluation function is chosen as $J_{r_i}(t) = r_i(t)$).*

*However, the feasible FDI times by using our methodology are generally less than the above methodology. Evaluation of residual evaluation function for the considered case studies above (especially, speed, heading, and depth subsystems) are demonstrated in Figures 2.9, 2.10, and 2.11, respectively.*

*The analysis of the residual evaluation functions in Figures 2.9–2.11 reveals that the fault $f(t)$ can be detected in each scenario as follows:*

- ***For the speed subsystem:*** *fault detection is achieved in 0.029 s after its occurrence, whereas by using our approach, this fault is detected in 0.01 s after its occurrence.*
- ***For the heading subsystem:*** *fault detection is achieved in 0.057 s after its occurrence, whereas by using our approach, this fault is detected in 0.012 s after its occurrence.*
- ***For the depth subsystem:*** *fault detection is achieved in 0.73 s after its occurrence, whereas by using our approach, this fault is detected in 0.484 s after its occurrence.*

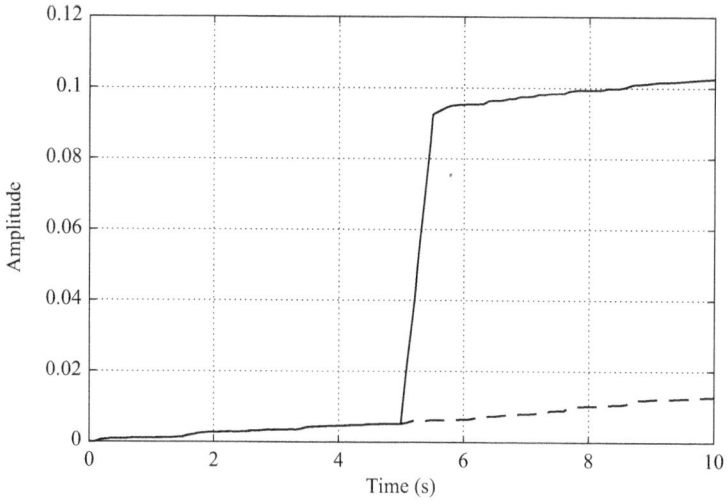

*Figure 2.10   Residual evaluation function for the heading subsystem; the solid line denotes the residual evaluation signal and the dash line denotes the threshold signal*

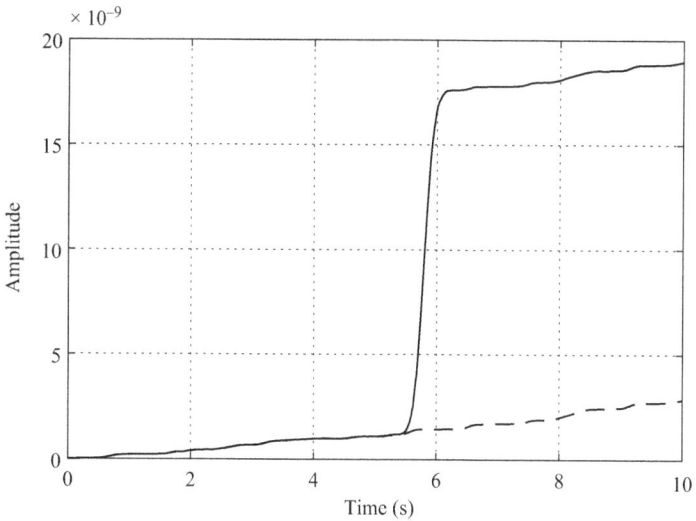

*Figure 2.11   Residual evaluation function for the depth subsystem; the solid line denotes the residual evaluation signal and the dash line denotes the threshold signal*

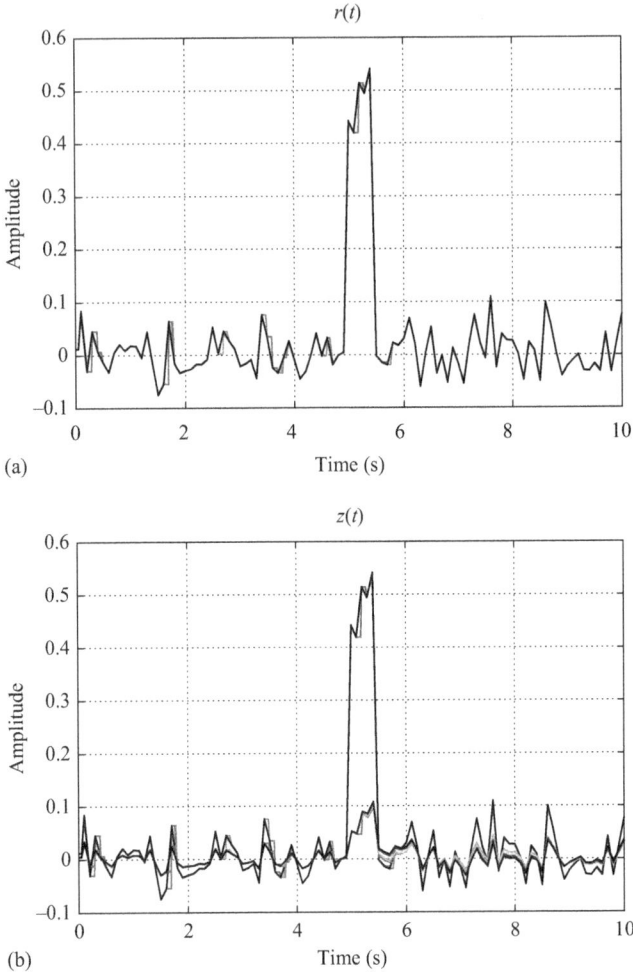

*Figure 2.12   The IFDC results for the speed subsystem: (a) the residual signal r(t), and (b) the regulated output z(t)*

*Furthermore, from Figures 2.9–2.11, it is observed that there is a significant gap between the residual evaluation functions and the thresholds, which reduces the possibility of occurrence of false alarms in comparison with our proposed methodology. Therefore, in general, there is a trade-off between decreasing the number of false alarms and increasing the FDI detection times.*

To apply the proposed IFDC methodology to the AUV, the parameter values of the AUV model must be known. However, in practice, their actual values are different from their nominal values. Therefore, the designed IFDC module must be robust against the uncertainty in the system parameters in order to maintain its nominal performance.

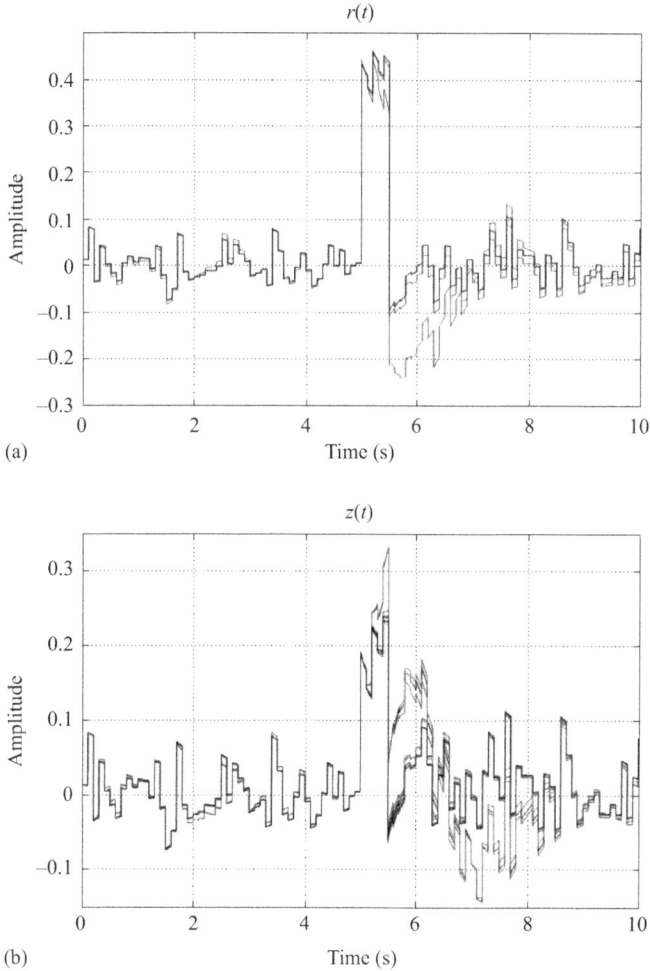

*Figure 2.13   The IFDC results for the heading subsystem: (a) the residual signal r(t) and (b) the regulated output z(t)*

On the other hand, our proposed methodology has intrinsic robustness with respect to parametric uncertainties. Indeed, we expect that the proposed IFDC design methodology should show some degrees of robustness against the parametric uncertainties.

To examine the robustness of the proposed scheme, it is assumed that the dynamic parameters of the AUV are uncertain but bounded. Indeed, the IFDC module is designed by using *only* the nominal parameters of the AUV ($p^*$), while the actual values of the parameters are randomly selected in the bound $[p^* - \varepsilon p^*, p^* + \varepsilon p^*]$. To be more specific, the following parametric uncertainties are considered:

- **For the speed subsystem:** It is assumed that the entries $A[1, 1]$ and $B_1[1, 1]$ are uncertain with $\varepsilon = 0.8$.

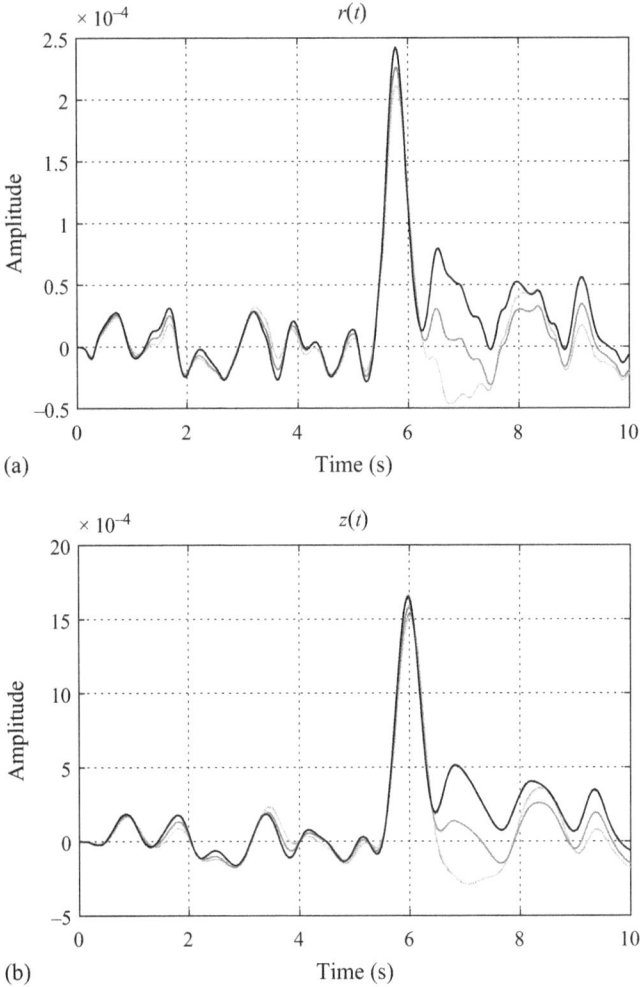

*Figure 2.14   The IFDC results for the depth subsystem: (a) the residual signal r(t) and (b) the regulated output z(t)*

- **For the heading subsystem:** It is assumed that the entries $A[3,2]$ and $B_1[1,1]$ are uncertain with $\varepsilon = 0.2$.
- **For the depth subsystem:** It is assumed that the entries $A[2,1]$ and $B_1[1,1]$ are uncertain with $\varepsilon = 0.2$.

The simulation results for these three subsystems are shown in Figures 2.12, 2.13 and 2.14, respectively. In each case, simulations are run for nine sets of the parameters. It can be concluded that the proposed IFDC methodology is quite robust against parametric uncertainties. Indeed, from observing the residual signals, it readily follows that the robustness against disturbances and the fault sensitivity have been achieved, and the fault is satisfactorily discriminated from the disturbances. Moreover, from the regulated outputs it follows that the effects of disturbances and the fault on the regulated outputs have been attenuated.

## 2.5 Conclusion

In this chapter, we have considered a mixed $H_2/H_\infty$ formulation of the IFDC problem using dynamic observer detector and state feedback controller. Dynamic observer was proposed to overcome the disadvantages of other filters in the design of the IFDC. In the proposed IFDC methodology, the proposed strategy enjoys major advantages as opposed to the previous results where the developed conditions have been obtained in term of the LMIs. Two numerical examples have been provided to demonstrate the effectiveness and capabilities of the proposed approach.

*Chapter 3*

# A single dynamic observer-based module for design of integrated fault detection, isolation, and tracking control scheme

In this chapter, the problem of integrated fault detection, isolation, and tracking (IFDIT) control is addressed for linear systems subject to both bounded energy and bounded peak disturbances. A dynamic observer is proposed and implemented by using the $H_\infty/H_-/L_1$ formulation of the IFDIT problem. A single dynamic observer module is designed that generates the residuals as well as the control signals.

The objective of the IFDIT module is to ensure that simultaneously the effects of disturbances and control signals on the residual signals are minimized (in order to accomplish the fault detection goal) subject to the constraint that the transfer matrix from the faults to the residuals is equal to a preassigned diagonal transfer matrix (in order to accomplish the fault isolation goal), while the effects of disturbances, reference inputs, and faults on the specified control outputs are minimized (in order to accomplish the fault-tolerant and tracking control goals).

A set of linear matrix inequality (LMI) feasibility conditions are derived to ensure solvability of the problem. We employ the extended LMI approach to reduce the conservativeness in the problem solution by introducing additional matrix variables to eliminate the couplings of the Lyapunov matrices with the system matrices. In order to illustrate and demonstrate the effectiveness of our proposed design methodology, the developed and proposed schemes are applied to an autonomous underwater vehicle (AUV). The work presented in this chapter has partially appeared in [264,265].

This chapter is organized as follows. We begin with a brief literature review. In Section 3.2, we formulate the IFDIT problem for linear systems subject to both bounded energy and bounded peak external disturbances. In Section 3.3, a systematic LMI approach for solving the proposed IFDIT problem and generating a set of residual and control signals is presented. In Section 3.3.1, the residual evaluation and threshold functions that are required for detection and isolation of the occurred faults in the system are discussed. In Section 3.4, the proposed approach is applied to the problem of integrated faults detection and isolation, and tracking control of the Subzero II AUV.

## 3.1 Introduction

The current literature in the domain of IFDC suffers from the following drawbacks and limitations:

1. First, most considered problems of IFDC in the literature can achieve only the "regulation" control objective and none consider the "tracking" control problem in the IFDC design.

2. Second, although most current IFDC schemes can achieve acceptable fault detection goal [36,38–45,48], they cannot accomplish and perform the fault isolation task. It should be noted that the fault isolation objective in most of the current literature on FDI is accomplished through the use of a bank of observers (see, e.g., [266–270]). For implementation of these methodologies, one requires to have a bank of observers where each observer's dimension is usually equal to the dimension of the entire system, and this puts a heavy computational burden on the developed FDI system. Therefore, reducing the dimension or the number of observers is of great significance and importance.

3. Third, a common assumption in almost all IFDC research is that the considered system is driven by bounded energy disturbances. However, in practical applications, the considered system is disturbed by unknown persistent signals bounded in magnitude. The $L_1$ performance index, which minimizes the worst case peak-to-peak gain of the system, provides a suitable framework for characterizing the effects of such signals [271].

As pointed out above, none of the work in the literature is capable of simultaneously accomplishing detecting and isolating faults as well as tracking control of the closed-loop system in presence of both bounded energy and bounded peak disturbances. Therefore, in this chapter, we propose an $H_\infty/H_-/L_1$ formulation of the IFDIT control problem by employing dynamic observer detector and state estimated feedback controller for linear systems. It is shown that by applying a single dynamic observer-based module whose order is equal to twice the order of the system, the IFDIT problem can be solved based on strict LMI feasibility conditions.

Consequently, the computational complexity as far as the number and dimension of the required observers are significantly reduced when compared to the existing techniques in the literature. Moreover, our proposed approach can now ensure that not only one will be able to detect and isolate a fault but also will be capable of tracking a given reference signal in presence of both bounded energy and bounded peak disturbances. Moreover, our developed scheme can also handle isolation of simultaneous faults.

To summarize, the main contributions of this work in comparison with other methodologies in the FD, FDI, and IFDC domains are stated in Table 3.1.

Finally, it should be pointed out that the IFDIT problem has a multiobjective framework since detection, isolation, and tracking control objectives need to be satisfied simultaneously. A multiobjective IFDIT problem that constraints the Lyapunov matrices that are involved to be the same leads to a conservative solution. Over the past few years, extensive effort has been dedicated to reducing the conservatism of

*Table 3.1   Comparison of the IFDIT scheme with other methodologies*

| Approach | Detection | Isolation | | Control | Tracking | TCD | | NO |
|---|---|---|---|---|---|---|---|---|
| | | SF | MF | | | BE | BP | |
| *FD* [27,28] | Yes | No | No | No | No | Yes | No | 1 |
| *FDI (SO)* [272,273] | Yes | Yes | Yes | No | No | Yes | No | 1 |
| *FDI (BGO)* [268,270,274] | Yes | Yes | No | No | No | Yes | No | $n_f$ |
| *FDI (BDO)* [269,272,274] | Yes | Yes | Yes | No | No | Yes | No | $n_f$ |
| *IFDC* [36,38–45] | Yes | No | No | Yes | No | Yes | No | 1 |
| *IFDIT* | Yes | Yes | Yes | Yes | Yes | Yes | Yes | 1 |

SO denotes the "single observer," BGO denotes the "bank of general observers," BDO denotes the "bank of dedicated observers," NO denotes the "number of observers," SF denotes the "single fault," MF denotes the "multiple fault," TCD denotes the "type of considered disturbances," BF denotes the "bounded energy," BP denotes the "bounded peak," and $n_f$ denotes the number of faults in the system.

the solutions through utilizing multiobjective, performance analysis, and robust stability. The work of [275,276] have shown how one can employ an *extended LMI* characterization of the problem to drastically reduce the conservatism in the obtained solutions. This is achieved by introducing additional matrix variables that eliminate the Lyapunov variables couplings with the original system matrices.

The extended LMI characterization was considered in [277] for ensuring stability and performance of linear systems. Motivated by this work, LMI feasibility conditions for solving our IFDIT problem are developed in this chapter such that no products of Lyapunov matrices with the system matrices are present. This will ensure to reduce the conservatism in obtaining a solution to our multiobjective problem. In order to illustrate and demonstrate the effectiveness of our proposed IFDIT strategy, the methodology is applied to a linear model of an AUV.

## 3.2   System description and problem formulation

In this section, we first describe the system under investigation and the corresponding IFDIT module governing equations. The IFDIT problem is then formulated for linear continuous-time systems.

### 3.2.1   System description

Consider the following linear time-invariant system:

$$\dot{x}(t) = Ax(t) + B_1u(t) + B_2d(t) + B_3w(t) + B_4f(t), \; x(0) = x_0,$$
$$y(t) = Cx(t) + D_1u(t) + D_2d(t) + D_3w(t) + D_4f(t),$$

(3.1)

where $x(t) \in \mathbb{R}^n$ denotes the state vector, $u(t) \in \mathbb{R}^{n_u}$ denotes the control input, $y(t) \in \mathbb{R}^{n_y}$ denotes the measured output, $d(t) \in \mathbb{R}^{n_d}$ denotes an $\mathbb{L}_2$-norm bounded external disturbance, $w(t) \in \mathbb{R}^{n_w}$ denotes an $\mathbb{L}_\infty$ disturbance signal satisfying $||w||_\infty \leq 1$, and the unknown input $f(t) \in \mathbb{R}^{n_f}$ denotes a possible additive fault input signal.

The matrices $A \in \mathbb{R}^{n \times n}$, $B_1 \in \mathbb{R}^{n \times n_u}$, $B_2 \in \mathbb{R}^{n \times n_d}$, $B_3 \in \mathbb{R}^{n \times n_w}$, $B_4 \in \mathbb{R}^{n \times n_f}$, $C \in \mathbb{R}^{n_y \times n}$, $D_1 \in \mathbb{R}^{n_y \times n_u}$, $D_2 \in \mathbb{R}^{n_y \times n_d}$, $D_3 \in \mathbb{R}^{n_y \times n_w}$, and $D_4 \in \mathbb{R}^{n_y \times n_f}$ are assumed to be known and constant of appropriate dimensions. In this work, it is assumed that we only consider either multiple actuator faults or multiple sensor faults at a time in the system (i.e., we do not consider occurrence of simultaneous actuator and sensor faults).

The reference model for accomplishing the tracking control objective is described as follows:

$$\dot{x}_r(t) = A_r x_r(t) + B_r u_r(t), \ x_r(0) = x_{r0},$$
$$y_r(t) = C_r x_r(t), \tag{3.2}$$

where $x_r(t) \in \mathbb{R}^n$ denotes the desired reference state vector, $u_r(t) \in \mathbb{R}^{n_{ur}}$ denotes the energy bounded reference input vector, and $y_r(t) \in \mathbb{R}^{n_y}$ denotes the desired reference output vector, and $A_r$, $B_r$, and $C_r$ are constant matrices with appropriate dimensions. It is assumed that $A_r$ is Hurwitz and $x_r(t)$ is available for use in the control implementation.

The main problem considered in this chapter is to design a fault detection and isolation filter and an output feedback tracking controller for the system (3.1) that will simultaneously accomplish these goals and objectives. The IFDIT module that is proposed for solving the above problem is specified as follows:

$$\dot{\hat{x}}(t) = A\hat{x}(t) + B_1 u(t) + n(t), \ \hat{x}(0) = \hat{x}_0,$$
$$\hat{y}(t) = C\hat{x}(t) + D_1 u(t),$$
$$u(t) = -K(\hat{x}(t) - x_r(t)),$$
$$r(t) = y(t) - \hat{y}(t), \tag{3.3}$$

where $\hat{x}(t) \in \mathbb{R}^n$ denotes the estimate of $x(t)$, $\hat{y}(t) \in \mathbb{R}^{n_y}$ denotes the observer output, $K \in \mathbb{R}^{n_u \times n}$ denotes the controller gain, $r(t) \in \mathbb{R}^{n_y}$ denotes the residual signal, and $n(t) \in \mathbb{R}^n$ denotes the correction signal, the dynamics of which is given by

$$\dot{x}_d(t) = A_d x_d(t) + B_d r(t), \ x_d(0) = x_{d0},$$
$$n(t) = C_d x_d(t) + D_d r(t), \tag{3.4}$$

with $x_d(t) \in \mathbb{R}^n$ denotes an auxiliary state vector, and the constant matrices $A_d$, $B_d$, $C_d$, $D_d$ denote the observer gains to be selected subsequently.

**Remark 3.1.** *It is assumed that the number of outputs and faults are the same (that is, $n_y = n_f$). In case that $n_y > n_f$, the residual signal is defined as $r(t) = H(y(t) - \hat{y}(t))$, where $H \in \mathbb{R}^{n_f \times n_y}$ is a predefined constant matrix allowing one to formulate and solve the proposed problem. Therefore, even when the number of outputs is not equal, but greater than the number of faults, the IFDIT problem is still solvable.*

By substituting the IFDIT module (3.3) and (3.4) into the system (3.1), and defining the controlled output $z(t) = y(t) - y_r(t)$ (to evaluate if we are achieving the tracking control objective), the following closed-loop system dynamics is obtained:

$$\dot{\xi}(t) = \overline{A}\xi(t) + \overline{B}_d d(t) + \overline{B}_w w(t) + \overline{B}_f f(t) + \overline{B}_r u_r(t),$$
$$r(t) = \overline{C}_1 \xi(t) + D_2 d(t) + D_3 w(t) + D_4 f(t), \qquad (3.5)$$
$$z(t) = \overline{C}_2 \xi(t) + D_2 d(t) + D_3 w(t) + D_4 f(t),$$

where $\xi^{\mathrm{T}}(t) = \begin{bmatrix} x_r^{\mathrm{T}}(t) & \hat{x}^{\mathrm{T}}(t) & e^{\mathrm{T}}(t) & x_d^{\mathrm{T}}(t) \end{bmatrix}$, $e(t) = x(t) - \hat{x}(t)$, $\overline{B}_r^{\mathrm{T}} = \begin{bmatrix} B_r^{\mathrm{T}} & 0 & 0 & 0 \end{bmatrix}$, $\overline{C}_1 = \begin{bmatrix} 0 & 0 & C & 0 \end{bmatrix}$, $\overline{C}_2 = \begin{bmatrix} D_1 K - C_r & C - D_1 K & C & 0 \end{bmatrix}$, and

$$\overline{A} = \begin{bmatrix} A_r & 0 & 0 & 0 \\ B_1 K & A - B_1 K & D_d C & C_d \\ 0 & 0 & A - D_d C & -C_d \\ 0 & 0 & B_d C & A_d \end{bmatrix}, \quad \overline{B}_d = \begin{bmatrix} 0 \\ D_d D_2 \\ B_2 - D_d D_2 \\ B_d D_2 \end{bmatrix},$$

$$\overline{B}_w = \begin{bmatrix} 0 \\ D_d D_3 \\ B_3 - D_d D_3 \\ B_d D_3 \end{bmatrix}, \quad \overline{B}_f = \begin{bmatrix} 0 \\ D_d D_4 \\ B_4 - D_d D_4 \\ B_d D_4 \end{bmatrix}.$$

In the next subsection, the IFDIT design problem is first formally introduced and is then reformulated as an $H_\infty / H_- / L_1$ optimization problem.

### 3.2.2 Problem formulation

The IFDIT problem that is addressed and solved in this chapter can be formally stated as follows:

**The IFDIT problem**: The problem is to design and specify for the system (3.1) the IFDIT module (3.3) and (3.4) such that the following requirements are satisfied simultaneously, namely,

- the augmented system (3.5) is stable,
- the effects of disturbances $(d(t), w(t))$ and reference input $(u_r(t))$ on the residual $r(t)$ are minimized and the effects of faults on the residual $r(t)$ are maximized (in order to accomplish the fault detection task),
- each element of the residual $r(t)$ is only sensitive to a specified potential fault (in order to accomplish the fault isolation task), and
- the effects of the bounded energy and bounded peak disturbances, fault, and reference input on the controlled output $z(t)$ are minimized (in order to accomplish the control performance and tracking tasks).

We now state and reformulate the above problem alternatively as follows. Specifically, our goal is to design the IFDIT block (3.3) such that the augmented system (3.5) is stable and the following optimization problem is solved, namely,

$$\text{minimize} \quad \sum_{i=1}^{8} \beta_i \gamma_i$$

subject to:

$$(\text{I}) \quad ||T_{zd}(s)||_\infty < \gamma_1, \quad (\text{II}) \ ||T_{zf}(s)||_\infty < \gamma_2,$$

$$(\text{III}) \quad ||T_{zu_r}(s)||_\infty < \gamma_3, \quad (\text{IV}) \ ||T_{rd}(s)||_\infty < \gamma_4,$$

$$(\text{V}) \quad ||T_{ru_r}(s)||_\infty < \gamma_5, \quad (\text{VI}) \ ||T_{rf}(s) - J||_\infty < \gamma_6,$$

$$(\text{VII}) \quad ||J||_- \geq 1, \quad\quad\quad (\text{VIII}) \ ||T_{zw}(s)||_1 < \gamma_7,$$

$$(\text{IX}) \quad ||T_{rw}(s)||_1 < \gamma_8,$$

(3.6)

where $J$ is selected as having the structure:

$$J = \text{diag}(j_1,\ldots,j_{n_f}) \in \mathbb{R}^{n_f \times n_f}, j_i > 0, \forall i \in \{1,\ldots,n_f\},$$

(3.7)

which is the simplest choice of $J$ (namely, a fixed diagonal matrix).

Refer to [278] for more details on other choices for $J$. It should be noted that the requirement of $J$ being diagonal is made to ensure that we can accomplish the fault isolability requirement in case when multiple faults do occur simultaneously. The positive constant weights $\beta_1,\ldots,\beta_8$ can be selected by the designer for a trade-off analysis among the performance indices (I)–(IX). Indeed, they can be used to emphasize the relative importance of the performance metrics among each other.

The $H_\infty$ and $L_1$ performance indices (I), (II), (III), and (VIII) are used to attenuate the effects of the bounded energy and bounded peak disturbances, as well as fault and reference inputs on the controlled output $z(t)$. This is a common methodology in the literature on tracking controller design, commonly known as "$H_\infty$ ($H_\infty/L_1$) tracking control problem" [279].

However, by implementing this methodology, one cannot guarantee zero tracking error, and it is only possible to ensure that the tracking objective is satisfied with a bounded tracking error. The $H_\infty$ and $L_1$ performance indices (IV), (V), and (IX) are used to attenuate the effects of disturbances ($d(t)$ and $w(t)$) and reference input on the residual. The performance indices (VI) and (VII) are used to guarantee a minimum level of sensitivity of residuals to the fault signals and to ensure fault isolation.

**Remark 3.2.** *In some approaches available in FD, FDI, and IFDC literature, a common assumption that is made is that $D_4 = 0$ and/or $CB_4$ has full rank (refer, e.g., to [280]). Moreover, in other approaches, the assumption that $D_4$ has a full rank is widely used (refer, e.g., to [48,281]). In this work, none of the above assumptions are imposed on our proposed methodology. However, it should be noted that if $D_4$ is not full column rank, for example, if $D_4 = 0$, this will have an adverse effect on the minimum* *values of* $\gamma_6$, *since* $\gamma_6 > ||T_{rf}(s) - J||_\infty \geq ||T_{rf}(\infty) - J|| = ||D_4 - J|| = ||J|| \geq 1$. *In other words, under the rank deficiency condition of* $D_4$ *the overall performance of our solution, which is measured by the objective function* $\sum_{i=1}^{8} \beta_i \gamma_i$, $i = 1, \ldots, 8$, *will be negatively impacted.*

In the next section, the introduced IFDIT problem is now solved for the closed-loop system (3.5) subject to the conditions that are specified by (3.6).

## 3.3  Main results

A total of nine performance indices, namely, (I)–(IX) must be simultaneously satisfied to solve the IFDIT problem for the closed-loop system (3.5). Note that it is straightforward to express the $H_-$ performance index (VII) as a matrix inequality, namely,

$$I - J < 0. \tag{3.8}$$

In the following discussion, at first in Theorem 3.1, sufficient conditions for the $H_\infty$ performance indices (I)–(VI) are formulated as LMI feasibility problem. Then, sufficient LMI conditions for the $L_1$ performance indices (VIII)–(IX) are obtained in Theorem 3.2. Finally, in Corollary 3.1, a feasible solution to the IFDIT problem is obtained by simultaneously considering all the indices (I)–(IX).

In other works, as in [282], an iterative LMI algorithm that requires a higher computational cost as compared to the standard LMI methods is applied to reduce the overall conservatism of the solutions to the fault diagnosis problems. It should be noted that performance of iterative LMI algorithms is affected by the selected initial conditions, and the schemes are generally not globally convergent.

Furthermore, the IFDC problem was studied in [41] where bilinear matrix inequality conditions are obtained. Consequently, to avoid the above drawbacks and limitations, the projection lemma (Lemma 1.1) is employed in this chapter to reduce the conservativeness of the IFDIT solution by introducing additional matrix variables $(X_1, X_2)$ to avoid the couplings of the Lyapunov matrices with the system state-space matrices. The main results of this chapter are now stated in Theorem 3.1.

**Theorem 3.1.** *For a predefined positive scalar* $\lambda$, *the closed-loop system* (3.5) *is stable and the* $H_\infty$ *performance indices* (I)–(VI) *are guaranteed with performance levels* $\gamma_1, \ldots, \gamma_6$ *if there exist symmetric positive definite matrices* $T_1, \ldots, T_6$ *and matrices* $X_{11}, Y_{11}, Q, S, \tilde{A}, \tilde{B}, \tilde{C}, \tilde{D}, M, J$, *such that the following LMIs hold*

$$\begin{bmatrix} E_{11} & T_1 + E_{12} & E_{13} & E_{14} \\ * & E_{22} & \lambda E_{13} & 0 \\ * & * & -\gamma_1^2 I & D_2^T \\ * & * & * & -I \end{bmatrix} < 0, \qquad \begin{bmatrix} E_{11} & T_2 + E_{12} & \Xi_{13} & E_{14} \\ * & E_{22} & \lambda \Xi_{13} & 0 \\ * & * & -\gamma_2^2 I & D_3^T \\ * & * & * & -I \end{bmatrix} < 0,$$

$$
\begin{bmatrix}
E_{11} & T_3 + E_{12} & \mathcal{E}_{13} & E_{14} \\
* & E_{22} & \lambda\mathcal{E}_{13} & 0 \\
* & * & -\gamma_3^2 I & 0 \\
* & * & * & -I
\end{bmatrix} < 0,
\qquad
\begin{bmatrix}
E_{11} & T_4 + E_{12} & E_{13} & \Xi_{14} \\
* & E_{22} & \lambda E_{13} & 0 \\
* & * & -\gamma_4^2 I & D_2^T \\
* & * & * & -I
\end{bmatrix} < 0,
$$

$$
\begin{bmatrix}
E_{11} & T_5 + E_{12} & \mathcal{E}_{13} & \Xi_{14} \\
* & E_{22} & \lambda\mathcal{E}_{13} & 0 \\
* & * & -\gamma_5^2 I & 0 \\
* & * & * & -I
\end{bmatrix} < 0,
\tag{3.9}
$$

$$
\begin{bmatrix}
E_{11} & T_6 + E_{12} & \Xi_{13} & \Xi_{14} \\
* & E_{22} & \lambda\Xi_{13} & 0 \\
* & * & -\gamma_6^2 I & (D_3 - J)^T \\
* & * & * & -I
\end{bmatrix} < 0,
$$

*where*

$$
E_{11} = Herm(\Omega_1),\ E_{12} = \lambda\Omega_1 - \Omega_2^T, E_{22} = -\lambda Herm(\Omega_2),
$$

$$
\Omega_1 =
\begin{bmatrix}
Q^T A_r^T & M^T B_1^T & 0 & 0 \\
0 & Q^T A^T - M^T B_1^T & 0 & 0 \\
0 & \tilde{C}^T & Y_{11}^T A^T - \tilde{C}^T & \tilde{A}^T \\
0 & C^T \tilde{D}^T & A^T - C^T \tilde{D}^T & A^T X_{11} - C^T \tilde{B}^T
\end{bmatrix},
$$

$$
\Omega_2 =
\begin{bmatrix}
Q^T & 0 & 0 & 0 \\
0 & Q^T & 0 & 0 \\
0 & 0 & Y_{11}^T & S \\
0 & 0 & I & X_{11}
\end{bmatrix},\quad
\Xi_{13} =
\begin{bmatrix}
0 \\
\tilde{D}D_4 \\
B_4 - \tilde{D}D_4 \\
X_{11}^T B_4 - \tilde{B}D_4
\end{bmatrix},\quad
\mathcal{E}_{13} =
\begin{bmatrix}
B_r \\
0 \\
0 \\
0
\end{bmatrix},\tag{3.10}
$$

$$
E_{13} =
\begin{bmatrix}
0 \\
\tilde{D}D_2 \\
B_2 - \tilde{D}D_2 \\
X_{11}^T B_2 - \tilde{B}D_2
\end{bmatrix},\quad
E_{14} =
\begin{bmatrix}
M^T D_1^T - Q^T C_r^T \\
Q^T C^T - M^T D_1^T \\
Y_{11}^T C^T \\
C^T
\end{bmatrix},\quad
\Xi_{14} =
\begin{bmatrix}
0 \\
0 \\
Y_{11}^T C^T \\
C^T
\end{bmatrix}.
$$

*Moreover, the control gain matrix $K$ and the dynamic observer parameters $A_d$, $B_d$, $C_d$, $D_d$ are selected according to*

$$
A_d = (X_{21}^T)^{-1}\left(\tilde{A} - X_{11}^T(A - D_d C)Y_{11} - X_{21}^T B_d C Y_{11} + X_{11}^T C_d Y_{21}\right)(Y_{21})^{-1},
$$

$$
B_d = (X_{21}^T)^{-1}(X_{11}^T D_d - \tilde{B}),\quad K = MQ^{-1},
\tag{3.11}
$$

$$
C_d = (\tilde{C} - D_d C Y_{11})(Y_{21})^{-1},\quad D_d = \tilde{D},
$$

where $X_{21}$ and $Y_{21}$ are invertible matrices that satisfy

$$Y_{21}^{\mathrm{T}}X_{21} = S - Y_{11}^{\mathrm{T}}X_{11}. \tag{3.12}$$

*Proof.* The closed-loop system (3.5) is stable and satisfies the performance index (I), if there exists a Lyapunov candidate function $V(\xi(t)) = \xi^{\mathrm{T}}(t)P_1\xi(t)$, where $P_1$ is a positive definite symmetric matrix, such that

$$\dot{V}(t) \leq -z^{\mathrm{T}}(t)z(t) + \gamma_1^2 d^{\mathrm{T}}(t)d(t). \tag{3.13}$$

From (3.13), the following condition is obtained:

$$\begin{bmatrix} \xi(t) \\ d(t) \end{bmatrix}^{\mathrm{T}} \begin{bmatrix} \overline{A}^{\mathrm{T}}P_1 + P_1\overline{A}^{\mathrm{T}} + \overline{C}_2^{\mathrm{T}}\overline{C}_2 & P_1\overline{B}_{\mathrm{d}} + \overline{C}_2^{\mathrm{T}}D_2 \\ * & D_2^{\mathrm{T}}D_2 - \gamma_1^2 I \end{bmatrix}^{\mathrm{T}} \begin{bmatrix} \xi(t) \\ d(t) \end{bmatrix} < 0. \tag{3.14}$$

The following inequality is sufficient to imply (3.14):

$$\begin{bmatrix} I & 0 \\ \overline{A} & \overline{B}_{\mathrm{d}} \end{bmatrix}^{\mathrm{T}} \begin{bmatrix} 0 & P_1 \\ P_1 & 0 \end{bmatrix} \begin{bmatrix} I & 0 \\ \overline{A} & \overline{B}_{\mathrm{d}} \end{bmatrix} + \begin{bmatrix} 0 & I \\ \overline{C}_2 & D_2 \end{bmatrix}^{\mathrm{T}} \begin{bmatrix} -\gamma_1^2 I & 0 \\ 0 & I \end{bmatrix} \begin{bmatrix} 0 & I \\ \overline{C}_2 & D_2 \end{bmatrix} < 0. \tag{3.15}$$

Note that in the inequality (3.15), the Lyapunov matrices and the system matrices are coupled with one another. When the LMI conditions involve product of Lyapunov matrices with the system matrices, then by selecting identical Lyapunov matrices one would end up with a conservative solution. Consequently, the Projection Lemma is now used to reduce the conservativeness of the IFDIT solution. Specifically, the matrix inequality (3.15) is now reformulated as

$$N_U^{\mathrm{T}}ZN_U < 0, \tag{3.16}$$

where $N_U$ and $Z$ are defined according to

$$Z = \begin{bmatrix} \overline{C}_2^{\mathrm{T}}\overline{C}_2 & P_1 & \overline{C}_2^{\mathrm{T}}D_2 \\ * & 0 & 0 \\ * & * & D_2^{\mathrm{T}}D_2 - \gamma_1^2 I \end{bmatrix}, \quad N_U = \begin{bmatrix} I & 0 \\ \overline{A} & \overline{B}_{\mathrm{d}} \\ 0 & I \end{bmatrix}. \tag{3.17}$$

If we choose $N_V$ in the inequality (1.9b) as follows:

$$N_V = \begin{bmatrix} \lambda I & 0 \\ -I & 0 \\ 0 & I \end{bmatrix} \rightarrow V = \begin{bmatrix} I & \lambda I & 0 \end{bmatrix}, \tag{3.18}$$

and then invoke Lemma 1.1, it can be shown that the inequality (3.16) becomes

$$Z + \begin{bmatrix} \overline{A}^{\mathrm{T}} \\ -I \\ \overline{B}_{\mathrm{d}}^{\mathrm{T}} \end{bmatrix} \begin{bmatrix} X & \lambda X & 0 \end{bmatrix} + \begin{bmatrix} X^{\mathrm{T}} \\ \lambda X^{\mathrm{T}} \\ 0 \end{bmatrix} \begin{bmatrix} \overline{A} & -I & \overline{B}_{\mathrm{d}} \end{bmatrix} < 0. \tag{3.19}$$

In order to transform the matrix inequality (3.19) into an LMI condition, the matrix $X \in \mathbb{R}^{4n \times 4n}$ is partitioned as follows:

$$X_{(4n \times 4n)} = \begin{bmatrix} X_{1(n \times n)} & 0 & 0 \\ 0 & X_{1(n \times n)} & 0 \\ 0 & 0 & X_{2(2n \times 2n)} \end{bmatrix}, \quad X_2 = \begin{bmatrix} X_{11} & X_{12} \\ X_{21} & X_{22} \end{bmatrix}. \tag{3.20}$$

Consequently, by using the Schur complement [283], the following inequality is obtained:

$$\left[ \begin{array}{c|c} \Delta \left| P_1 + \lambda \overline{A}^{\mathrm{T}} \begin{bmatrix} X_1 & 0 & 0 \\ 0 & X_1 & 0 \\ 0 & 0 & X_2 \end{bmatrix} - \begin{bmatrix} X_1^{\mathrm{T}} & 0 & 0 \\ 0 & X_1^{\mathrm{T}} & 0 \\ 0 & 0 & X_2^{\mathrm{T}} \end{bmatrix} \right. & \begin{bmatrix} X_1^{\mathrm{T}} & 0 & 0 \\ 0 & X_1^{\mathrm{T}} & 0 \\ 0 & 0 & X_2^{\mathrm{T}} \end{bmatrix} \overline{B}_{\mathrm{d}} \; \overline{C}_2^{\mathrm{T}} \\ \hline * & \lambda \begin{bmatrix} -X_1 - X_1^{\mathrm{T}} & 0 & 0 \\ 0 & -X_1 - X_1^{\mathrm{T}} & 0 \\ 0 & 0 & -X_2 - X_2^{\mathrm{T}} \end{bmatrix} \; \lambda \begin{bmatrix} X_1^{\mathrm{T}} & 0 & 0 \\ 0 & X_1^{\mathrm{T}} & 0 \\ 0 & 0 & X_2^{\mathrm{T}} \end{bmatrix} \overline{B}_{\mathrm{d}} \; 0 \\ \hline * & * & -\gamma_1^2 I \; D_2^{\mathrm{T}} \\ \hline * & * & * \; -I \end{array} \right] < 0, \tag{3.21}$$

where $\Delta = \mathrm{Herm}\left( \overline{A}^{\mathrm{T}} \begin{bmatrix} X_1 & 0 & 0 \\ 0 & X_1 & 0 \\ 0 & 0 & X_2 \end{bmatrix} \right)$.

Now, let us define new matrices $Y$, $Q$, $\Pi_1$, $\Pi_2$, and $\tilde{\Pi}_1$ as follows:

$$Y = X_2^{-1} = \begin{bmatrix} Y_{11} & Y_{12} \\ Y_{21} & Y_{22} \end{bmatrix}, \quad \Pi_1 = \begin{bmatrix} Y_{11} & I \\ Y_{21} & 0 \end{bmatrix}, \quad \Pi_2 = X_2 \Pi_1 = \begin{bmatrix} I & X_{11} \\ 0 & X_{21} \end{bmatrix},$$

$$Q = X_1^{-1}, \quad \tilde{\Pi}_1 = \mathrm{diag}(Q, Q, \Pi_1).$$

$$\tag{3.22}$$

Note that from (3.21) it follows that $X_1 + X_1^{\mathrm{T}} > 0$, $X_2 + X_2^{\mathrm{T}} > 0$, and therefore, $Y + Y^{\mathrm{T}} > 0$, $X_{11} + X_{11}^{\mathrm{T}} > 0$, $X_{22} + X_{22}^{\mathrm{T}} > 0$, and $Y_{11} + Y_{11}^{\mathrm{T}} > 0$, which imply the non-singularity of $X_1, X_2, X_{11}, X_{22}, Y_{11}$. Also, without loss of any generality, we assume that $X_{21}$ and $Y_{21}$ are nonsingular matrices. Therefore, $\tilde{\Pi}_1$ is a nonsingular matrix. It should be noted that the nonsingularity of $X_{21}$ is guaranteed by simply taking $X_{21} = I$. It also follows that $Y_{21} = -X_{22}^{-1} X_{21} Y_{11}$, which is nonsingular since the matrices $X_{21}$, $X_{22}$ and $Y_{11}$ are invertible.

Now, if we perform congruence transformation with $\mathrm{diag}(\tilde{\Pi}_1^{\mathrm{T}}, \tilde{\Pi}_1^{\mathrm{T}}, I, I)$ on (3.21), we obtain

$$\begin{bmatrix} E_{11} & T_1 + E_{12} & E_{13} & E_{14} \\ * & E_{22} & \lambda E_{13} & 0 \\ * & * & -\gamma_1^2 I & D_2^{\mathrm{T}} \\ * & * & * & -I \end{bmatrix} < 0, \tag{3.23}$$

where $E_{11}$, $E_{12}$, $E_{13}$, $E_{14}$, and $E_{22}$ are defined in (3.10) and

$$\tilde{A} = A_{11}^{\mathrm{T}}((A - D_{\mathrm{d}}C)Y_{11} - C_{\mathrm{d}}Y_{21}) + X_{21}^{\mathrm{T}}(B_{\mathrm{d}}CY_{11} + A_{\mathrm{d}}Y_{21}),$$

$$\tilde{B} = X_{11}^{\mathrm{T}}D_{\mathrm{d}} - X_{21}^{\mathrm{T}}B_{\mathrm{d}}, \quad \tilde{C} = D_{\mathrm{d}}CY_{11} + C_{\mathrm{d}}Y_{21}, \tag{3.24}$$

$$\tilde{D} = D_{\mathrm{d}}, \quad M = KQ, \quad T_1 = \tilde{\Pi}_1^{\mathrm{T}}P_1\tilde{\Pi}_1.$$

Therefore, the first inequality in the conditions in (3.9) is now obtained. Note that since $P_1 > 0$, it follows that $T_1$ must be a positive definite matrix.

The performance indices (II)–(VI) can also be shown by employing the same procedure and following along the same lines as in the derivation of the performance index (I). In order to validate this claim, a brief proof of the performance index (VI) is provided below, and the details for the performance indices (II)–(V) are omitted for sake of brevity.

The performance index (VI) is equivalent to the following condition:

$$\left\| \left[ \begin{array}{c|c} \overline{A} & \overline{B}_{\mathrm{f}} \\ \hline \overline{C}_1 & \underbrace{D_4 - J}_{\hat{D}_4} \end{array} \right] \right\|_\infty < \gamma_6. \tag{3.25}$$

To guarantee stability and the performance index (3.25), the following condition must be satisfied, namely,

$$\begin{bmatrix} I & 0 \\ \overline{A} & \overline{B}_{\mathrm{f}} \end{bmatrix}^{\mathrm{T}} \begin{bmatrix} 0 & P_6 \\ P_6 & 0 \end{bmatrix} \begin{bmatrix} I & 0 \\ \overline{A} & \overline{B}_{\mathrm{f}} \end{bmatrix} + \begin{bmatrix} 0 & I \\ \overline{C}_1 & \hat{D}_4 \end{bmatrix}^{\mathrm{T}} \begin{bmatrix} -\gamma_6^2 I & 0 \\ 0 & I \end{bmatrix} \begin{bmatrix} 0 & I \\ \overline{C}_1 & \hat{D}_4 \end{bmatrix} < 0, \tag{3.26}$$

where $P_6$ is a positive definite symmetric Lyapunov matrix.

If we choose $N_V$ in the inequality (1.9b) as in (3.18) and the $N_U$ and $Z$ in the inequality (1.9a) as follows:

$$Z = \begin{bmatrix} \overline{C}_1^{\mathrm{T}}\overline{C}_1 & P_6 & \overline{C}_1^{\mathrm{T}}\hat{D}_4 \\ * & 0 & 0 \\ * & * & \hat{D}_4^{\mathrm{T}}\hat{D}_4 - \gamma_6^2 I \end{bmatrix}, \quad N_U = \begin{bmatrix} I & 0 \\ \overline{A} & \overline{B}_{\mathrm{f}} \\ 0 & I \end{bmatrix}, \tag{3.27}$$

and invoke Lemma 1.1, then it can be shown that (3.26) becomes

$$Z + \begin{bmatrix} \overline{A}^{\mathrm{T}} \\ -I \\ \overline{B}_{\mathrm{f}}^{\mathrm{T}} \end{bmatrix} \begin{bmatrix} X & \lambda X & 0 \end{bmatrix} + \begin{bmatrix} X^{\mathrm{T}} \\ \lambda X^{\mathrm{T}} \\ 0 \end{bmatrix} \begin{bmatrix} \overline{A} & -I & \overline{B}_{\mathrm{f}} \end{bmatrix} < 0. \tag{3.28}$$

By partitioning $X \in \mathbb{R}^{4n \times 4n}$ as in (3.20) and by using the Schur complement, the following inequality is obtained:

$$
\left[
\begin{array}{c|c|c|c}
\Delta \quad P_6 + \lambda \overline{A}^{\mathrm{T}} \begin{bmatrix} X_1 & 0 & 0 \\ 0 & X_1 & 0 \\ 0 & 0 & X_2 \end{bmatrix} - X^{\mathrm{T}} & \begin{bmatrix} X_1^{\mathrm{T}} & 0 & 0 \\ 0 & X_1^{\mathrm{T}} & 0 \\ 0 & 0 & X_2^{\mathrm{T}} \end{bmatrix} \overline{B}_{\mathrm{f}} & \overline{C}_1^{\mathrm{T}} \\
\hline
* \quad \lambda \begin{bmatrix} -X_1-X_1^{\mathrm{T}} & 0 & 0 \\ 0 & -X_1-X_1^{\mathrm{T}} & 0 \\ 0 & 0 & -X_2-X_2^{\mathrm{T}} \end{bmatrix} & \lambda \begin{bmatrix} X_1^{\mathrm{T}} & 0 & 0 \\ 0 & X_1^{\mathrm{T}} & 0 \\ 0 & 0 & X_2^{\mathrm{T}} \end{bmatrix} \overline{B}_{\mathrm{f}} & 0 \\
\hline
* \qquad\qquad * & -\gamma_6^2 I & (D_4 - J)^{\mathrm{T}} \\
\hline
* \qquad\qquad * & * & -I
\end{array}
\right] < 0. \quad (3.29)
$$

Now, if we perform congruence transformation with $\mathrm{diag}(\tilde{\Pi}_1^{\mathrm{T}}, \tilde{\Pi}_1^{\mathrm{T}}, I, I)$ on (3.29), where $\tilde{\Pi}_1$ is defined in (3.22), the sixth inequality in the conditions in (3.9) is obtained.

Finally, the control gain matrix $K$ and the dynamic observer parameters $A_{\mathrm{d}}$, $B_{\mathrm{d}}$, $C_{\mathrm{d}}$, $D_{\mathrm{d}}$ in (3.11) are obtained according to the expressions in (3.24). This completes the proof of the theorem. $\qquad\square$

The following theorem now provides the LMI constraints for the $L_1$ performance indices (VIII) and (IX).

**Theorem 3.2.** *For given predefined positive scalars $\lambda$, $\eta$, and $\mu$, the closed-loop system (3.5) is stable and the $L_1$ performance indices (VIII) and (IX) are guaranteed with performance levels $\gamma_7$ and $\gamma_8$ if there exist symmetric positive definite matrices $T_7$, $T_8$ and matrices $X_{11}$, $Y_{11}$, $Q$, $S$, $\tilde{A}$, $\tilde{B}$, $\tilde{C}$, $\tilde{D}$, $M$, such that the following LMIs hold*

$$
\begin{bmatrix}
E_{11} + \eta T_7 & T_7 + E_{12} & \mathscr{E}_{13} \\
* & E_{22} & \lambda \mathscr{E}_{13} \\
* & * & -\mu I
\end{bmatrix} < 0,
$$

$$
\begin{bmatrix}
-\eta T_7 & 0 & E_{14} \\
* & (\mu - \gamma_7)I & D_3^{\mathrm{T}} \\
* & * & -\gamma_7 I
\end{bmatrix} < 0,
$$

$$
\begin{bmatrix}
E_{11} + \eta T_8 & T_8 + E_{12} & \mathscr{E}_{13} \\
* & E_{22} & \lambda \mathscr{E}_{13} \\
* & * & -\mu I
\end{bmatrix} < 0,
$$

$$
\begin{bmatrix}
-\eta T_8 & 0 & \Xi_{14} \\
* & (\mu - \gamma_8)I & D_3^{\mathrm{T}} \\
* & * & -\gamma_8 I
\end{bmatrix} < 0,
$$

$$(3.30)$$

where $\mathscr{E}_{13} = \begin{bmatrix} 0 & D_3^{\mathrm{T}}\tilde{D}^{\mathrm{T}} & B_3^{\mathrm{T}} - D_3^{\mathrm{T}}\tilde{D}^{\mathrm{T}} & B_3^{\mathrm{T}}X_{11} - D_3^{\mathrm{T}}\tilde{B}^{\mathrm{T}} \end{bmatrix}^{\mathrm{T}}$, $E_{11}$, $E_{12}$, $E_{22}$, $E_{14}$, and $\Xi_{14}$ matrices are defined as in Theorem 3.1. The control gain matrix $K$ and the dynamic observer parameters $A_d$, $B_d$, $C_d$, $D_d$ are then given by (3.11).

*Proof.* Consider the Lyapunov function candidate $V(\xi(t)) = \xi^{\mathrm{T}}(t)P_7\xi(t)$, where $P_7$ is a positive definite matrix. According to the first inequality in (3.30), by substituting $\tilde{A}$, $\tilde{B}$, $\tilde{C}$, $\tilde{D}$, $N$ with (3.24) and $T_7 = \tilde{\Pi}_1^{\mathrm{T}}P_7\tilde{\Pi}_1$, and performing the congruence transformation with $\mathrm{diag}(\tilde{\Pi}_1^{-\mathrm{T}}, \tilde{\Pi}_1^{-\mathrm{T}}, I)$, one can conclude that

$$\begin{bmatrix} \mathrm{Herm}(\bar{A}^{\mathrm{T}}X) + \eta P_7 & P_7 + \lambda \bar{A}^{\mathrm{T}}X - X^{\mathrm{T}} & X^{\mathrm{T}}\bar{B}_{\mathrm{w}} \\ * & -\lambda(X + X^{\mathrm{T}}) & \lambda X^{\mathrm{T}}\bar{B}_{\mathrm{w}} \\ * & * & -\mu I \end{bmatrix} < 0, \tag{3.31}$$

Using Lemma 1.1, the previous inequality holds if and only if the following LMIs are satisfied, namely,

$$\begin{bmatrix} I & 0 \\ \bar{A} & \bar{B}_{\mathrm{w}} \\ 0 & I \end{bmatrix}^{\mathrm{T}} \begin{bmatrix} \eta P_7 & P_7 & 0 \\ P_7 & 0 & 0 \\ 0 & 0 & -\mu I \end{bmatrix} \begin{bmatrix} I & 0 \\ \bar{A} & \bar{B}_{\mathrm{w}} \\ 0 & I \end{bmatrix} < 0, \tag{3.32}$$

$$\begin{bmatrix} \lambda I & 0 \\ -I & 0 \\ 0 & I \end{bmatrix}^{\mathrm{T}} \begin{bmatrix} \eta P_7 & P_7 & 0 \\ P_7 & 0 & 0 \\ 0 & 0 & -\mu I \end{bmatrix} \begin{bmatrix} \lambda I & 0 \\ -I & 0 \\ 0 & I \end{bmatrix} < 0. \tag{3.33}$$

By multiplying both sides of the inequality (3.32) by $[\xi^{\mathrm{T}}(t)\ w^{\mathrm{T}}(t)]$ from the left and $[\xi^{\mathrm{T}}(t)\ w^{\mathrm{T}}(t)]^{\mathrm{T}}$ from the right, it follows that

$$\dot{V}(t) + \eta \xi^{\mathrm{T}}(t)P_7\xi(t) - \mu w^{\mathrm{T}}(t)w(t) < 0. \tag{3.34}$$

For $w(t) = 0$, the inequality (3.34) leads to $\dot{V}(t) < 0$, that guarantees the stability of the system. Based on the fact that $w(t)^{\mathrm{T}}w(t) \leq 1$, then $\dot{V}(t) < 0$ holds whenever $V(\xi(t)) > \mu/\eta$. Since $V(0) = 0$, this implies that $V(\xi(t))$ cannot exceed the value $\mu/\eta$, and therefore $V(\xi(t)) < \mu/\eta$, for $t \geq 0$ [271].

On the other hand, by considering $T_7 = \tilde{\Pi}_1^{\mathrm{T}}P_7\tilde{\Pi}_1$ and substituting $\tilde{A}$, $\tilde{B}$, $\tilde{C}$, $\tilde{D}$, $N$ with (3.24) in the second inequality in (3.30), and performing the congruence transformation with $\mathrm{diag}(\tilde{\Pi}_1^{-\mathrm{T}}, I, I)$, one can conclude that

$$\begin{bmatrix} -\eta P_7 & 0 & \bar{C}_2^{\mathrm{T}} \\ * & (\mu - \gamma_7)I & D_3^{\mathrm{T}} \\ * & * & -\gamma_7 I \end{bmatrix} < 0. \tag{3.35}$$

Now, by multiplying both sides of the inequality (3.35) by $[\xi^{\mathrm{T}}(t)\ w^{\mathrm{T}}(t)]$ from the left and $[\xi^{\mathrm{T}}(t)\ w^{\mathrm{T}}(t)]^{\mathrm{T}}$ from the right, we get

$$z^{\mathrm{T}}(t)z(t) < \gamma_7[\eta V(\xi(t)) + (\gamma_7 - \mu)w^{\mathrm{T}}(t)w(t)]. \tag{3.36}$$

Since $w(t)^T w(t) \leq 1$ and $V(\xi(t)) < \mu/\eta$, hence $z^T(t)z(t) < \gamma_7^2$. Therefore, the $L_1$ norm (VIII) is satisfied. The $L_1$ performance index (IX) can also be shown by employing the same procedure as in the derivation of the performance index (VIII). These details are omitted for sake of brevity. This completes the proof of the theorem. □

Based on the above results, the following corollary is now proposed to solve the proposed IFDIT problem (3.6).

**Corollary 3.1.** *For the given positive constants $\lambda$, $\eta$, $\mu$, $\beta_1,...,\beta_8$, a feasible solution to the IFDIT control problem given by (3.6) is obtained by solving the following convex optimization problem:*

$$\min_{T_i, X_{11}, Y_{11}, Q, S, \tilde{A}, \tilde{B}, \tilde{C}, \tilde{D}, M, J} \sum_{i=1}^{8} \beta_i \gamma_i \tag{3.37}$$

*subject to the inequalities (3.8), (3.9), and (3.30).*

*Proof.* The proof can be easily derived by considering the IFDIT problem (3.6), the inequality (3.8) and by invoking Theorems 3.1 and 3.2 and is not included here for sake of brevity. □

**Remark 3.3.** *Note that in Corollary 3.1, the solution to the IFDIT problem is formulated as strict LMI conditions that allows one to easily carry out the optimization solution. The main advantage of the proposed LMI formulation is that it is convex and can therefore be solved effectively and efficiently by using interior-point methods. Moreover, the values of $(\beta_1, \beta_2, \beta_3, \beta_7)$ and $(\beta_4, \beta_5, \beta_6, \beta_8)$ are used to emphasize the tracking control and the FDI objectives, respectively. Therefore, our LMI-based solution enables one to explore the trade-offs between the tracking control and the FDI objectives and analyze the limits of performance and feasibility in the design process.*

The following algorithm summarizes the required steps that one needs to follow for designing the IFDIT module parameters.

**Algorithm 3.1:** *Designing the IFDIT problem.*

1. *Select the positive weights $\beta_1,...,\beta_8$ according to the level of importance or priority among various objectives in the IFDIT problem.*
2. *Set the positive scalars $\lambda$, $\eta$, and $\mu$.*
3. *Solve the optimization problem (3.37) for $T_i, X_{11}, Y_{11}, Q, S, \tilde{A}, \tilde{B}, \tilde{C}, \tilde{D}, M, J$ and obtain the minimum values of $\gamma_i$ for $i = 1,...,8$.*
4. *If the problem becomes infeasible, change the scalars $\lambda$, $\eta$, and $\mu$ and go back to step 3; else go to step 5.*

5. *Solve the equation* $Y_{21}^{\mathrm{T}} X_{21} = S - Y_{11}^{\mathrm{T}} X_{11}$ *for* $X_{21}$ *and* $Y_{21}$.
6. *Obtain the control gain matrix* $K$ *and the dynamic observer parameters* $A_d$, $B_d$, $C_d$, $D_d$ *from the expressions in* (3.11).
7. *Construct the IFDIT module according to* (3.3) *and* (3.4).

We are now in a position to present the evaluation criterion that is employed for performing the FDI task.

### 3.3.1 Residual evaluation criterion

Following the construction of the residuals $r_i(t) \in \mathbb{R}, \forall i \in \{1, \ldots, n_y\}$, the final step in accomplishing the IFDIT strategy is to determine the threshold functions $J_{\mathrm{th}_i}$ and the evaluation functions $J_{r_i}(t)$. Various evaluation functions can be considered as presented in [30].

The upper and lower threshold values are selected as $J_{\mathrm{th}_i}^u = \sup_{f=0, u_r, d \in \mathbb{L}_2, w \in \mathbb{L}_\infty} r_i(t)$ and $J_{\mathrm{th}_i}^l = \inf_{f=0, u_r, d \in \mathbb{L}_2, w \in \mathbb{L}_\infty} r_i(t)$, respectively. Based on the selected thresholds and the evaluation function taken as $J_{r_i}(t) = r_i(t)$, the occurrence of a fault can then be detected and isolated in our proposed IFDIT strategy by using the following decision logic: if $r_i(t) > J_{\mathrm{th}_i}^u$ or $r_i(t) < J_{\mathrm{th}_i}^l \implies f_i \neq 0$.

Note that according to the above decision logic, the occurred faults in the system are automatically isolated. Indeed, based on the proposed optimization problem (3.6) [in particular the performance indices (VI) and (VII)], the transfer matrix from the faults to the residuals has a diagonal structure, and therefore, the designed IFDIT module generates a set of dedicated structured residuals $r_i(t)$ ("one residual-one fault" scheme) that leads to the fault isolation objective. However, the problem of fault analysis or identification which deals with determining the type, magnitude, and severity of the fault is beyond the scope of this chapter and is left as a topic of future research.

## 3.4 A case study: an autonomous underwater vehicle

In this section, we present the application of our proposed IFDIT methodology to a linearized longitudinal model of the Subzero II AUV [284] to illustrate and demonstrate the effectiveness of our proposed solution. The linearized longitudinal model of the Subzero II AUV is given as follows [284]:

$$\dot{x}(t) = Ax(t) + B_1 u(t) + B_2 d(t) + B_3 w(t) + B_4 f(t),$$
$$y(t) = Cx(t) + D_1 u(t) + D_2 d(t) + D_3 w(t) + D_4 f(t),$$

where $x = \begin{bmatrix} u & w & q & \theta & z & n & \delta s \end{bmatrix}^{\mathrm{T}}$, $u = \begin{bmatrix} M_d & \delta s_d \end{bmatrix}^{\mathrm{T}}$, $y = \begin{bmatrix} u & z \end{bmatrix}^{\mathrm{T}}$, with u denoting the forward speed, w denoting the vertical speed, q denoting the pitch rate, z denoting the depth, $\theta$ denoting the pitch angle, n denoting the propeller rotation speed, $\delta s$ denoting the control surface deflection, $M_d$ denoting the motor command, and $\delta s_d$ denoting the control surface command. Moreover,

$$A = \begin{bmatrix} -0.5558 & 0.0474 & 0.0516 & 0.0038 & 0 & 0.0582 & 0.0797 \\ 0.001 & -2.1258 & -0.3734 & -0.0175 & 0 & -0.0001 & -1.8211 \\ -0.0365 & -7.9966 & -8.7065 & -0.6474 & 0 & -0.0038 & -13.4438 \\ 0 & 0 & 1 & 0 & 0 & 0 & 0 \\ 0 & 1 & 0 & -1.3009 & 0 & 0 & 0 \\ 7.2126 & 0 & 0 & 0 & 0 & -4.0813 & 0 \\ 0 & 0 & 0 & 0 & 0 & 0 & -11.53 \end{bmatrix},$$

$$B_1 = \begin{bmatrix} 0 & 0 \\ 0 & 0 \\ 0 & 0 \\ 0 & 0 \\ 0 & 0 \\ 0.0492 & 0 \\ 0 & 10.377 \end{bmatrix}, \quad C = \begin{bmatrix} 1 & 0 & 0 & 0 & 0 & 0 & 0 \\ 0 & 0 & 0 & 0 & 1 & 0 & 0 \end{bmatrix}, \quad D_1 = \begin{bmatrix} 0 & 0 \\ 0 & 0 \end{bmatrix}.$$

The bounded energy disturbance matrices $B_2$ and $D_2$ are selected as $B_2 = 0$ and $D_2 = 0.1 I_{2 \times 2}$, which represent the occurrence of sensor noise in the AUV. The bounded peak disturbance matrices $B_3$ and $D_3$ are selected as $B_3 = 0.1 B_1$ and $D_3 = 0$, as these are common in industrial settings where disturbances enter the system dynamics by corrupting the input channel. The fault matrices $B_4$ and $D_4$ are selected as $B_4 = B_1$ and $D_4 = D_1$, which represent the occurrence of actuator faults in the system.

It is expected and desired that the AUV IFDIT system detects and isolates the occurrence of different types of faults $f(t) \in \mathbb{R}^2$ (corresponding to various fault severities and different locations of fault occurrences) in presence of disturbances $d(t) \in \mathbb{R}^2$, $w(t) \in \mathbb{R}^2$ and reference inputs $u_r(t) \in \mathbb{R}^2$ and the IFDIT system should also simultaneously track the desired output $y_r(t) \in \mathbb{R}^2$. Note that the problem of tracking a set point for the Subzero II and REMUS AUVs is solved by using $H_\infty$ robust control and state-dependent Riccati equation methods in [284,285], respectively. Nevertheless, it is important for us to design a controller that makes the AUV track a desired time-varying trajectory given complex missions and scenarios.

In our case study, it is assumed that due to certain physical obstacles, the AUV has to track a damped sinusoidal trajectory with a constant speed. Therefore, the reference input $u_{r_1}(t) \in \mathbb{R}$ is selected as a rectangular pulsed signal with an amplitude of 1 that is applied in the time intervals 0–80 s. Moreover, the reference input $u_{r_2}(t) \in \mathbb{R}$ is selected as $u_{r_2}(t) = -1.6013 e^{-0.05t} \sin(0.3122t)$.

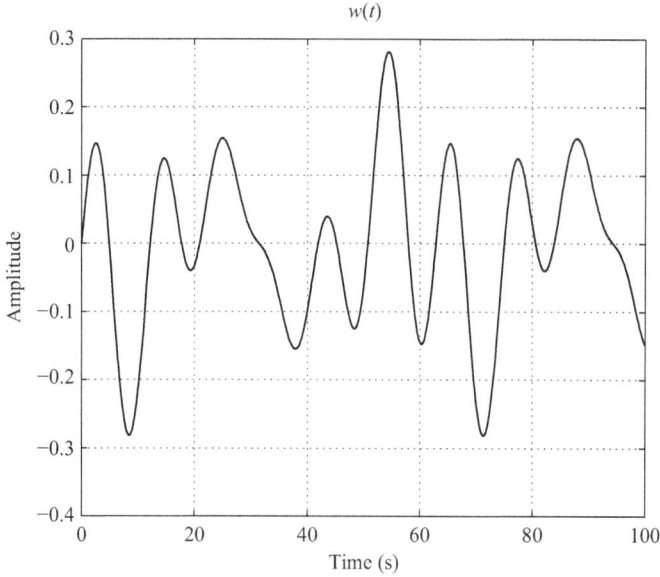

*Figure 3.1   The bounded peak disturbance w(t)*

The reference model matrices for achieving the above tracking objective are arbitrarily selected as

$$A_r = \text{diag}(-3, -3, -4, -4, -5, -6, -7), \ C_r = C,$$

$$B_r = \begin{bmatrix} 1 & 0 & 0 & 0 & 0 & 27.9 & 0 \\ 0 & 0 & 0 & 0 & 0.5 & 0 & 0 \end{bmatrix}^{\text{T}}.$$

Note that selection of the reference model matrices is not unique and multiple choices are available in general, although each model will lead to a different tracking error characteristic. Therefore, based on and according to the specifications of the Subzero II AUV model, after certain trials and errors we have selected the above matrices for the reference model.

It should be pointed out that waves and ocean currents are the most important sources of environmental disturbances for marine vehicles [262]. The current velocity can be modeled by a first order Gauss–Markov process as given by $\dot{V}_c(t) + \rho V_c(t) = v(t)$, where $\rho \geq 0$ and $v(t)$, $t \geq 0$ is a Gaussian white noise. For $\rho = 0$, the model becomes a random walk. Hence, the injected disturbance $d(t) \in \mathbb{R}^2$ to the AUV is a random walk.

Moreover, the waves can be modeled by sinusoidal signals or a combination of them as a pseudo random signal. Therefore, the bounded peak disturbance $w(t)$ is taken as $w(t) = 0.1(\sin(0.5t) - \sin(0.2t) + \sin(0.6t))$, which is shown in Figure 3.1. The fault signal $f_1(t)$ is represented as a rectangular pulse with an amplitude of 6, and which is applied during the time interval 35–45 s, and the fault signal $f_2(t)$ is simulated as a soft bias (slope = 0.04) and which is applied during the time interval 12–17 s.

The positive constant weights $\beta_1,\ldots,\beta_8$ are assumed to be selected as equal to one, which implies that the designer imposes the same level of importance and priority to the fault detection, isolation, and tracking objectives [the objectives (I)–(IX)]. Moreover, the positive constants $\lambda$, $\eta$, and $\mu$ are selected as $\lambda = 0.1$, $\eta = 1$, and $\mu = 0.08$. According to step 3 of Algorithm 3.1, the optimization problem (3.37) for $T_i, X_{11}, Y_{11}, Q, S, \tilde{A}, \tilde{B}, \tilde{C}, \tilde{D}, M, J$ is solved and the values of these matrices are obtained. Moreover, the optimization parameters $\gamma_1,\ldots,\gamma_8$ are obtained as $\gamma_1 = 0.4143$, $\gamma_2 = 0.6827$, $\gamma_3 = 0.2014$, $\gamma_4 = 0.3923$, $\gamma_5 = 0.1791$, $\gamma_6 = 1.6551$, $\gamma_7 = 1.7313$, and $\gamma_8 = 0.0802$. Following step 5 of Algorithm 3.1, the values of $X_{21}$ and $Y_{21}$ matrices are obtained. The IFDIT module matrices ($A_d$, $B_d$, $C_d$, $D_d$, $K$) are then obtained from (3.11), and finally, the IFDIT module is constructed according to (3.3) and (3.4).

It is worth noting that by considering the worst case analysis of the residuals corresponding to the healthy operation of the AUV system that is subjected to various disturbances, the threshold values $J^u_{th_i}$ and $J^l_{th_i}$ are selected. The residual signals $r_1(t)$ and $r_2(t)$ are shown in Figure 3.2, where the robustness against (i) the disturbances, (ii) the reference inputs, and (iii) the enhanced fault sensitivity can be confirmed. Note that by using our threshold evaluation logic, the faults $f(t) \in \mathbb{R}^2$ are effectively detected and isolated. Moreover, the fault detection times for the faults $f_1(t)$ and $f_2(t)$ are within 1.15 and 2 s, respectively.

The system output $y_1(t)$ and the reference output $y_{r_1}(t)$ are shown in Figure 3.3(a). The AUV system and the reference outputs $y_2(t)$ and $y_{r_2}(t)$ are shown in Figure 3.3(b). In this figure, the black circles depict and designate eight physical obstacles that lead to the selection of the desired time-varying trajectory for the tracking problem. From Figure 3.3, it can be concluded that the AUV system can effectively track the reference outputs in the presence of disturbances and fault signals.

The tracking errors $y_i(t) - y_{r_i}(t)$, $i = 1, 2$ are shown in Figure 3.4. The source of these errors is due to the formulation of the IFDIT problem in condition (3.6). In this condition, the performance indices (I), (II), (III), and (VIII) are used to attenuate the effects of disturbances $(d(t), w(t))$, reference input $(u_r(t))$, and the fault signal $(f(t))$ on the regulated output $z(t)$.

The physical significance of the performance indices (I)–(III) is that the effects of any $d(t), f(t), u_r(t)$ on the controlled output $z(t)$ must be attenuated below the desired levels $\gamma_1, \gamma_2, \gamma_3$, from the energy viewpoint, no matter what $d(t), f(t), u_r(t)$ are, i.e., the $L_2$ gains from $d(t), f(t), u_r(t)$ to $z(t)$ must be equal to or less than the prescribed values $\gamma_1, \gamma_2, \gamma_3$. Moreover, the $L_1$ performance index (VIII) guarantees that the controlled output has a bounded peak ($\|z(t)\|_\infty$) less than the desired level $\gamma_7$.

Consequently, the disturbances, reference input, and fault signals attenuations on the controlled output $z(t)$ are given by $\gamma_1 = 0.4143$, $\gamma_2 = 0.6827$, $\gamma_3 = 0.2014$, and $\gamma_7 = 1.7313$, respectively. The above results demonstrate and illustrate that our proposed IFDIT methodology simultaneously satisfies the fault detection, isolation, and tracking problem requirements.

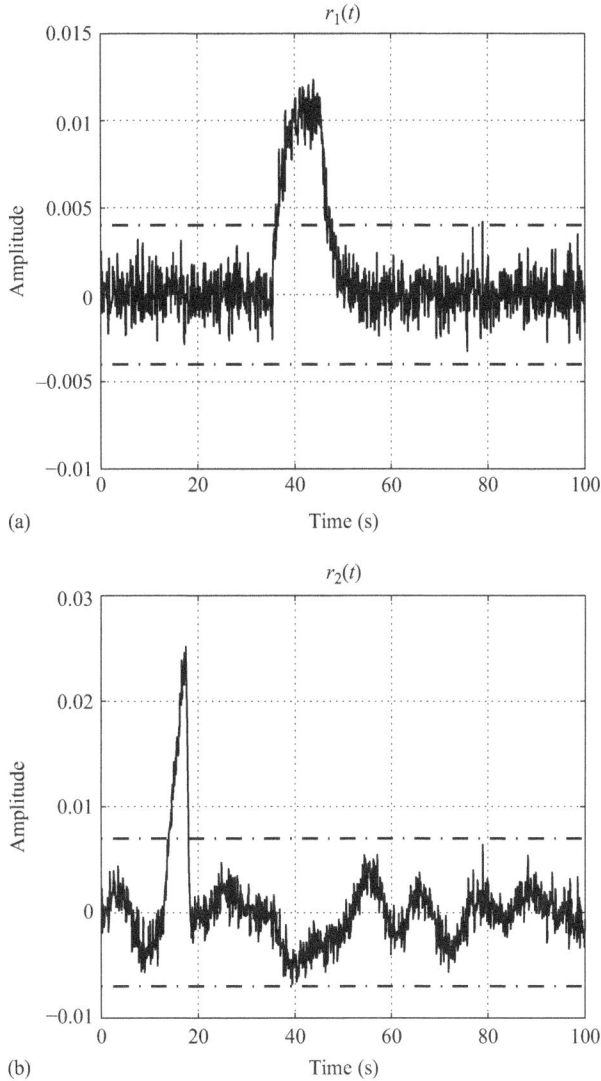

*Figure 3.2   The residual signals of (a) $r_1(t)$ and (b) $r_2(t)$, where the solid lines denote the residual signals and the dash-dot lines denote the residual upper and lower thresholds*

**Remark 3.4.** *In simulations conducted above, the effectiveness and capabilities of our methodology were demonstrated by providing a case study corresponding to an important industrial system, namely, the Subzero II AUV. However, it should be noted that the application of our proposed methodology is not restricted to this system and can be applied to and extended for numerous other classes of engineering systems,*

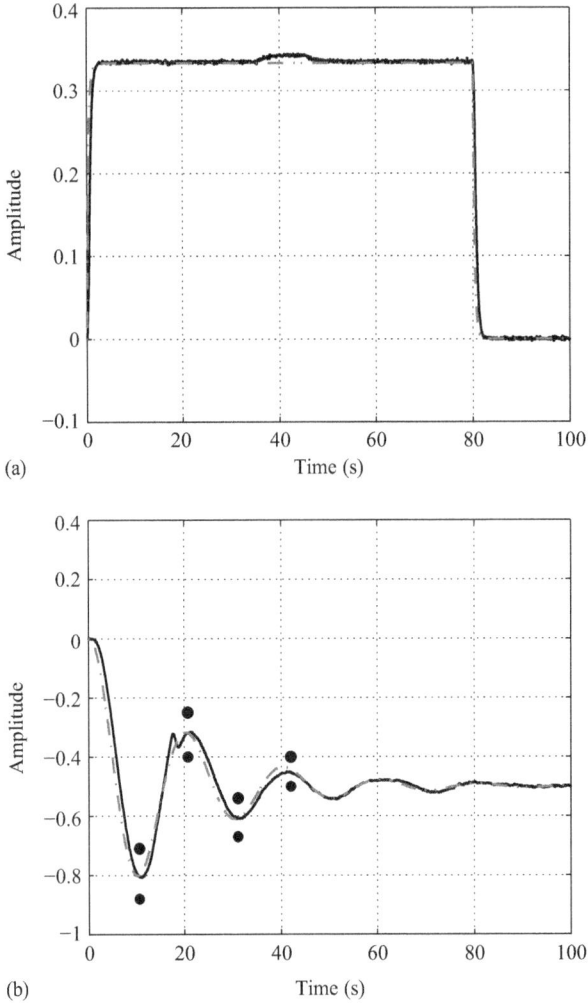

*Figure 3.3    The system and reference outputs (a) $y_1(t)$ and $y_{r_1}(t)$, (b) $y_2(t)$ and $y_{r_2}(t)$, where the solid lines denote the system outputs and the dash-dot lines denote the reference outputs*

such as aero engines, chemical processes, manufacturing systems, power networks, wind energy conversion systems, oil and gas systems, and industrial electronic equipment. Indeed, the considered system in (3.1) is quite general and the linearized model of many industrial and engineering systems in presence of external disturbances and faults can be expressed according to (3.1). Hence, our methodology, which is based on the general model in (3.1), can be considered and applied to a wide range of industrial applications.

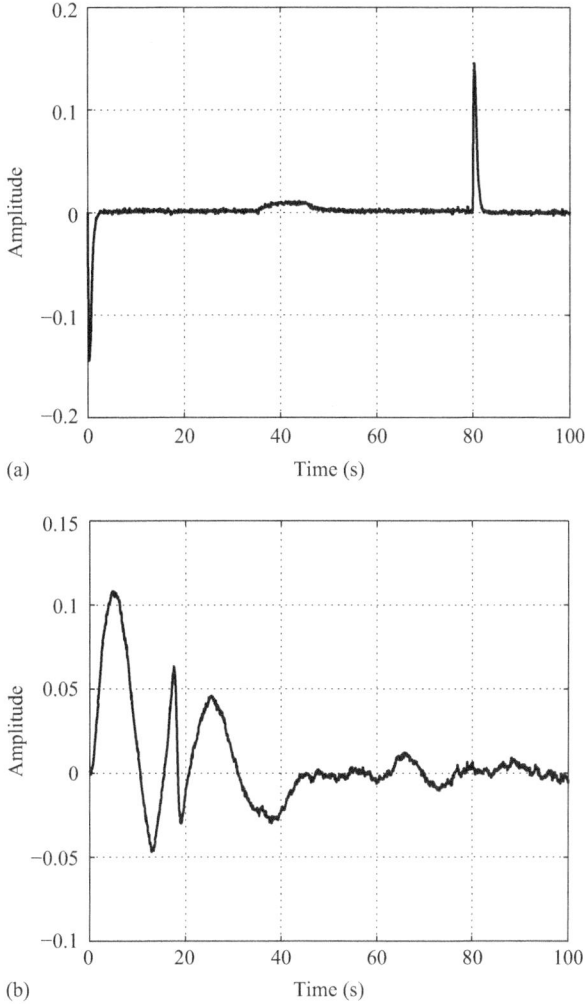

Figure 3.4 *The tracking errors (a) $y_1(t) - y_{r_1}(t)$ and (b) $y_2(t) - y_{r_2}(t)$*

Below we provide comparisons with other possible methodologies in order to further illustrate the effectiveness of our proposed design methodology.

**Comparison with the methodology that uses a common Lyapunov function**

By applying the methodology that is proposed in [264] which uses identical Lyapunov matrices that are coupled with the system matrices, the weights $\gamma_1, \ldots, \gamma_8$ are now obtained as $\gamma_1 = 0.7104$, $\gamma_2 = 0.924$, $\gamma_3 = 0.2602$, $\gamma_4 = 0.418$, $\gamma_5 = 0.225$, $\gamma_6 = 1.692$, $\gamma_7 = 2.015$, and $\gamma_8 = 0.138$. Note that by using the scheme proposed in this chapter, smaller values of $\gamma_1,\ldots,\gamma_8$ are obtained, implying a better achievable performance and a lower cost that is associated with our methodology. Consequently, by considering identical Lyapunov functions for all the performance indices (I)–(IX) can indeed impose more conservativeness in the final design solution.

**Comparison with a separate design**

Another set of simulations are now conducted to compare our proposed methodology with a separate design of the controller and the FDI filters. At first, a dynamic observer-based feedback controller using the performance indices (I), (II), (III), and (VIII) is obtained. The optimization parameters $\gamma_1$, $\gamma_2$, $\gamma_3$, and $\gamma_7$ are obtained as $\gamma_1 = 0.4133$, $\gamma_2 = 0.685$, $\gamma_3 = 0.2025$, and $\gamma_7 = 1.64$, respectively. An observer-based FDI filter is then designed for the system by using the performance indices (IV), (V), (VI), (VII), and (IX). The optimization parameters $\gamma_4$, $\gamma_5$, $\gamma_6$, $\gamma_8$ are obtained as $\gamma_4 = 0.289$, $\gamma_5 = 0.173$, $\gamma_6 = 1.6503$, and $\gamma_8 = 0.138$, respectively.

Although the optimization parameters $\gamma_1,\ldots,\gamma_8$ for both separate and integrated design scenarios are almost of the same order of magnitude, it should be noted that the separate approach does not yield an optimal design, since the methodology we proposed leads to a lower computational complexity in comparison to the separate design. Indeed, in case of the separate design, the order of the resulting detector/controller module is $4n$. However, our proposed methodology yields a detector/controller module with a lower complexity (the IFDIT module has an order of $2n$). Moreover, it should be pointed out that the separate design of the controller and the FDI modules does not take into account the rather significant interactions and couplings that exist between these modules.

## 3.5    Conclusion

In this chapter, we developed and presented an $H_\infty/H_-/L_1$ methodology for the problem of IFDIT control of linear systems using a dynamic observer. An LMI approach for the IFDIT design was introduced that simultaneously stabilizes the closed-loop system while guaranteeing that the fault detection, isolation, and also tracking control objectives are accomplished. The LMI conditions were derived where products of Lyapunov matrices with system matrices were not involved. This results in a significant reduction in the conservatism of the IFDIT solution. Moreover, the IFDIT problem was solved in such a manner that each element of the residual vector is only sensitive to a specified fault, and therefore, occurrence of simultaneous faults in the system can also be handled. Application of our methodology to a linearized longitudinal model of the Subzero II AUV was also presented to demonstrate and illustrate the effectiveness of our proposed diagnostic and control approach.

*Chapter 4*
# Integrated design of fault detection, isolation, and control for continuous-time Markovian jump systems

In this chapter, the problem of integrated fault detection, isolation, and control (IFDIC) design of continuous-time Markovian jump linear systems with uncertain transition probabilities and subject to both energy bounded and peak bounded disturbances is introduced and addressed. A single Markovian jump module designated as the IFDIC under a mixed robust $H_\infty/H_-/L_1$ framework is considered to simultaneously achieve the desired detection, isolation, and control objectives. Conventional mixed robust $H_\infty/H_-/L_1$ approaches to the fault detection and isolation (FDI) problem lead to conservative results due to the selection of identical Lyapunov matrices.

Consequently, the extended linear matrix inequalities (LMIs) methodology is utilized in this work to reduce the conservativeness of standard approaches by introducing additional matrix variables so that the coupling of Lyapunov matrices with the system matrices is eliminated. Simulation results for an application to the GE F-404 aircraft engine system illustrate the effectiveness and capabilities of our proposed design methodologies. Comparisons with relevant work in the literature are also provided to demonstrate the advantages of our proposed solutions. The work presented in this chapter has partially appeared in [286,287].

This chapter is organized as follows. We begin with a brief literature review in Section 4.1. We describe the system and the IFDIC module governing equations in Section 4.2.1. The IFDIC problem is then formulated for the Markovian jump linear system with uncertain transition probabilities in Section 4.2.2. In essence, the IFDIC design problem is transformed into a mixed $H_\infty/H_-/L_1$ optimization problem. The main results of this chapter, that is, the solution of the proposed IFDIC problem, are provided in Section 4.3. In Section 4.4, the proposed approach is applied to the problem of IFDIC design for the GE F-404 aircraft engine system.

## 4.1 Introduction

Markovian jump systems belong to an important class of hybrid systems and have recently received a great deal of attention by the control research community [288,289]. This family of systems is generally modeled by a set of linear models with transitions among them that are determined by a Markov chain taking values in a finite set. Markovian jump systems are popular in modeling a number of practical systems subject to random failures and structural changes, such as electric power systems, communication systems, aircraft flight control systems, control of nuclear power plants, and manufacturing systems [290].

In order to guarantee system safety and reliability, model-based FDI methodologies have been developed for Markovian jump systems. In [291], a fault detection (FD) filter for a discrete-time Markovian jump linear system with partially known transition probabilities is developed based on an $H_\infty$ filtering framework. In [292], the FD problem for discrete-time Markovian jump linear systems using a recursive mode independent Kalman filter is considered. In [293], the FD problem for a class of nonlinear stochastic systems with Markovian switching and mixed time-delays is addressed.

In [281], the FD problem for a class of nonhomogeneous Markovian jump systems based on a delta operator is considered and an LMI approach is developed for solving the problem. In [294], the FD problem for discrete-time Markovian jump systems with sensor saturation and randomly varying nonlinearities is addressed. In view of the above, the current literature in fault diagnosis of Markovian jump systems suffers from the following limitations and drawbacks.

First, in almost all the published work in fault diagnosis of Markovian jump systems, an open-loop model of the process is considered, and/or it is assumed that the controller maintains stability of the closed-loop system upon the failure. Indeed, most of the current literatures consider the controller design and the fault diagnosis units separately. Consequently, it is of great interest to study and develop an integrated design of FDI and controller modules for Markovian jump systems.

Moreover, most of the current references in the field of Markovian jump systems FDI only consider the detection objective, and the methods cannot isolate occurrence of system faults [281,290–293]. The problem of FDI of continuous-time and discrete-time Markovian jump linear systems using a geometric approach was considered in [295] and [296], respectively. Second, the fault isolation objective in [295,296] and also in most of the current FDI literature (see, e.g., [266–268,270]) is accomplished through the use of a bank of observers. For implementation of these methodologies, one requires to have a bank of observers, and this puts a heavy computational burden on design of the FDI module. Therefore, reducing the number of observers is of great significance and importance.

Third, a common assumption in almost all Markovian jump systems FDI research [281,291–293,295,296] is that the considered system is driven by bounded energy disturbances. However, in practical applications, the considered system is disturbed by unknown persistent signals bounded in magnitude. The $L_1$ performance index, which minimizes the worst case peak-to-peak gain of the system, provides a suitable framework for characterizing the effects of such signals. The problems of robust $L_1$ filtering and robust $L_1$ control design of Markovian jump systems have been studied in [297] and [298], respectively. In [299], the problem of robust FDI design for discrete-time linear time-invariant systems is solved by using robust $L_1$ estimation and a bank of observers.

Based on the above discussion, proposing a new and simple methodology to address the problem of IFDIC of Markovian-jump systems that are disturbed by both energy and peak bounded disturbances is of great importance. In this work, the IFDIC problem under a mixed $H_\infty/H_-/L_1$ framework for continuous-time Markovian jump linear systems with uncertain transition probabilities and subject to both energy bounded and peak bounded external disturbances is considered.

It should be pointed out that in practice only estimated mode transition rates are available, and estimation errors, referred to as switching probability uncertainties, may lead to instability or at least degraded system performance that is similar to uncertainties in the system model [300]. Hence, in this work, the element-wise uncertainties in the mode transition rate matrix are considered, and an LMI-based methodology for designing the IFDIC module subject to the uncertain Markovian jump system is proposed.

Our contributions in this chapter can therefore be stated as follows:

- first, we consider Markovian jump systems with uncertain transition probabilities that are disturbed by both bounded energy and bounded peak disturbances (to address the third limitation above);
- second, we propose a single observer-based module that is designated as the IFDIC module with an order that is the same as that of the system order (to address the second limitation above); and
- third, we formulate the IFDIC problem under a mixed $H_\infty/H_-/L_1$ framework and solve the proposed problem based on the LMI feasibility conditions (to address the first limitation above).

Indeed, a single IFDIC module is proposed that can produce two signals, namely, the residual and the control. The parameters of the IFDIC module are then designed such that the effects of energy bounded and peak bounded disturbances on the residual signals are minimized (for accomplishing the FD objective) subject to the constraint that the mapping matrix function from the faults to the residuals is equal to a pre-assigned diagonal mapping matrix (for accomplishing the fault isolation objective), while the effects of the energy bounded and peak bounded disturbances and faults on the specified control outputs are minimized (for accomplishing the fault-tolerant control (FTC) objective).

Consequently, the computational complexity in terms of the number and the dimension of the required observer model is significantly reduced in comparison with all the existing methodologies in the literature. Moreover, our proposed methodology can also handle isolation of simultaneous faults.

Motivated by [277], the LMI feasibility conditions for solving our IFDIC problem are developed such that products of Lyapunov matrices with the system state space matrices are not involved and the multiobjective solution conservativeness is reduced. To summarize, the main capabilities and advantages of our proposed IFDIC solution in comparison with other relevant work in the literature on Markovian jump systems are listed in Table 4.1.

*Table 4.1   Comparison of the IFDIC scheme with other methodologies where FI denotes the "fault isolation," TCD denotes the "type of considered disturbances," NO denotes the "number of observers," and $n_f$ denotes the "number of faults" in the system*

| Approach | FD feature | FI feature | Control feature | TCD | | NO |
| --- | --- | --- | --- | --- | --- | --- |
| | | | | Energy bounded | Peak bounded | |
| [240,301] | No | No | Yes | Yes | No | – |
| [297] | No | No | Yes | Yes | Yes | – |
| [281,290–294,302–304] | Yes | No | No | Yes | No | 1 |
| [295,296] | Yes | Yes | No | Yes | No | $n_f$ |
| IFDIC | Yes | Yes | Yes | Yes | Yes | 1 |

The following definitions will be used throughout this chapter.

**Definition 4.1.** *(H₋ index) [48]. For the operator $T_{yx}$, the $H_-$ index is defined as $||T_{yx}||_- = \inf( (||y||_2) / (||x||_2) )$.*

**Definition 4.2.** *(Stochastic $H_\infty$ norm) [293]. For the operator $T_{yx}$, the stochastic $H_\infty$ norm is defined as $||T_{yx}||_\infty^{\mathbb{E}} = \sup( (||y||_2^{\mathbb{E}}) / (||x||_2) )$.*

**Definition 4.3.** *(Stochastic $L_1$ norm) [297]. For the operator $T_{yx}$, the stochastic $L_1$ norm is defined as $||T_{yx}||_1^{\mathbb{E}} = \sup( (||y||_\infty^{\mathbb{E}}) / (||x||_\infty) )$.*

## 4.2   System description and problem formulation

In this section, we first describe the system and the IFDIC module governing equations. The IFDIC problem is then formulated for the Markovian jump linear systems with uncertain transition probabilities.

### 4.2.1   System description

Consider the following continuous-time Markovian jump linear system defined on a complete probability space $(\Omega, \mathfrak{F}, \mathrm{P})$:

$$
\begin{cases}
\dot{x}(t) = A_{\lambda_t} x(t) + B_{\mathrm{u}\lambda_t} u(t) + B_{\mathrm{d}\lambda_t} d(t) + B_{\mathrm{w}\lambda_t} w(t) + B_{\mathrm{f}\lambda_t} f(t) \\
y(t) = C_{\lambda_t} x(t) + D_{\mathrm{u}\lambda_t} u(t) + D_{\mathrm{d}\lambda_t} d(t) + D_{\mathrm{w}\lambda_t} w(t) + D_{\mathrm{f}\lambda_t} f(t) \ , \\
z(t) = E_{\lambda_t} x(t) + F_{\mathrm{d}\lambda_t} d(t) + F_{\mathrm{w}\lambda_t} w(t) + F_{\mathrm{f}\lambda_t} f(t)
\end{cases}
\tag{4.1}
$$

where $x(t) \in \mathbb{R}^n$ denotes the state vector, $u(t) \in \mathbb{R}^{n_u}$ denotes the control input, $d(t) \in \mathbb{R}^{n_d}$ denotes an $\mathbb{L}_2$-norm bounded external disturbance, $w(t) \in \mathbb{R}^{n_w}$ denotes an $\mathbb{L}_\infty$

disturbance signal satisfying $||w||_\infty \leq 1$, the unknown input $f(t) \in \mathbb{R}^{n_f}$ denotes a possible fault, $y(t) \in \mathbb{R}^{n_y}$ with $n_y \geq n_f$ denotes the measured output, and $z(t) \in \mathbb{R}^{n_z}$ denotes the regulated output. Without loss of generality, it is assumed that $f(t)$ is $\mathbb{L}_2$-norm bounded. Note that the fault matrices $B_{f\lambda_t}$ and $D_{f\lambda_t}$ are specified according to the faults that are to be detected in the components, actuators, or sensors.

The jumping mode process $\lambda_t, t \geq 0$ is a continuous-time, discrete state homogeneous Markov process on the probability space $(\Omega, \mathfrak{F}, P)$ that takes on values in a finite state space $\mathfrak{L} \triangleq \{1, 2, \ldots, s\}$ and is governed by the mode transition probabilities:

$$\text{Prob}(\lambda_{t+\Delta t} = j \mid \lambda_t = i) = \begin{cases} \pi_{ij}\Delta t + o(\Delta t) & \text{if } j \neq i \\ 1 + \pi_{ii}\Delta t + o(\Delta t) & \text{if } j = i \end{cases}, \tag{4.2}$$

where $\Delta t > 0$, $\lim_{\Delta t \to 0} o(\Delta t)/\Delta t = 0$, and $\pi_{ij} \geq 0$ is the transition rate from mode $i$ at time $t$ to mode $j \neq i$ at time $t + \Delta t$ and $\pi_{ii} = -\sum_{j=1, j \neq i}^s \pi_{ij}$.

It is further assumed that the mode transition rate matrix $\Pi \triangleq [\pi_{ij}]$ is not precisely known *a priori*, but belongs to the following admissible uncertainty domain:

$$\mathfrak{D}_\pi \triangleq \{\hat{\Pi} + \Delta\Pi : |\Delta\pi_{ij}| \leq \varepsilon_{ij}, \varepsilon_{ij} \geq 0 \text{ for all } i, j \in \mathfrak{L}, j \neq i\}, \tag{4.3}$$

where the matrix $\hat{\Pi} \triangleq [\hat{\pi}_{ij}]$ is a known constant real matrix, and $\Delta\Pi \triangleq (\Delta\pi_{ij})$ denotes the uncertainty in the mode transition rate matrix. For all $i, j \in \mathfrak{L}, j \neq i$, $\hat{\pi}_{ij}(\geq 0)$ denotes the estimated value of $\pi_{ij}$, and the error between them is designated by $\Delta\pi_{ij}$ that can take on any value in $[-\varepsilon_{ij}, \varepsilon_{ij}]$. For all $i \in \mathfrak{L}$, $\hat{\pi}_{ii} \triangleq -\sum_{j=1, j \neq i}^s \hat{\pi}_{ij}$ and $\Delta\pi_{ii} \triangleq -\sum_{j=1, j \neq i}^s \Delta\pi_{ij}$.

For simplicity, the model uncertainties are assumed to be recast as disturbances, although other types of uncertainties can also be formally investigated. For simplicity in the presentation, the following notations are adopted, namely, $A_i \triangleq A_{\lambda_t}$, $B_{ui} \triangleq B_{u\lambda_t}$, $B_{di} \triangleq B_{d\lambda_t}$, $B_{wi} \triangleq B_{w\lambda_t}$, $B_{fi} \triangleq B_{f\lambda_t}$, $C_i \triangleq C_{\lambda_t}$, $D_{ui} \triangleq D_{u\lambda_t}$, $D_{di} \triangleq D_{d\lambda_t}$, $D_{wi} \triangleq D_{w\lambda_t}$, $D_{fi} \triangleq D_{f\lambda_t}$, $E_i \triangleq E_{\lambda_t}$, $F_{di} \triangleq F_{d\lambda_t}$, $F_{wi} \triangleq F_{w\lambda_t}$, and $F_{fi} \triangleq F_{f\lambda_t}$ whenever $\lambda_t = i, i \in \mathfrak{L}$.

Our proposed IFDIC module is now described as follows:

$$\begin{cases} \dot{\hat{x}}(t) = A_{\lambda_t}\hat{x}(t) + B_{u\lambda_t}u(t) + L_{\lambda_t}(y(t) - \hat{y}(t)) \\ \hat{y}(t) = C_{\lambda_t}\hat{x}(t) + D_{u\lambda_t}u(t) \\ u(t) = K_{\lambda_t}\hat{x}(t) \\ r(t) = V_{\lambda_t}(y(t) - \hat{y}(t)) \end{cases}, \tag{4.4}$$

where $\hat{x}(t) \in \mathbb{R}^n$ denotes the estimate of $x(t)$, $\hat{y}(t) \in \mathbb{R}^{n_y}$ denotes the observer output, $r(t) \in \mathbb{R}^{n_f}$ denotes the residual signal, $K_{\lambda_t} \in \mathbb{R}^{n_u \times n}$ denotes the controller gain and the constant matrices $L_{\lambda_t} \in \mathbb{R}^{n \times n_y}$, and $V_{\lambda_t} \in \mathbb{R}^{n_f \times n_y}$ denotes the observer parameters to be designed subsequently. For $\lambda_t = i \in \mathfrak{L}$, the matrices $K_{\lambda_t}$, $L_{\lambda_t}$, and $V_{\lambda_t}$ are denoted by $K_i = K_{\lambda_t}$, $L_i = L_{\lambda_t}$, and $V_i = V_{\lambda_t}$, respectively.

By substituting the IFDIC module (4.4) into the system (4.1), the following closed-loop dynamics is obtained:

$$\begin{cases} \dot{x}_{cl}(t) = A_{cl\lambda_t}x_{cl}(t) + B_{dcl\lambda_t}d(t) + B_{wcl\lambda_t}w(t) + B_{fcl\lambda_t}f(t) \\ r(t) = C_{cl\lambda_t}x_{cl}(t) + D_{dcl\lambda_t}d(t) + D_{wcl\lambda_t}w(t) + D_{fcl\lambda_t}f(t) \quad , \\ z(t) = E_{cl\lambda_t}x_{cl}(t) + F_{d\lambda_t}d(t) + F_{w\lambda_t}w(t) + F_{f\lambda_t}f(t) \end{cases} \quad (4.5)$$

where $x_{cl}(t) = [x^{\mathrm{T}}(t) \quad e^{\mathrm{T}}(t)]^{\mathrm{T}}$, $e(t) = x(t) - \hat{x}(t)$, and

$$A_{cli} = \begin{bmatrix} A_i + B_{ui}K_i & -B_{ui}K_i \\ 0 & A_i - L_iC_i \end{bmatrix}, \quad C_{cli} = \begin{bmatrix} 0 & V_iC_i \end{bmatrix},$$

$$B_{fcli} = \begin{bmatrix} B_{fi} \\ B_{fi} - L_iD_{fi} \end{bmatrix}, \quad B_{dcli} = \begin{bmatrix} B_{di} \\ B_{di} - L_iD_{di} \end{bmatrix}, \quad B_{wcli} = \begin{bmatrix} B_{wi} \\ B_{wi} - L_iD_{wi} \end{bmatrix},$$

$$E_{cli} = \begin{bmatrix} E_i & 0 \end{bmatrix}, \quad D_{dcli} = V_iD_{di}, \quad D_{wcli} = V_iD_{wi}, \quad D_{fcli} = V_iD_{fi}.$$

In the next subsection, the IFDIC design problem is transformed into a mixed $H_\infty/H_-/L_1$ optimization problem.

## 4.2.2 *The IFDIC design problem formulation*

The IFDIC problem that is addressed in this chapter is now formally stated as follows.

***The IFDIC problem***:

Our objective is to design the IFDIC module (4.4) for the system (4.1) such that

- the augmented closed-loop system (4.5) is mean-square stable over all admissible uncertainty domain in (4.3),
- the effects of disturbances $(d(k), w(k))$ on the residual $r(t)$ are minimized and the effects of faults on the residual $r(t)$ are maximized (in order to accomplish the FD goal),
- each element of the residual $r(t)$ is only sensitive to a specified potential fault (in order to accomplish the fault isolation goal), and
- the effects of bounded energy and bounded peak disturbances and faults on the regulated output $z(t)$ are minimized (in order to accomplish the control performance goal).

Stated alternatively, our goal is to design the IFDIC module (4.4) such that the augmented closed-loop system (4.5) is mean-square stable and the following optimization problem is satisfied:

$$\text{minimize} \quad \beta_1 \gamma_1 + \beta_2 \gamma_2 + \beta_3 \gamma_3 + \beta_4 \gamma_4 + \beta_5 \gamma_5 + \beta_6 \gamma_6$$

subject to

(I) $||T_{zd}||_\infty^{\mathbb{E}} < \gamma_1,$        (II) $||T_{zf}||_\infty^{\mathbb{E}} < \gamma_2,$

(III) $||T_{rf} - T_{ref}||_\infty^{\mathbb{E}} < \gamma_3,$    (IV) $||T_{rd}||_\infty^{\mathbb{E}} < \gamma_4,$      (4.6)

(V) $||T_{ref}||_- \geq 1,$          (VI) $||T_{zw}||_1^{\mathbb{E}} < \gamma_5,$

(VII) $||T_{rw}||_1^{\mathbb{E}} < \gamma_6,$

where $r_e(t) = Qf(t)$ and $Q$ is selected to have the following structure:

$$Q = \text{diag}(q_1, \ldots, q_{n_f}) \in \mathbb{R}^{n_f \times n_f}, q_l > 0, \forall l \in \{1, \ldots, n_f\}. \tag{4.7}$$

The positive constant weights $\beta_1$–$\beta_6$ can be used by the designer as a trade-off analysis among the objectives (I)–(VII). The $H_\infty$ and $L_1$ performance indices (I), (II), and (VI) are used to attenuate the effects of the disturbances ($d(t)$ and $w(t)$) and faults on the regulated output. The $H_\infty$ and $L_1$ performance indices (IV) and (VII) are used to attenuate the effects of bounded energy and bounded peak disturbances on the residual, respectively. The $H_\infty$ and $H_-$ performance indices (III) and (V) are used to guarantee a minimum level of sensitivity of residuals to the fault signals and to ensure the fault isolation goal.

**Remark 4.1.** *It should be noted that in this chapter we have selected the simplest choice of $T_{ref}$ (viz., a fixed diagonal matrix). Refer to [278] for more detail on other choices for $T_{ref}$. It should also be noted that the requirement of $T_{ref}$ being diagonal is made to ensure the fault isolability in case that multiple faults occur simultaneously in the system.*

In the next section, the IFDIC problem is solved for the closed-loop system (4.5) subject to the conditions in (4.6).

## 4.3 Main results

There are seven performance indices (I)–(VII) that must be satisfied simultaneously for solving the IFDIC problem for the closed-loop system (4.5). Note that it is straightforward to express the inequality (V) as a matrix inequality. Indeed, from Definition 4.1 it follows that

$$||T_{ref}||_- \geq 1 \Leftrightarrow \left\{ \int_0^{+\infty} ||r_e(t)||^2 dt \right\} \geq \left\{ \int_0^{+\infty} ||f(t)||^2 dt \right\}. \tag{4.8}$$

It can easily be shown that the above inequality can be converted to an LMI condition, namely,

$$I - Q \leq 0. \tag{4.9}$$

In the following discussion, at first in Theorem 4.1, sufficient conditions for the $H_\infty$ performance indices (I)–(IV) are formulated as LMI feasibility problems. Then, the sufficient LMI conditions for the $L_1$ performance indices (VI) and (VII) are obtained in Theorem 4.2. Finally, in Corollary 4.1, a feasible solution to the IFDIC problem is obtained by simultaneously considering all the indices (I)–(VII).

The extended LMI conditions are provided that do not involve products of Lyapunov matrices and the system matrices. Consequently, the conservativeness in the solution will be reduced as the system Lyapunov matrices can be selected to be different from one another.

**Theorem 4.1.** *The closed-loop system (4.5) is mean-square stable and the performance indices (I)–(IV) are guaranteed if there exist symmetric positive-definite matrices $P_{1i}$, $P_{2i}$, $P_{3i}$, $P_{4i}$, diagonal matrix $Q$, matrices $X_{1i}$, $X_{2i}$, $\hat{X}_{1i}$, $M_i$, $N_i$, $V_i$, $i \in \mathfrak{L}$, and positive scalars $\alpha$, $\gamma_1$, $\gamma_2$, $\gamma_3$, $\gamma_4$, $\theta_{kij}$, $i,j \in \mathfrak{L}, i \neq j$, $k = 1, \ldots, 4$, such that*

$$
\begin{bmatrix}
E_{11i} & E_{2i}+P_{1i} & E_{3i} & E_{4i} & \Phi_{1i} \\
* & E_{5i} & \alpha E_{3i} & 0 & 0 \\
* & * & -\gamma_1^2 I & F_{di}^{\mathrm{T}} & 0 \\
* & * & * & -I & 0 \\
* & * & * & * & \Theta_{1i}
\end{bmatrix} < 0,
$$

$$
\begin{bmatrix}
E_{21i} & E_{2i}+P_{2i} & \Xi_{3i} & E_{4i} & \Phi_{2i} \\
* & E_{5i} & \alpha \Xi_{3i} & 0 & 0 \\
* & * & -\gamma_2^2 I & F_{fi}^{\mathrm{T}} & 0 \\
* & * & * & -I & 0 \\
* & * & * & * & \Theta_{2i}
\end{bmatrix} < 0,
$$

(4.10)

$$
\begin{bmatrix}
E_{31i} & E_{2i}+P_{3i} & \Xi_{3i} & \Xi_{4i} & \Phi_{3i} \\
* & E_{5i} & \alpha \Xi_{3i} & 0 & 0 \\
* & * & -\gamma_3^2 I & \Xi_{6i} & 0 \\
* & * & * & -I & 0 \\
* & * & * & * & \Theta_{3i}
\end{bmatrix} < 0,
$$

$$
\begin{bmatrix}
E_{41i} & E_{2i}+P_{4i} & E_{3i} & \Xi_{4i} & \Phi_{4i} \\
* & E_{5i} & \alpha E_{3i} & 0 & 0 \\
* & * & -\gamma_4^2 I & E_{6i} & 0 \\
* & * & * & -I & 0 \\
* & * & * & * & \Theta_{4i}
\end{bmatrix} < 0,
$$

$$B_{ui}^{\mathrm{T}} X_{1i} = \hat{X}_{1i} B_{ui}^{\mathrm{T}},$$

*for $i \in \mathfrak{L}$, where $E_{k1i} = Herm(\Omega_i) + \sum_{j=1,j\neq i}^{s} (\theta_{kij}/4)\varepsilon_{ij}^2 I + \sum_{j=1}^{s} \hat{\pi}_{ij} P_{kj}$, $E_{2i} = \lambda\Omega_i - X_i^{\mathrm{T}}$, $E_{5i} = -\lambda(X_i + X_i^{\mathrm{T}})$, $\Phi_{ki} = [P_{ki} - P_{k1}, \ldots, P_{ki} - P_{k(i-1)}, P_{ki} - P_{k(i+1)}, \ldots, P_{ki} - P_{ks}]$, $\Theta_{ki} = diag(\theta_{ki1} I, \ldots, \theta_{ki(i-1)} I, \theta_{ki(i+1)} I, \ldots, \theta_{kis} I)$, $X_i = diag(X_{1i}, X_{2i})$, and*

$$\Omega_i = \begin{bmatrix} A_i^\mathrm{T} X_{1i} + M_i^\mathrm{T} B_{ui}^\mathrm{T} & 0 \\ -M_i^\mathrm{T} B_{ui}^\mathrm{T} & A_i^\mathrm{T} X_{2i} - C_i^\mathrm{T} N_i^\mathrm{T} \end{bmatrix},$$

$$E_{3i} = \begin{bmatrix} X_{1i}^\mathrm{T} B_{di} \\ X_{2i}^\mathrm{T} B_{di} - N_i^\mathrm{T} D_{di} \end{bmatrix}, \quad E_{4i} = \begin{bmatrix} E_i^\mathrm{T} \\ 0 \end{bmatrix},$$

$$\Xi_{3i} = \begin{bmatrix} X_{1i}^\mathrm{T} B_{fi} \\ X_{2i}^\mathrm{T} B_{fi} - N_i^\mathrm{T} D_{fi} \end{bmatrix}, \quad \Xi_{4i} = \begin{bmatrix} 0 \\ C_i^\mathrm{T} V_i^\mathrm{T} \end{bmatrix},$$

$$E_{6i} = D_{di}^\mathrm{T} V_i^\mathrm{T}, \quad \Xi_{6i} = D_{fi}^\mathrm{T} V_i^\mathrm{T} - Q^\mathrm{T}.$$

*The filter parameters $L_i$ and $K_i$ are then given by*

$$K_i = (M_i^\mathrm{T} \hat{X}_{1i}^{-1})^\mathrm{T}, \quad L_i = (N_i^\mathrm{T} X_{2i}^{-1})^\mathrm{T}. \tag{4.11}$$

*Proof.* To guarantee the mean-square stability and the performance index ($I$) for the closed-loop system (4.5), the following condition must be satisfied for $i \in \mathcal{L}$ [295], namely,

$$\begin{bmatrix} I & 0 \\ A_{\mathrm{cl}i} & B_{\mathrm{dcl}i} \end{bmatrix}^\mathrm{T} \begin{bmatrix} \sum_{j=1}^s \pi_{ij} P_{1j} & P_{1i} \\ P_{1i} & 0 \end{bmatrix} \begin{bmatrix} I & 0 \\ A_{\mathrm{cl}i} & B_{\mathrm{dcl}i} \end{bmatrix}$$
$$+ \begin{bmatrix} 0 & I \\ E_{\mathrm{cl}i} & F_{di} \end{bmatrix}^\mathrm{T} \begin{bmatrix} -\gamma_1^2 I & 0 \\ 0 & I \end{bmatrix} \begin{bmatrix} 0 & I \\ E_{\mathrm{cl}i} & F_{di} \end{bmatrix} < 0. \tag{4.12}$$

Note that in inequality (4.12), the Lyapunov matrices and the system matrices are coupled with one another. If the LMI conditions involve the product of Lyapunov matrices and the system state-space matrices, then by selecting identical Lyapunov matrices, one would end up with a conservative solution. Consequently, the Projection Lemma [238] (as stated in Lemma 1.1) is used here to reduce the conservativeness in the solution to the IFDIC problem.

Specifically, the inequality (4.12) is now reformulated as

$$N_{U_i}^\mathrm{T} Z_i N_{U_i} < 0, \tag{4.13}$$

where $N_{U_i}$ and $Z_i$ are given by

$$Z_i = \begin{bmatrix} \sum_{j=1}^s \pi_{ij} P_{1j} + E_{\mathrm{cl}i}^\mathrm{T} E_{\mathrm{cl}i} & P_{1i} & E_{\mathrm{cl}i}^\mathrm{T} F_{di} \\ * & 0 & 0 \\ * & * & F_{di}^\mathrm{T} F_{di} - \gamma_1^2 I \end{bmatrix},$$

$$N_{U_i} = \begin{bmatrix} I & 0 \\ A_{\mathrm{cl}i} & B_{\mathrm{dcl}i} \\ 0 & I \end{bmatrix} \rightarrow U_i = \begin{bmatrix} A_{\mathrm{cl}i} & -I & B_{\mathrm{dcl}i} \end{bmatrix}. \tag{4.14}$$

If one chooses the matrix $N_{V_i}$ in (1.9b) as

$$N_{V_i} = \begin{bmatrix} \alpha I & 0 \\ -I & 0 \\ 0 & I \end{bmatrix} \rightarrow V_i = \begin{bmatrix} I & \alpha I & 0 \end{bmatrix}, \tag{4.15}$$

and uses Lemma 1.1, then it can be concluded that there exists a matrix $X_i$ such that the inequality (4.13) is equivalent to

$$Z_i + \mathrm{Herm} \left\{ \begin{bmatrix} A_{cli}^{\mathrm{T}} \\ -I \\ B_{dcli}^{\mathrm{T}} \end{bmatrix} \begin{bmatrix} X_i & \alpha X_i & 0 \end{bmatrix} \right\} < 0. \tag{4.16}$$

By partitioning $X_i$ into

$$X_i = \begin{bmatrix} X_{1i(n \times n)} & 0 \\ 0 & X_{2i(n \times n)} \end{bmatrix}, \tag{4.17}$$

and using (4.5), (4.14), (4.16), (4.17) and also the Schur complement, it follows that

$$\begin{bmatrix} \varphi_i & P_{1i} + \alpha \varphi_{1i} - X_i^{\mathrm{T}} & \varphi_{2i} & \varphi_{3i} \\ * & -\alpha(X_i + X_i^{\mathrm{T}}) & \alpha \varphi_{2i} & 0 \\ * & * & -\gamma_1^2 I & F_{di}^{\mathrm{T}} \\ * & * & * & -I \end{bmatrix} < 0, \tag{4.18}$$

where $\varphi_i = \mathrm{Herm}(\varphi_{1i}) + \sum_{j=1}^{s} \pi_{ij} P_{1j}$ and

$$\varphi_{1i} = \begin{bmatrix} A_i^{\mathrm{T}} X_{1i} + K_i^{\mathrm{T}} B_{ui}^{\mathrm{T}} X_{1i} & 0 \\ -K_i^{\mathrm{T}} B_{ui}^{\mathrm{T}} X_{1i} & A_i^{\mathrm{T}} X_{2i} - C_i^{\mathrm{T}} L_i^{\mathrm{T}} X_{2i} \end{bmatrix},$$

$$\varphi_{2i} = \begin{bmatrix} X_{1i}^{\mathrm{T}} B_{di} \\ X_{2i}^{\mathrm{T}} B_{di} - X_{2i}^{\mathrm{T}} L_i D_{di} \end{bmatrix}, \quad \varphi_{3i} = \begin{bmatrix} E_i^{\mathrm{T}} \\ 0 \end{bmatrix}. \tag{4.19}$$

Note that $\sum_{j=1}^{s} \pi_{ij} P_{1j}$ in (4.18) can be further written as

$$\sum_{j=1}^{s} \hat{\pi}_{ij} P_{1j} + \sum_{j=1, j \neq i}^{s} \left[ \frac{1}{2} \Delta \pi_{ij} (P_{1j} - P_{1i}) + \frac{1}{2} \Delta \pi_{ij} (P_{1j} - P_{1i}) \right]. \tag{4.20}$$

By applying Lemma 1.2 and using (4.20), the matrix inequality (4.18) holds for all $|\Delta \pi_{ij}| \leq \varepsilon_{ij}$ if there exist $\theta_{1ij} > 0$, $i, j \in \mathfrak{L}$, $i \neq j$ such that

$$\begin{bmatrix} \Upsilon_i & P_{1i} + \alpha \varphi_{1i} - X_i^{\mathrm{T}} & \varphi_{2i} & \varphi_{3i} \\ * & -\alpha(X_i + X_i^{\mathrm{T}}) & \alpha \varphi_{2i} & 0 \\ * & * & -\gamma_1^2 I & F_{di}^{\mathrm{T}} \\ * & * & * & -I \end{bmatrix} < 0, \tag{4.21}$$

where $\Upsilon_i = \mathrm{Herm}(\varphi_{1i}) + \sum_{j=1}^{s} \hat{\pi}_{ij} P_{1j} + \sum_{j=1,j\neq i}^{s} [(\theta_{1ij}/4)\varepsilon_{ij}^2 I + \frac{1}{\theta_{1ij}}(P_{1j} - P_{1i})^2]$ and $\varphi_{1i}, \varphi_{2i}$, and $\varphi_{3i}$ are defined in (4.19).

Since (4.21) is not in the form of an LMI, by using the Schur complement and the equality constraint $B_{ui}^{T} X_{1i} = \hat{X}_{1i} B_{ui}^{T}$ and substituting $N_i^{T} = L_i^{T} X_{2i}$ and $M_i^{T} = K_i^{T} \hat{X}_{1i}$ into (4.21), it follows that the first inequality in (4.10) is obtained. Note that, according to (4.21), $-\alpha(X_i + X_i^{T}) < 0$, and this guarantees the invertibility of the matrices $X_{1i}$ and $X_{2i}$. Moreover, if the matrix $B_{ui}$ is full column rank ($\mathrm{rank}(B_{ui}) = n_u$), then the matrix equation $B_{ui}^{T} X_{1i} = \hat{X}_{1i} B_{ui}^{T}$ implies

$$
\begin{aligned}
n_u &\geq \mathrm{rank}(\hat{X}_{1i}) \geq \mathrm{rank}(\hat{X}_{1i} B_{ui}^{T}) = \mathrm{rank}(B_{ui}^{T} X_{1i}) \\
&\geq \mathrm{rank}((B_{ui}^{T} X_{1i}) X_{1i}^{-1}) = \mathrm{rank}(B_{ui}^{T}) = n_u.
\end{aligned}
\tag{4.22}
$$

In other words, $\hat{X}_{1i}$ is nonsingular. Therefore, the filter parameters $L_i$ and $K_i$ can be obtained from $K_i = (M_i^{T} \hat{X}_{1i}^{-1})^{T}$ and $L_i = (N_i^{T} X_{2i}^{-1})^{T}$.

The performance indices (II)–(IV) can also be shown to hold by following along the same lines as in the derivation of the performance index (I). These details are not included here for brevity. This now completes the proof of the theorem. $\qquad \square$

The following theorem provides the LMI constraints for the $L_1$ performance indices (VI) and (VII).

**Theorem 4.2.** *The closed-loop system (4.5) is mean-square stable and the $L_1$ performance indices (VI) and (VII) are guaranteed if there exist symmetric positive-definite matrices $P_{5i}, P_{6i}$, matrices $X_{1i}, X_{2i}, \hat{X}_{1i}, M_i, N_i, V_i, i \in \mathfrak{L}$ and positive scalars $\alpha, \mu_i, \sigma_i, \gamma_5, \gamma_6, \theta_{kij}, i,j \in \mathfrak{L}, i \neq j, k = 5,6$, such that*

$$
\begin{bmatrix}
E_{51i} + \sigma_i P_{5i} & E_{2i} + P_{5i} & \mathscr{E}_{3i} & \Phi_{5i} \\
* & E_{5i} & \alpha\mathscr{E}_{3i} & 0 \\
* & * & -\mu_i I & 0 \\
* & * & * & \Theta_{5i}
\end{bmatrix} < 0,
$$

$$
\begin{bmatrix}
-\sigma_i P_{5i} & 0 & E_{4i} \\
* & -(\gamma_5 - \mu_i) I & F_{wi}^{T} \\
* & * & -\gamma_5 I
\end{bmatrix} < 0,
$$

$$
\begin{bmatrix}
E_{61i} + \sigma_i P_{6i} & E_{2i} + P_{6i} & \mathscr{E}_{3i} & \Phi_{6i} \\
* & E_{5i} & \alpha\mathscr{E}_{3i} & 0 \\
* & * & -\mu_i I & 0 \\
* & * & * & \Theta_{6i}
\end{bmatrix} < 0,
\tag{4.23}
$$

$$
\begin{bmatrix}
-\sigma_i P_{6i} & 0 & \Xi_{4i} \\
* & -(\gamma_6 - \mu_i) I & D_{wi}^{T} V_i^{T} \\
* & * & -\gamma_6 I
\end{bmatrix} < 0,
$$

$$
B_{ui}^{T} X_{1i} = \hat{X}_{1i} B_{ui}^{T},
$$

*for $i \in \mathcal{L}$, where $\mathscr{E}_{3i}^{\mathrm{T}} = [B_{wi}^{\mathrm{T}} X_{1i}\ B_{wi}^{\mathrm{T}} X_{2i} - D_{wi}^{\mathrm{T}} N_i]$, and $E_{k1i}$, $E_{2i}$, $E_{5i}$, $\Phi_{ki}$, $\Theta_{ki}$ are defined in Theorem 4.1. The filter parameters $L_i$ and $K_i$ are then given by (4.11).*

*Proof.* Let the mode at time $t$ be $i$, that is $\lambda_t = i, i \in \mathcal{L}$. Consider the Lyapunov function $V(x_{\mathrm{cl}}(t), \lambda_t = i, t \geq 0) \equiv V(x_{\mathrm{cl}}(t), i) = x_{\mathrm{cl}}^{\mathrm{T}}(t) P_{5i} x_{\mathrm{cl}}(t)$, where $P_{5i}$ is a symmetric positive-definite matrix for each modes $i \in \mathcal{L}$. Now, according to the first inequality in (4.23), by substituting $M_i$, $N_i$ with $M_i = \hat{X}_{1i}^{\mathrm{T}} K_i$, $N_i = X_{2i}^{\mathrm{T}} L_i$, and applying the Schur complement and Lemma 1.2, one can conclude that

$$\begin{bmatrix} \varphi_i & P_{5i} + \alpha \varphi_{1i} - X_i^{\mathrm{T}} & \varphi_{2i} \\ * & -\alpha(X_i + X_i^{\mathrm{T}}) & \alpha \varphi_{2i} \\ * & * & -\mu_i I \end{bmatrix} < 0, \tag{4.24}$$

where $\varphi_i = \mathrm{Herm}(\varphi_{1i}) + \sum_{j=1}^{s} \pi_{ij} P_{5j} + \sigma_i P_{5i}$, and $\varphi_{1i}$ is defined in (4.19) and

$$\varphi_{2i} = \begin{bmatrix} X_{1i}^{\mathrm{T}} B_{wi} \\ X_{2i}^{\mathrm{T}} B_{wi} - X_{2i}^{\mathrm{T}} L_i D_{wi} \end{bmatrix}. \tag{4.25}$$

Using Lemma 1.1, the previous inequality holds if and only if the following LMIs are satisfied, namely,

$$\begin{bmatrix} I & 0 \\ A_{\mathrm{cl}i} & B_{\mathrm{wcl}i} \\ 0 & I \end{bmatrix}^{\mathrm{T}} \begin{bmatrix} \sum_{j=1}^{s} \pi_{ij} P_{5j} + \sigma_i P_{5i} & P_{5i} & 0 \\ P_{5i} & 0 & 0 \\ 0 & 0 & -\mu_i I \end{bmatrix} \begin{bmatrix} I & 0 \\ A_{\mathrm{cl}i} & B_{\mathrm{wcl}i} \\ 0 & I \end{bmatrix} < 0, \tag{4.26}$$

$$\begin{bmatrix} \alpha I & 0 \\ -I & 0 \\ 0 & I \end{bmatrix}^{\mathrm{T}} \begin{bmatrix} \sum_{j=1}^{s} \pi_{ij} P_{5j} + \sigma_i P_{5i} & P_{5i} & 0 \\ P_{5i} & 0 & 0 \\ 0 & 0 & -\mu_i I \end{bmatrix} \begin{bmatrix} \alpha I & 0 \\ -I & 0 \\ 0 & I \end{bmatrix} < 0. \tag{4.27}$$

By multiplying both sides of (4.26) by $[x_{\mathrm{cl}}^{\mathrm{T}}(t)\ w^{\mathrm{T}}(t)]$ from the left and $[x_{\mathrm{cl}}^{\mathrm{T}}(t)\ w^{\mathrm{T}}(t)]^{\mathrm{T}}$ from the right, it follows that

$$\Gamma V(x_{\mathrm{cl}}(t), i) + \sigma_i x_{\mathrm{cl}}^{\mathrm{T}}(t) P_{5i} x_{\mathrm{cl}}(t) - \mu_i w^{\mathrm{T}}(t) w(t) < 0, \tag{4.28}$$

where $\Gamma$ is the weak infinitesimal operator of the stochastic process $\{x_{\mathrm{cl}}(t), \lambda_t = i\}$, $t \geq 0$ which is defined as

$$\Gamma V(x_{\mathrm{cl}}(t), \lambda_t = i) = \lim_{\Delta t \to 0} \frac{1}{\Delta t} [\mathbb{E}\{V(x_{\mathrm{cl}}(t + \Delta t), \lambda_{t+\Delta t} | x_{\mathrm{cl}}(t), \lambda_t = i)\} - V(x_{\mathrm{cl}}(t), \lambda_t = i)]$$

$$= \frac{\partial}{\partial t} V(x_{\mathrm{cl}}(t), i) + \frac{\partial}{\partial x_{\mathrm{cl}}} V(x_{\mathrm{cl}}(t), i) \dot{x}_{\mathrm{cl}}(t, i) + \sum_{j=1}^{s} \pi_{ij} V(x_{\mathrm{cl}}(t), j). \tag{4.29}$$

For $w(t) = 0$, (4.28) leads to $\Gamma V(x_{\mathrm{cl}}(t), i) < 0$ that guarantees the asymptotic stability of the system. Moreover, from the results in [301] and [271], it follows that

$$\sigma_i x_{\mathrm{cl}}^{\mathrm{T}}(t) P_{5i} x_{\mathrm{cl}}(t) < \mu_i w(t)^{\mathrm{T}} w(t) \quad \text{for all } t \geq 0. \tag{4.30}$$

Based on the fact that $w(t)^\mathrm{T} w(t) \leq 1$, one can conclude that $V(x_\mathrm{cl}(t), i) < \mu_i/\sigma_i$, for $t \geq 0$.

By substituting $M_i$, $N_i$ with $M_i = \hat{X}_{1i}^\mathrm{T} K_i$, $N_i = X_{2i}^\mathrm{T} L_i$, respectively, in the second inequality in (4.23), and applying the Schur complement it follows that

$$\gamma_5^{-1} \begin{bmatrix} E_{4i} \\ F_{wi}^\mathrm{T} \end{bmatrix} \begin{bmatrix} E_{4i} & F_{wi} \end{bmatrix} < \begin{bmatrix} -\sigma_i P_{5i} & 0 \\ * & -(\gamma_5 - \mu_i)I \end{bmatrix}. \tag{4.31}$$

Now, by multiplying both sides of (4.31) by $[x_\mathrm{cl}^\mathrm{T}(t) \quad w^\mathrm{T}(t)]$ from the left and $[x_\mathrm{cl}^\mathrm{T}(t) \quad w^\mathrm{T}(t)]^\mathrm{T}$ from the right, one can conclude that

$$\mathbb{E}\{z^\mathrm{T}(t)z(t)\} < \mathbb{E}\{\gamma_5 \left(\sigma_i x_\mathrm{cl}^\mathrm{T}(t) P_{5i} x_\mathrm{cl}(t) + (\gamma_5 - \mu_i) w^\mathrm{T}(t)w(t)\right)\}. \tag{4.32}$$

Since $w^\mathrm{T}(t)w(t) \leq 1$ and $V(x_\mathrm{cl}(t), i) < \mu_i/\sigma_i$, therefore $\mathbb{E}\{z^\mathrm{T}(t)z(t)\} < \gamma_5^2$. Consequently, the $L_1$ norm (VI) is satisfied. The $L_1$ performance index (VII) can also be shown to hold by employing the same procedure as in the derivation of the performance index (VI). These details are not included for sake of brevity. This completes the proof of the theorem. □

**Remark 4.2.** *Note that the second and the fourth inequalities in (4.23) are only linear if $\sigma_i$ is fixed. Determining the best bound guaranteed by these inequalities requires that one performs a line search over $\sigma_i > 0$.*

It should be pointed out that the $L_1$ performance index has been used in [297] and [298] to consider the filtering and control problems for Markovian jump systems, respectively. However, in this chapter, the $L_1$ performance index is used in the IFDIC problem for Markovian jump linear systems with uncertain transition probabilities. Moreover, unlike the LMI conditions in [297] and [298], in this work based on the projection lemma, the LMI feasibility conditions for solving the IFDIC problem are developed such that products of Lyapunov matrices with the system state space matrices are not involved that will lead to less conservative results.

Based on the above results, the following corollary is now proposed to solve the proposed IFDIC problem (4.6).

**Corollary 4.1.** *For the given positive constants $\beta_1, \ldots, \beta_6$, a feasible solution to the problem of IFDIC filter design is obtained by solving the following convex optimization problem for $i \in \mathfrak{L}$, $k = 1, \ldots, 6$,*

$$\min_{P_{ki}, X_i, \hat{X}_{1i}, M_i, N_i, V_i, Q} \beta_1\gamma_1 + \beta_2\gamma_2 + \beta_3\gamma_3 + \beta_4\gamma_4 + \beta_5\gamma_5 + \beta_6\gamma_6 \tag{4.33}$$

*subject to the inequalities (4.9), (4.10), and (4.23).*

*Proof.* The proof can be easily obtained by considering the IFDIC problem (4.6), the inequalities (4.8), (4.9), and by invoking Theorems 4.1 and 4.2 and is not included here for sake of brevity. □

**Remark 4.3.** *In the previous work in literature (e.g., [282]), an iterative LMI algorithm that requires a higher computational time than other LMI methods (e.g., [37,43]) is applied to reduce the overall conservatism of the solution for the fault diagnosis problem. It should be noted that the performance of iterative LMI algorithms is affected by initial conditions and the schemes are not globally convergent. Furthermore, the IFDC problem has been studied in [41] where bilinear matrix inequality conditions are proposed. To avoid these drawbacks, we have utilized the advantages of extended LMIs to develop LMI conditions for solving the IFDIC problem. Indeed as can be obtained from Theorems 4.1 and 4.2, the conservativeness in our solutions is reduced by introducing additional matrix variables ($X_{1i}$, $X_{2i}$) and the coupling of the Lyapunov matrices with the system state space matrices are eliminated.*

### Residual evaluation criterion:

Following construction of the residuals $r_i(t) \in \mathbb{R}$, $\forall i \in \{1, \ldots, n_f\}$ from (4.4), the final step in the IFDIC scheme is to determine the thresholds $J_{\text{th}_i}$ and the evaluation functions $J_{r_i}(t)$. Various evaluation functions can be considered [30].

The upper and lower threshold bands are selected as $J_{\text{th}_i}^u = \sup_{f=0, d \in \mathbb{L}_2, w \in \mathbb{L}_\infty} r_i(t)$ and $J_{\text{th}_i}^l = \inf_{f=0, d \in \mathbb{L}_2, w \in \mathbb{L}_\infty} r_i(t)$, respectively. Based on the selected thresholds and the evaluation function taken as $J_{r_i}(t) = r_i(t)$, the occurrence of a fault can then be detected and isolated by using the following decision logic: if $r_i(t) > J_{\text{th}_i}^u$ or $r_i(t) < J_{\text{th}_i}^l \implies f_i \neq 0$.

Given that our proposed IFDIC methodology is a passive FTC methodology, the FD information is not explicitly used in the controller design. However, the generated diagnostic information can be used subsequently in the framework of active FTC system design.

## 4.4 Case study

Application of our proposed methodology to a linearized model of the GE F-404 aircraft engine [293] is now presented in this section to illustrate its effectiveness and capabilities. By linearizing the model of the GE F-404 aircraft engine, the nominal system matrix $A_{\lambda_t}$ and the measurement output matrix $C_{\lambda_t}$ are obtained as follows [293,305,306]:

$$A_{\lambda_t} = \begin{bmatrix} -1.46 & 0 & 2.428 \\ 0.1643 + 0.5\delta(\lambda_t) & -0.4 - 2\delta(\lambda_t) & -0.3788 \\ 0.3107 & 0 & -2.23 \end{bmatrix},$$

$$C_{\lambda_t} = \begin{bmatrix} 1 & 0 & 0 \\ 0 & 1 & 0 \end{bmatrix},$$

where the three system states are defined as follows, with $x_1(t)$ denotes the sideslip angle, $x_2(t)$ denotes the roll rate, and $x_3(t)$ denotes the yaw rate. Let $A_{\lambda_t}$ be subject to a Markov process $\lambda_t$ changes in the set $\mathfrak{L} = \{1, 2\}$ with the transition rate

$$\hat{\Pi} = \begin{bmatrix} -3 & 3 \\ 4 & -4 \end{bmatrix}.$$

The uncertainties in the mode transition rate matrix $\pi_{ij}$ are such that $\Delta\pi_{12} \leq \varepsilon_{12}$ with $\varepsilon_{12} \triangleq \hat{\pi}_{12}/2$, $\Delta\pi_{21} \leq \varepsilon_{21}$ with $\varepsilon_{21} \triangleq \hat{\pi}_{21}/2$. The uncertainty $\delta(\lambda_t)$ is assumed to be 0.4 when $\lambda_t = 1$ and 0.2 when $\lambda_t = 2$. Hence, we have

$$A_1 = \begin{bmatrix} -1.46 & 0 & 2.428 \\ 0.3643 & -1.2 & -0.3788 \\ 0.3107 & 0 & -2.23 \end{bmatrix}, \quad A_2 = \begin{bmatrix} -1.46 & 0 & 2.428 \\ 0.2643 & -0.8 & -0.3788 \\ 0.3107 & 0 & -2.23 \end{bmatrix}.$$

In addition to the main system matrices $A_i$ and $C_i$, we set the other matrices as $E_1 = E_2 = C_1, D_{u1} = D_{u2} = 0_{2\times2}, F_{w1} = F_{w2} = 0_{2\times2}$ and

$$B_{u1} = B_{u2} = \begin{bmatrix} -0.07 & 0.08 \\ -0.05 & 0.11 \\ 0.09 & -0.06 \end{bmatrix}, \quad F_{d1} = F_{f1} = \begin{bmatrix} 0 & 0.005 \\ 0.1 & 0 \end{bmatrix},$$

$$F_{d2} = F_{f2} = \begin{bmatrix} 0.1 & 0.05 \\ 0 & 0.1 \end{bmatrix}.$$

The fault matrices $B_{fi}$ and $D_{fi}$ are selected as

$$B_{f1} = B_{f2} = \begin{bmatrix} 0 & 0.08 \\ 0 & 0.11 \\ 0 & -0.06 \end{bmatrix}, \quad D_{f1} = D_{f2} = \begin{bmatrix} 1 & 0 \\ 0 & 0 \end{bmatrix},$$

which represent the occurrence of an actuator fault in the second actuator and a sensor fault in the first sensor of the system.

According to [307], the disturbances produced by external environmental factors, such as wind gusts, gravity gradients, and sensor and actuator noise, may enter the aircraft engine in different ways. It is, therefore, reasonable to take disturbances into account when the aircraft engine is modeled. Hence, the bounded energy disturbance matrices $B_{di}$ and $D_{di}$ are selected as $B_{d1} = B_{d2} = B_{f1}$ and $D_{d1} = D_{d2} = 0.1I_{2\times2}$, which represent the occurrence of disturbances in the actuator and sensor noise in the system, respectively. Moreover, the bounded peak disturbance matrices $B_{wi}$ and $D_{wi}$ are selected as $B_{w1} = B_{w2} = B_{u1}$ and $D_{w1} = D_{w2} = 0_{2\times2}$, which represent the occurrence of disturbance in the system actuator.

In the following, the performance of our proposed IFDIC approach is demonstrated through considering concurrent abrupt faults with different severities in the first sensor and the second actuator of the system. In simulations that are conducted, and without loss of generality, the bounded energy disturbance $d(t) \in \mathbb{R}^2$ that is injected to the system is band-limited white noise with the power of 0.0001. Moreover, the bounded peak disturbance $w(t) \in \mathbb{R}^2$ is taken as $0.35 \sin(0.5t)$ [308,309].

The sensor fault signal $f_1(t)$ is simulated as a rectangular pulsed signal with an amplitude of 0.05 that is applied during the interval 20–30 s and the actuator fault signal $f_2(t)$ is considered as a rectangular pulsed signal with an amplitude of $-2.5$ that is applied during the interval 25–35 s. It is desired that the system detects and isolates the occurrence of the fault $f(t) \in \mathbb{R}^2$ in presence of disturbances $d(t), w(t) \in \mathbb{R}^2$ and also attenuates the effects of fault $f(t) \in \mathbb{R}^2$ and disturbances $d(t), w(t) \in \mathbb{R}^2$ on the regulated output $z(t) \in \mathbb{R}^2$.

To illustrate the trade-offs between the control and FDI objectives, three scenarios are considered by having different values of $\beta_i$, $i = 1, \ldots, 6$. In the first scenario, positive constant weights $\beta_1, \ldots, \beta_6$ are assumed to be equal to one, implying the same level of importance to FD, isolation, and control objectives [the objectives (I)–(VII)]. For a given $\alpha = 0.1$, $\theta_{kij} = 0.01$, $k = 1, \ldots, 4$, $i, j \in \mathfrak{L}, i \neq j$, $\sigma_1 = \sigma_2 = 0.5$, the optimization problem of Corollary 4.1 is solved and the parameters $\mu_1, \mu_2$ and the optimization parameters $\gamma_1, \ldots, \gamma_6$ are obtained as $\mu_1 = \mu_2 = 0.0818$, $\gamma_1 = 0.0287$, $\gamma_2 = 0.0489$, $\gamma_3 = 1.0182$, $\gamma_4 = 0.0226$, $\gamma_5 = 0.1727$, and $\gamma_6 = 0.0818$. Moreover, it should be pointed out that by conducting Monte Carlo simulations and considering the worst case analysis of the residuals corresponding to the healthy operation of the system subject to various disturbances, the threshold values $J_{\mathrm{th}_i}^u$ and $J_{\mathrm{th}_i}^l$ for $i = 1, 2$ are selected as $J_{\mathrm{th}_1}^l = -0.009$, $J_{\mathrm{th}_2}^l = -0.01$, $J_{\mathrm{th}_1}^u = 0.009$, and $J_{\mathrm{th}_2}^u = 0.01$.

The closed-loop system regulated output $z(t)$ is shown in Figure 4.1, where it can be concluded that the effects of disturbances and faults on the regulated output have been attenuated. Note that the performance indices (I), (II), and (VI) in condition (4.6) are used to *attenuate* the effects of disturbances ($d(t), w(t)$) and fault signals on the regulated output $z(t)$.

Indeed, the $H_\infty$ performance indices (I) and (II) guarantee that the energy of the bounded energy disturbance and fault signals ($\|d(t)\|_2$ and $\|f(t)\|_2$) to the energy of the regulated output ($\|z(t)\|_2^{\mathbb{E}}$) are less than the satisfactory levels represented by $\gamma_1$ and $\gamma_2$, respectively. Moreover, the $L_1$ performance index (VI) guarantees that the regulated output has a bounded peak ($\|z(t)\|_\infty^{\mathbb{E}}$) less than the satisfactory level $\gamma_5$. Hence, the level of attenuation of disturbances and fault signals on the regulated output $z(t)$ are given by $\gamma_1 = 0.0287$, $\gamma_2 = 0.0489$, and $\gamma_5 = 0.1727$, respectively.

The residual signals are shown in Figure 4.2, where robustness against disturbances as well as enhanced fault sensitivity can be readily observed. Also the faults are satisfactorily discriminated from the disturbances. Note that by using our threshold test, although there exist false alarms during the time interval 34–35 s, the simultaneous faults $f(t) \in \mathbb{R}^2$ are both effectively detected and isolated.

The above simulation results demonstrate that our proposed IFDIC scheme does indeed simultaneously achieve the IFDIC requirements satisfactorily.

For the second scenario, the values of $\beta_i$, $i = 1, \ldots, 6$ are selected as $\beta_1 = 1$, $\beta_2 = 1$, $\beta_3 = 1,000$, $\beta_4 = 1,000$, $\beta_5 = 1$, and $\beta_6 = 1,000$, which indicate that the design interest and priority is in more diagnosis performance than in the control performance. The performance indices for this case are obtained as $\gamma_1 = 0.0458$, $\gamma_2 = 0.0554$, $\gamma_3 = 1.0007$, $\gamma_4 = 7.9581e - 04$, $\gamma_5 = 4.1609$, and $\gamma_6 = 0.0031$. Therefore, it is expected that in this case, the performance indices corresponding to the FDI objective ($\gamma_3, \gamma_4, \gamma_6$) and the control objective ($\gamma_1, \gamma_2, \gamma_5$) are improved and degraded, respectively. Moreover, the threshold values of $J_{\mathrm{th}_i}^u$ and $J_{\mathrm{th}_i}^l$ for $i = 1, 2$ are selected as

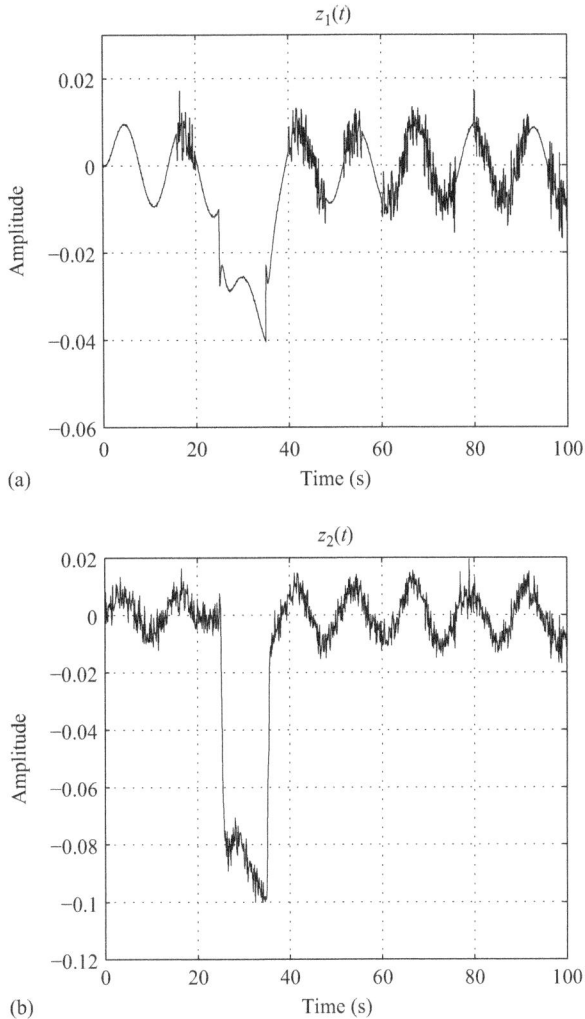

(a)

(b)

*Figure 4.1    The regulated outputs (a) $z_1(t)$ and (b) $z_2(t)$ corresponding to Scenario 1*

$J_{th_1}^l = -0.0003$, $J_{th_2}^l = -0.00024$, $J_{th_1}^u = 0.0003$, and $J_{th_2}^u = 0.00024$. The regulated output $z(t)$ of the closed-loop system is shown in Figure 4.3, where it can be concluded that the effects of disturbances and faults on the regulated output have been attenuated.

The residual signals are shown in Figure 4.4, where the robustness against disturbances and an enhanced fault sensitivity can be readily observed. Moreover, the faults are satisfactorily discriminated from disturbances. It follows that by using our developed threshold test, simultaneous faults $f(t) \in \mathbb{R}^2$ are both effectively detected and isolated.

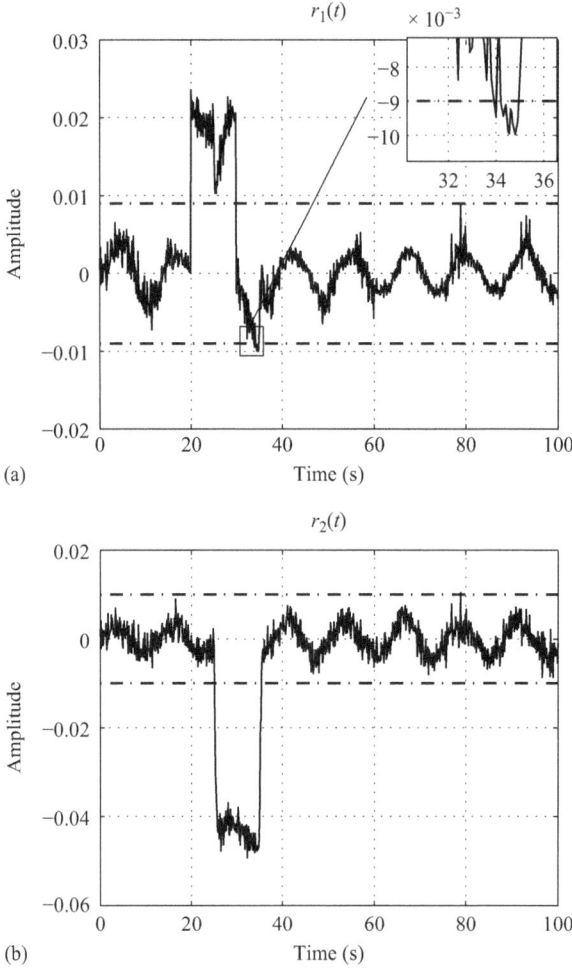

*Figure 4.2*   *The residual signals of (a) $r_1(t)$ and (b) $r_2(t)$ corresponding to Scenario 1, where the solid lines denote the residual signals and the dash-dot lines denote the residual upper and lower thresholds*

From Figures 4.3 and 4.4, and comparing them with Figures 4.1 and 4.2, it follows readily that in these simulations, the diagnosis and control objectives have been improved and degraded, respectively. Indeed, by comparing Figures 4.1 and 4.3, it can be concluded that compared with the latter simulation results, the effects of disturbances are more attenuated in the former results. On the other hand, from Figure 4.4, it can be concluded that the fault $f(t)$ has been effectively detected and isolated without any false alarms. In order to quantitatively document that the second scenario has better diagnosis performance in comparison with the first scenario, the minimum detectable faults for each scenario are provided in Table 4.2.

$$z_1(t)$$

(a)

$$z_2(t)$$

(b)

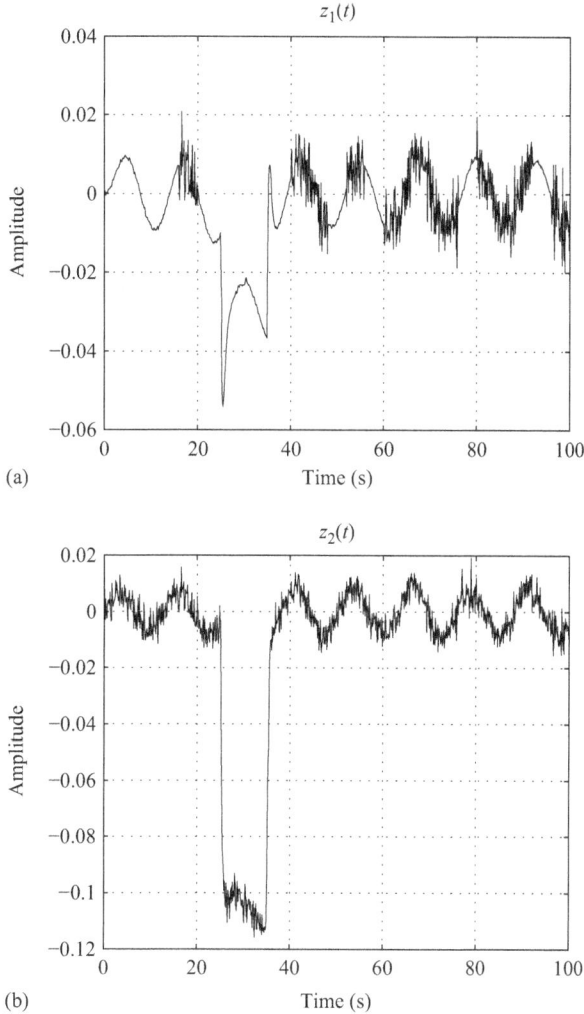

*Figure 4.3   The regulated outputs (a) $z_1(t)$ and (b) $z_2(t)$ corresponding to Scenario 2*

Finally, for the third scenario, the values of $\beta_i$, $i = 1, \ldots, 6$ are selected as $\beta_1 = 0.1$, $\beta_2 = 0.1$, $\beta_3 = 0.1$, $\beta_4 = 0.1$, $\beta_5 = 1,000$, and $\beta_6 = 0.1$, which implies that the design priority is in having more attenuation of bounded peak disturbance $w(t)$ on the regulated output $z(t)$ (the performance index $\gamma_5$) as opposed to the other performance indices $\gamma_1, \gamma_2, \gamma_3, \gamma_4,$ and $\gamma_6$. The performance indices for this case are obtained as $\gamma_1 = 17.0853$, $\gamma_2 = 17.0039$, $\gamma_3 = 17.0359$, $\gamma_4 = 17.0876$, $\gamma_5 = 0.0472$, and $\gamma_6 = 0.0994$. It should be pointed out that by considering the worst case analysis of the residuals corresponding to the healthy operation of the system being subject to various disturbances, the threshold values of $J_{\mathrm{th}_i}^u$ and $J_{\mathrm{th}_i}^l$ for $i = 1, 2$ are selected as

(a)

(b)

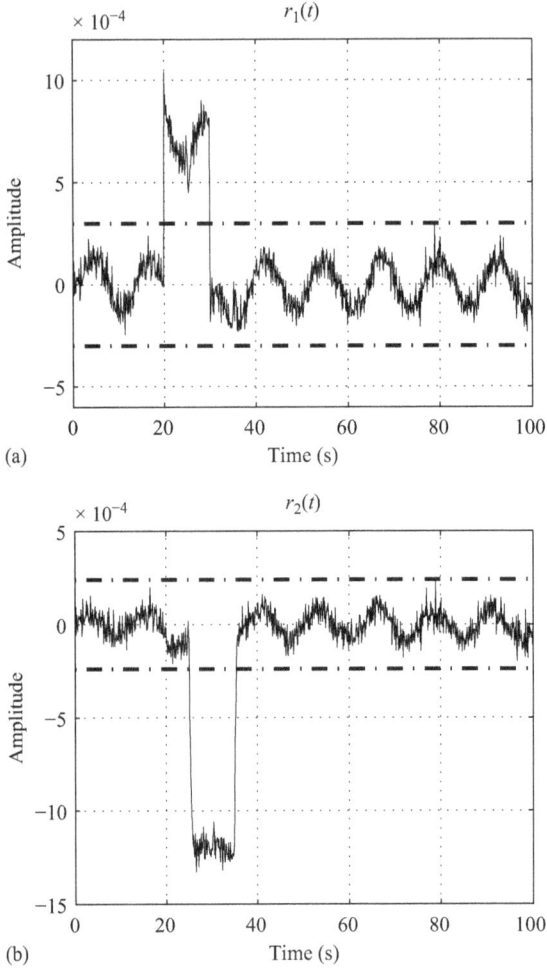

*Figure 4.4  The residual signals of (a) $r_1(t)$, (b) $r_2(t)$ corresponding to Scenario 2, where the solid lines denote the residual signals and the dash-dot lines denote the residual upper and lower thresholds*

*Table 4.2  Minimum detectable faults*

| Type of faults | Scenario 1 (same importance for all objectives) | Scenario 2 (more priority to the diagnosis objective vs. the control objective) |
|---|---|---|
| Actuator | 0.86 | 0.52 |
| Sensor | 0.035 | 0.0295 |

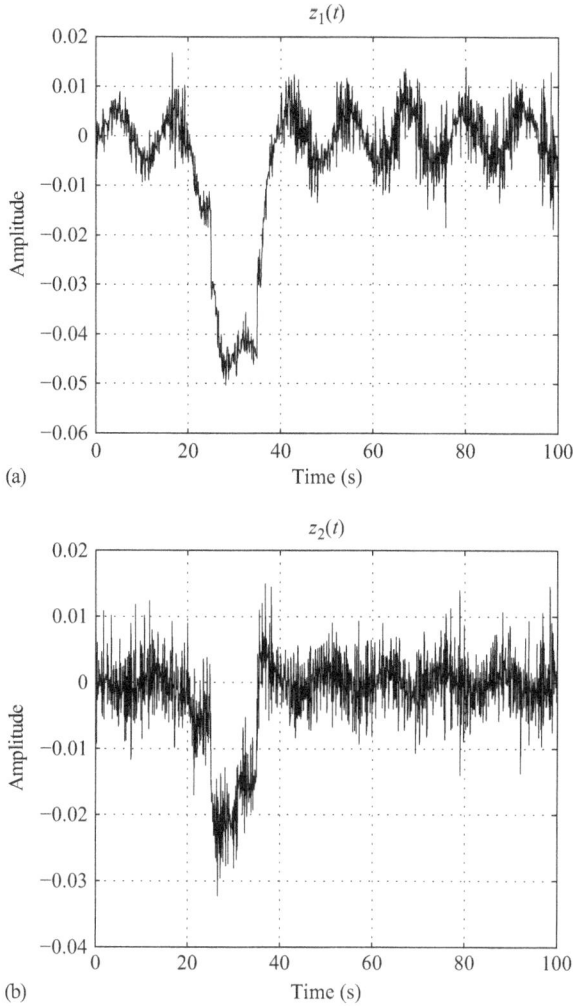

*Figure 4.5   The regulated outputs (a) $z_1(t)$ and (b) $z_2(t)$ corresponding to Scenario 3*

$J_{\mathrm{th}_1}^l = -0.016$, $J_{\mathrm{th}_2}^l = -0.0004$, $J_{\mathrm{th}_1}^u = 0.016$, and $J_{\mathrm{th}_2}^u = 0.0004$. The regulated output $z(t)$ and the residual signal $r(t)$ are shown in Figures 4.5 and 4.6, respectively.

From Figure 4.5 and evaluation of Figures 4.1 and 4.3, it can be concluded that as compared with the previous results, the effects of the bounded peak disturbance $w(t)$ on the regulated output $z(t)$ are more attenuated in this scenario (especially for the second regulated output $z_2(t)$). On the other hand, by comparing Figures 4.2, 4.4, and 4.6, it follows readily that the diagnosis objective has been degraded and the faults $f(t)$ have not been effectively detected and isolated.

Comparisons with other methodologies are now considered below to demonstrate the capabilities and effectiveness of our proposed design methodology.

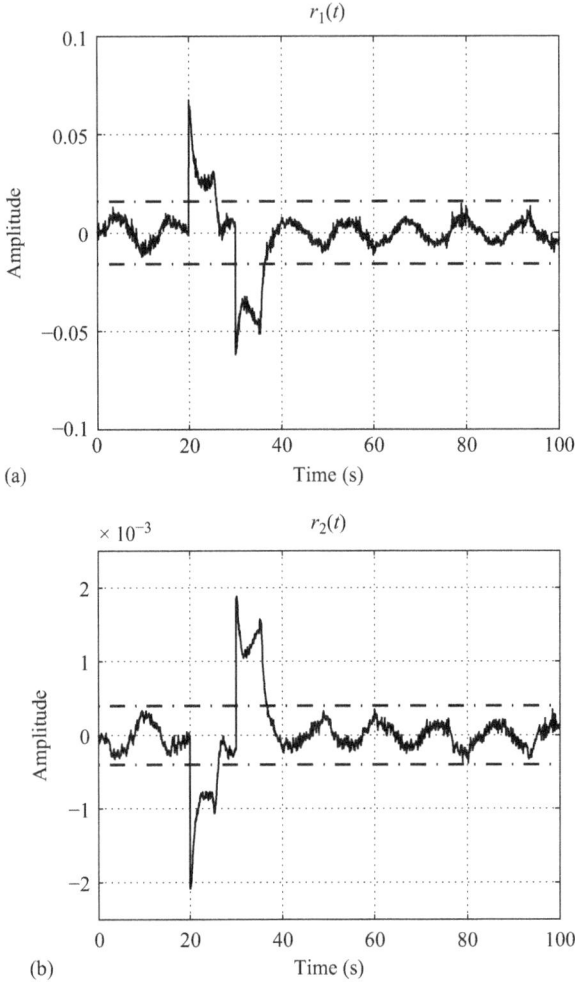

*Figure 4.6    The residual signals of (a) $r_1(t)$ and (b) $r_2(t)$ corresponding to Scenario 3, where the solid lines denote the residual signals and the dash-dot lines denote the residual upper and lower thresholds*

### Comparison with relevant work in FDI of continuous-time Markovian jump systems

Using the methodologies in [281,293], the occurrence of sensor faults in the Markovian jump system can be detected. Our methodology has certain advantages in comparison to [281,293]. Indeed, using our methodology, the occurrence of different types of faults (actuator, sensor, and component) can be simultaneously detected, isolated, and the desired control performance can be satisfied for the Markovian jump system. In [295], the occurrence of actuator faults in the Markovian jump systems is detected and isolated by using a bank of filters. However, our methodology can consider simultaneously the detection, isolation, and control objectives for different types of faults in the Markovian jump system by using only a single-observer-based module (IFDIC module). Moreover, another key advantage of our methodology lies in the fact that the IFDIC module states are practical variables as they represent estimates of the physical plant states. It follows that the observer states can be used to monitor (online or offline) the performance of the system. Note that this simple property does not hold for the general FDI representation that are developed in [281,293,295,303].

Another advantage of our methodology in comparison with [281,293,295,304] is the ease of its implementation. For example, in [304], a dynamic output feedback structure is used as a filter in the FD block. Therefore, four static gains should be designed and obtained to characterize the FD filter dynamics. However, in this work, in addition to the plant model, only three static gains (viz., $L_i$, $K_i$, $V_i$) characterize the entire IFDIC dynamics.

Finally, it should be pointed out that the proposed methodologies in [281,291–293,295,296,303,304] are limited to Markovian jump systems that are disturbed by only bounded energy disturbances. However, both bounded energy and bounded peak disturbances can be considered by using the results of this chapter.

### *Comparison with separate design*

Another set of simulations are finally conducted in order to compare our proposed methodology with the separate design of the controller and FDI filters. At first, an observer-based feedback controller using the performance indices (I), (II), and (VI) is obtained. The optimization parameters $\gamma_1$, $\gamma_2$, and $\gamma_5$ are obtained as $\gamma_1 = 0.0302$, $\gamma_2 = 0.0576$, and $\gamma_5 = 0.1259$, respectively. An observer-based FDI filter is then designed for the controlled system by using the performance indices (III), (IV), (V), and (VII). The optimization parameters $\gamma_3$, $\gamma_4$, $\gamma_6$ are obtained as $\gamma_3 = 1.0159$, $\gamma_4 = 0.0317$, and $\gamma_6 = 0.0801$, respectively.

Although, the optimization parameters $\gamma_1, \ldots, \gamma_6$ for both separate and integrated design cases are almost of the same order of magnitude; however, it should be noted that the separate approach does not yield an optimal design since our methodology leads to a lower computational complexity in comparison to the separate design. Indeed, in the case of separate design, the order of the resulting detector/controller module is $2n$. However, our methodology yields a detector/controller module of a lower complexity (the IFDIC module has an order of $n$). Moreover, it should be pointed out that the separate design of the controller and FDI modules does not take into account the rather significant interactions that occur between these modules.

## 4.5 Conclusion

In this chapter, we developed and presented a mixed $H_\infty/H_-/L_1$ formulation of the IFDIC problem for a continuous-time Markovian jump linear systems with uncertain transition probabilities. An LMI approach for solving the IFDIC design was introduced, which in addition to mean square stabilizing the closed-loop system, simultaneously guarantees FD, isolation and control objectives. LMI conditions were presented so that no products of Lyapunov matrices and system state space matrices were involved. This results in a significant reduction in the conservatism of the IFDIC solution. Moreover, the IFDIC problem was solved so that each element of the residual vector was only sensitive to a specified fault, and therefore, occurrence of simultaneous faults in the system can also be handled and resolved. Application of our methodology to a linearized model of the GE F-404 aircraft engine was also presented to demonstrate and illustrate the capabilities and effectiveness of our proposed approach.

*Chapter 5*
# Event-triggered multiobjective control and fault diagnosis: a unified framework

In this chapter, we deal with a new linear matrix inequality (LMI) approach to the problems of event-triggered multiobjective synthesis of feedback controllers and fault diagnosis filters through a unified framework. Toward this end, at first, we define a general problem known as "event-triggered integrated fault detection, isolation and control" (E-IFDIC). By utilizing a filter to represent, characterize, and specify the E-IFDIC module, we develop a multiobjective formulation of the problem based on $H_\infty$, $H_-$, $l_1$ and generalized $H_2$ performance criteria. It is shown that when an event-triggered strategy is applied to both the sensor and E-IFDIC module, the amount of data that is sent through the sensor-to-E-IFDIC module and E-IFDIC module-to-actuator channels are dramatically reduced.

A set of LMI feasibility conditions are derived to ensure the solvability of the problem as well as to simultaneously obtain the E-IFDIC module parameters and the event-triggered conditions. In the area of robust control, fault diagnosis, and fault tolerant control of linear systems, many fundamental problems can be recast as $H_\infty$, $l_1$ and generalized $H_2$ control frameworks leading to the so-called mixed norm or multiobjective optimization problems.

Consequently, it is shown that certain existing problems in the fields of time and event-trigged control and fault diagnosis can be considered as special cases of our proposed methodology. Finally, two industrial case studies corresponding to an unstable batch reactor process and an unmanned underwater vehicle are provided to illustrate and demonstrate the effectiveness of our proposed design methodology when compared to the available work in the literature. The work presented in this chapter has partially appeared in [310,311].

This chapter is organized as follows. After a brief overview of the relevant literature, we begin in Section 5.2 by presenting the statement of problem for E-IFDIC and the governing equations for the considered discrete-time LTI system under investigation. In Section 5.3, we develop and implement an LMI-based solution to our proposed E-IFDIC problem. In Section 5.4, different aspects and certain capabilities of our methodology for event-triggered multiobjective synthesis of feedback controllers and fault diagnosis filters are demonstrated by providing two industrial case studies under different scenarios.

## 5.1  Introduction

In the past few years, the problem of event-triggered feedback control design in which the control action is only driven when required and determined by a certain condition on the plant measurements has received considerable attention [312,313]. The condition, commonly referred to as event-triggering mechanism, triggers a new control task typically when the measurement signal becomes sufficiently large to exceed a prescribed threshold [314]. Therefore, an event-triggered control system drives the control action aperiodically when it is needed and is different from the conventional time-triggered systems in which the control action is executed periodically [315]. It follows that an event-triggered strategy can reduce the usage of computational and communication resources without degrading the overall desired closed-loop system performance capabilities [316].

### 5.1.1  Literature review on event-triggered control

Recently, there has been a growing interest on using event-triggered strategies for both control [3,317,318] and estimation [312,319] problems. Most of the existing event-triggered controllers are designed by emulation [320]. Indeed, at first, a stabilizing feedback control law is constructed in absence of the network, and then, a triggering condition is synthesized to preserve the closed-loop stability. The main limitation and drawback of this approach is that it is difficult to obtain an optimal design given that one is restricted by the initial choice of the feedback control law [321]. To overcome this drawback, codesign of feedback control laws and event-triggering conditions have been considered in [1,321].

One of the fundamental problems in feedback control design is the capability of the controller to reject (or attenuate) external disturbances and noise that occur in practical systems. It should be pointed out that most current work in *codesign of event detection and control* (C-EDC) do not consider the effects of noise and external disturbances [1,321].

The key difference between various controller design methodologies for disturbance attenuation can be traced to the modeling and treatment of system disturbances. For example, $H_\infty$ theory has been applied on a disturbance model consisting of bounded energy $l_2$ signals. The generalized $H_2$ which has been used in control, filtering, and fault detection problems [322,323] also provides a proper method to characterize the effects of such signals. Hence, $H_\infty$ and generalized $H_2$ performances quantify disturbance attenuation for signals with bounded energy.

The key difference between $H_\infty$ and generalized $H_2$ is that the $H_\infty$ theory can be used to capture the energy of certain output signals in a feedback control design problem, whereas the generalized $H_2$ might impose a bound on the peak value of certain outputs.

In practical applications such as persistent disturbance rejection and command following in presence of input saturation, performance criteria should be stated based on the bound on the peak amplitude of the signals as opposed to bound on their energy. This cannot be appropriately recast into the $H_\infty$ or generalized $H_2$ frameworks.

Hence, an alternate optimization framework, namely, the $l_1$ optimization (peak-to-peak nominal performance) has been formulated [271]. This framework is based on the $l_1$ norm which is the induced norm corresponding to peak bounded input/output signals. Although there exists some work on the problem of event-triggered $H_\infty$ control design [314,324], there are no published results that are reported in the literature on codesign schemes of event detector and $l_1$ (or generalized $H_2$) controllers.

### 5.1.2  Literature review on event-triggered fault diagnosis

Very limited work exists on event-triggered fault diagnosis (EFD) systems. In [325], a particular form of the Kalman filter was proposed for accomplishing fault isolation through the send-on-delta scheme, where data is sent if the difference between the current data and the previously transmitted data exceeds a certain predetermined threshold. An event-triggered fault detection filter for networked control systems subject to communication delays and nonlinear perturbations is proposed in [326].

Recently, the problem of event-triggered fault detection and isolation filter design for discrete-time linear time-invariant (LTI) systems is addressed in [327,328]. The methods [325–328] discussed above are developed by using an open-loop model of the process. However, in many practical control systems, the fault detection systems correspond to closed-loop feedback systems. In such cases, faults may become hidden by the control actions and the prompt detection of process faults becomes clearly more challenging [48].

The above has motivated researchers to consider the problem of integrated fault detection and control (IFDC), which has attracted a lot of interest in the past two decades [329]. However, there are no published results on event-triggered IFDC design in the literature.

### 5.1.3  Our contributions

Based on the above discussion, the main objectives of this chapter are to study through a unified framework the problems of *event-triggered multiobjective controllers design in presence of different disturbances* as well as *EFD* for LTI systems. To accomplish the above problems in a unified framework, the multiobjective problem of E-IFDIC is defined and studied for LTI systems.

Indeed, at first, a single E-IFDIC module based on a dynamic filter structure is designed that produces two signals, one representing the residual and the other representing the control signal. To reduce the communication rates, two event-triggered conditions, one on the sensor to the E-IFDIC module and the other on the E-IFDIC module to the actuator are also designed to determine whether the newly measured data or control output, respectively, should be transmitted or not. A generalized $H_2/H_\infty/H_-/l_1$ formulation of the problem is then developed and sufficient conditions for solvability of the problem are obtained in terms of LMI feasibility conditions.

The main advantage of the proposed LMI formulation is that it is a convex problem and can therefore be solved effectively by using interior-point methods. Moreover, our LMI-based solution enables one to explore trade-offs studies and analyze the limitations of the performance and feasibility in the design process.

The contributions of this work can therefore be summarized as follows:

1. The problem of E-IFDIC is defined and developed.
2. To solve the E-IFDIC problem, sufficient conditions for codesign of event detector and $H_\infty$, generalized $H_2$ and $l_1$ controllers are obtained in terms of LMI feasibility conditions. Note that these problems can be considered as alternatives to the time-triggered $H_\infty$, generalized $H_2$ and $l_1$ control frameworks to mitigate the over-provisioning of the computational and communication resources.
3. As opposed to the two-step schemes in [314], in this work, the controller gain and the event-triggering conditions are designed simultaneously to meet the $H_\infty$ performance requirement with respect to disturbances. Moreover, unlike [324, 330] that mediate the transmissions by considering event detectors at sensor nodes, in this work a more general framework is developed and applied. Indeed, we propose and construct event-triggered strategies for both the sensor and the E-IFDIC module nodes to reduce the communication load in the sensor-to-E-IFDIC module and E-IFDIC module-to-actuator channels, respectively.
4. It is shown that a number of existing problems in the fields of time and event-trigged control and fault diagnosis (e.g., [1,271,322,323,329]) can be considered as special cases of our proposed framework and are solvable by using our proposed methodologies. Hence, our proposed E-IFDIC problem is quite general and can cover a number of existing problems in a unified framework (refer to Remark 5.1).

## 5.2   System description and problem formulation

In this section, we present the statement of problem for E-IFDIC and the governing equations for the considered discrete-time LTI system under investigation.

### 5.2.1   System description

Consider the following discrete-time LTI system:

$$x(k+1) = Ax(k) + B_u\bar{u}(k) + B_dd(k) + B_ww(k) + B_ff(k),$$
$$y(k) = Cx(k) + D_dd(k) + D_ww(k) + D_ff(k), \qquad (5.1)$$
$$z(k) = Ex(k) + F_u\bar{u}(k) + F_dd(k) + F_ww(k) + F_ff(k),$$

where $x(k) \in \mathbb{R}^n$ denotes the state vector, $\bar{u}(k) \in \mathbb{R}^{n_u}$ denotes the last control signal that is transmitted from the controller to the actuator, $y(k) \in \mathbb{R}^{n_y}$ denotes the measured output, $z(k) \in \mathbb{R}^{n_z}$ denotes the regulated output, $d(k) \in \mathbb{R}^{n_d}$ denotes an $l_2$ external disturbance, $w(k) \in \mathbb{R}^{n_w}$ denotes an $l_\infty$ disturbance signal satisfying $||w(.)||_\infty \leq 1$, and the unknown input $f(k) \in \mathbb{R}^{n_f}$ denotes a possible fault.

Without loss of generality, assume that $f(k)$ is $l_2$-norm bounded. The matrices $A, B_u, B_d, B_w, B_f, C, D_d, D_w, D_f, E, F_u, F_d, F_w$, and $F_f$ are constant with appropriate dimensions. Note that the fault matrices $B_f, D_f$, and $F_f$ are specified according to faults that are to be detected and isolated in the components, actuators, or sensors, respectively.

Note that the LTI model (5.1) can be considered as a discretized model of a continuous-time system. Therefore, the time interval between two successive transmissions are at least lower bounded by the sampling time [1]. Consequently, the results of this work can be applied to both continuous-time and discrete-time systems.

The main objective of this work is to develop a unified framework that is capable of solving various challenges in event-triggered multiobjective control and fault diagnosis problems. Toward this end, the problem of event-triggered IFDIC is defined and solved for the discrete-time LTI system (5.1). Specifically, a single E-IFDIC module based on a dynamic filter structure is designed that generates residuals as well as the control signals. Using the control signal, the controlled output $z(k)$ is regulated in presence of faults and external disturbances. Using the residual signal, the occurred faults in the system can be detected and isolated. It is shown that a number of existing problems in the time and event-triggered control and fault diagnosis areas can be considered as special cases of the E-IFDIC problem.

### 5.2.2   E-IFDIC module and event detector description

The following module is now proposed for addressing the E-IFDIC problem:

$$
\begin{aligned}
x_c(k+1) &= A_c x_c(k) + L\bar{y}(k),\\
y_c(k) &= C x_c(k),\\
u(k) &= K x_c(k),\\
r(k) &= H(\bar{y}(k) - y_c(k)),
\end{aligned}
\tag{5.2}
$$

where $x_c(k) \in \mathbb{R}^n$ denotes the filter state, $y_c(k) \in \mathbb{R}^{n_y}$ denotes the filter output, $r(k) \in \mathbb{R}^{n_f}$ denotes the residual signal, $u(k) \in \mathbb{R}^{n_u}$ denotes the controller output, and the filter input $\bar{y}(k) \in \mathbb{R}^{n_y}$ denotes the last measurement that is transmitted from the sensor to the filter module. The constant matrices $A_c \in \mathbb{R}^{n \times n}$, $L \in \mathbb{R}^{n_u \times n}$, $K \in \mathbb{R}^{n \times n_y}$, $H \in \mathbb{R}^{n_f \times n_y}$ denote the filter parameters to be designed subsequently.

In an event-triggered implementation, the measured output of the plant ($y(k)$) is not transmitted to the E-IFDIC module at every sampling instant, rather this is done only at the transmission times that are denoted by $k_i^y$, $i \in \mathbb{N}$ with the initial time $k_0^y = 0$. Therefore, the input to the E-IFDIC module can be expressed as $\bar{y}(k) = y(k_i^y)$, for $k \in [k_i^y, k_{i+1}^y)$, $i, k \in \mathbb{N}$. Moreover, a very basic sample and hold strategy, i.e., $\bar{y}(k) = \bar{y}(k-1)$ is applied when no new output measurement is transmitted.

It is assumed that the system output is transmitted to the E-IFDIC module only when the following condition

$$
\sigma_y y^{\mathrm{T}}(k) y(k) - \varepsilon_y \delta_y^{\mathrm{T}}(k) \delta_y(k) > 0,
\tag{5.3}
$$

is not satisfied, where $\varepsilon_y > 0$ and $\sigma_y > 0$ are weighting constants to be designed and selected by the user, and $\delta_y(k) = \bar{y}(k) - y(k)$. Hence, in our framework, a sensor measurement is transmitted to the E-IFDIC module only when the difference between the latest transmitted value and the current sensor measurement is sufficiently large as compared to the current reading value.

Hence, in this setup unnecessary usage of the transmission bandwidth and data overloading is avoided. It also saves battery-powered sensors energy. It should be noted that the parameters $\varepsilon_y$ and $\sigma_y$ will affect the length of the interevent intervals and the communication rates. In other words, the smaller value of $\sigma_y/\varepsilon_y$ will lead to higher communication rates. Moreover, unlike [325], the signal variations are compared with their current values instead of a constant threshold, leading to a less conservative result that is achieved by using our approach.

It is also assumed that a similar event-triggering condition is implemented for the controller output ($u(k)$). Therefore, the controller output is sent to the actuator only when the following event condition is violated

$$\sigma_u u^T(k)u(k) - \varepsilon_u \delta_u^T(k)\delta_u(k) > 0, \tag{5.4}$$

where $\delta_u(k) = \bar{u}(k) - u(k)$, $\varepsilon_u > 0$ and $\sigma_u > 0$ are constant parameters to be designed subsequently. Hence, the control signal that is received by the actuator is given by $\bar{u}(k) = u(k_i^u)$, for $k \in [k_i^u, k_{i+1}^u)$, $i, k \in \mathbb{N}$ and $k_0^u = 0$. In addition, a similar sample and hold strategy, i.e., $\bar{u}(k) = \bar{u}(k-1)$ is implemented when no new control output is transmitted.

It should be pointed out that in this work we consider scenarios where the sensor event times ($k_i^y$) and the filter event times ($k_i^u$) are asynchronous.

By integrating the E-IFDIC module dynamics (5.2) with the system in (5.1), the following augmented system dynamics is obtained, namely,

$$x_{\mathrm{cl}}(k+1) = \tilde{A}x_{\mathrm{cl}}(k) + \tilde{B}_d d(k) + \tilde{B}_w w(k) + \tilde{B}_f f(k) + \tilde{B}_{\delta_u}\delta_u(k) + \tilde{B}_{\delta_y}\delta_y(k),$$

$$r(k) = \tilde{C}x_{\mathrm{cl}}(k) + \tilde{D}_d d(k) + \tilde{D}_w w(k) + \tilde{D}_f f(k) + \tilde{D}_{\delta_y}\delta_y(k), \tag{5.5}$$

$$z(k) = \tilde{E}x_{\mathrm{cl}}(k) + \tilde{F}_d d(k) + \tilde{F}_w w(k) + \tilde{F}_f f(k) + \tilde{F}_{\delta_u}\delta_u(k),$$

where $x_{\mathrm{cl}}(k) = [x(k)^T \, x_c(k)^T]^T$ and

$$\tilde{A} = \begin{bmatrix} A & B_u K \\ LC & A_c \end{bmatrix}, \quad \tilde{B}_d = \begin{bmatrix} B_d \\ LD_d \end{bmatrix}, \quad \tilde{B}_w = \begin{bmatrix} B_w \\ LD_w \end{bmatrix},$$

$$\tilde{B}_f = \begin{bmatrix} B_f \\ LD_f \end{bmatrix}, \quad \tilde{B}_{\delta_u} = \begin{bmatrix} B_u \\ 0 \end{bmatrix}, \quad \tilde{B}_{\delta_y} = \begin{bmatrix} 0 \\ L \end{bmatrix}, \quad \tilde{E} = \begin{bmatrix} E & F_u K \end{bmatrix},$$

$$\tilde{C} = \begin{bmatrix} HC & -HC \end{bmatrix}, \quad \tilde{D}_d = HD_d, \quad \tilde{D}_w = HD_w, \quad \tilde{D}_f = HD_f,$$

$$\tilde{D}_{\delta_y} = H, \quad \tilde{F}_d = F_d, \quad \tilde{F}_w = F_w, \quad \tilde{F}_f = F_f, \quad \tilde{F}_{\delta_u} = F_u.$$

The overall block diagram of an event-triggered IFDIC problem is shown in Figure 5.1. The "event detector" blocks [based on the event-triggered conditions (5.3) and (5.4)] are used to reduce the transmissions from the sensor to the E-IFDIC module and from the E-IFDIC module to the actuator. Indeed, in the sensor and E-IFDIC nodes, only part of the sensor data and E-IFDIC output will be transmitted to the E-IFDIC module and actuator, respectively. Therefore, the filter and actuator

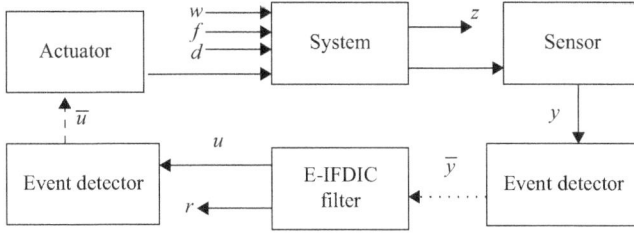

*Figure 5.1   The block diagram of the E-IFDIC problem*

inputs represent simply the last transmitted measurement signal $\bar{y}(k)$ and the control signal $\bar{u}(k)$, respectively.

By using this strategy, the burden on the network communication is reduced, and the network communication bandwidth will be saved as well. Consequently, it is possible to significantly reduce the usage of the communication resources for detection, isolation, and control tasks as compared to a conventional IFDIC approach. Note that, this is an important feature especially for wireless networks, since one will be able to save on the transmission energy and also increase the lifespan of the batteries at the nodes.

In the following, the E-IFDIC problem is transformed into a mixed generalized $H_2/H_\infty/H_-/l_1$ optimization problem in order to formally develop and design our proposed methodologies and solutions.

## 5.2.3   Problem formulation

The E-IFDIC problem is now formally stated as follows.

### The E-IFDIC problem:

The problem under investigation is to design for the system (5.1) an E-IFDIC module that is governed by (5.2) and subject to the event-triggered conditions (5.3) and (5.4) such that simultaneously the following properties hold, namely,

- the augmented system (5.5) is stable,
- the effects of disturbances $(d(k), w(k))$ on the residual signal $r(k)$ are minimized and the effects of faults on the residual signal $r(k)$ are maximized (in order to accomplish the fault detection task),
- each element of the residual signal $r(k)$ is only sensitive to a specified potential fault (in order to accomplish the fault isolation task), and
- the effects of disturbances and faults on the regulated output $z(k)$ are minimized (in order to accomplish the control performance task).

In other words, the main objective that is pursued in this work is to design the E-IFDIC block (5.2) and the event-triggered conditions (5.3) and (5.4) such that the augmented system (5.5) is stable and the following conditions are satisfied:

$$\text{(I)} \sup_{f,w=0} \frac{||r||_2}{||d||_2} < \gamma_1, \qquad \text{(II)} \sup_{d,w=0} \frac{||r-r_e||_2}{||f||_2} < \gamma_2,$$

$$\text{(III)} \sup_{f,w=0} \frac{||z||_2}{||d||_2} < \gamma_3, \qquad \text{(IV)} \sup_{d,w=0} \frac{||z||_2}{||f||_2} < \gamma_4,$$

$$\text{(V)} \inf_{d,w=0} \frac{||r_e||_2}{||f||_2} > 1, \qquad \text{(VI)} \sup_{w,f=0} \frac{||r||_\infty}{||d||_2} < \gamma_5,$$

$$\text{(VII)} \sup_{d,f=0} \frac{||r||_\infty}{||w||_\infty} < \gamma_6, \quad \text{(VIII)} \sup_{d,f=0} \frac{||z||_\infty}{||w||_\infty} < \gamma_7,$$

(5.6)

where $r_e(k) = Jf(k)$ and $J$ is selected to have the following structure:

$$J = \text{diag}(j_1, \ldots, j_{n_f}) \in \mathbb{R}^{n_f \times n_f}, \; j_i > 0, \; \forall i \in \{1, \ldots, n_f\}.$$

The $H_\infty$ and $l_1$ performance indices (I) and (VII) are used to attenuate the effects of the bounded energy and bounded peak disturbances on the residual signal, respectively. The $H_\infty$ and $H_-$ performance indices (II) and (V) are used to guarantee a minimum level of sensitivity of the residual signal to the fault signals and to ensure the fault isolation goal, respectively. The $H_\infty$ and $l_1$ performance indices (III), (IV), and (VIII) are used to attenuate the effects of disturbances ($d(k)$ and $w(k)$) and faults on the regulated output. The generalized $H_2$ performance index (VI) is used to improve the fault detection efficiency by guaranteeing that the residual signal remains below a threshold value corresponding to the healthy system or fault-free condition [37].

**Remark 5.1.** *By utilizing our proposed E-IFDIC problem (5.6), the following problems and methodologies within the conventional time-triggered systems can be considered and solved in the framework of event-triggered systems. Specifically,*

- **Problem 1.** *Codesign of the event detector and $l_1$ controller.*
- **Problem 2.** *Codesign of the event detector and generalized $H_2$ controller.*
- **Problem 3.** *Event-triggered mixed generalized $H_2/H_\infty/l_1$ (or $H_\infty/l_1$) controller design.*
- **Problem 4.** *Event-triggered IFDC design problem.*

*The problems above can have many applications in event-triggered systems, such as in peak (or energy) bounded disturbance attenuation, and command following in presence of input saturations, etc.*

In the next section, an LMI-based solution to our proposed E-IFDIC problem will be developed and implemented.

## 5.3  Main results

There are eight performance criteria (I)–(VIII) that must be satisfied *simultaneously* for designing the E-IFDIC problem for the augmented system (5.5). Note that the $H_-$ performance index (V) can be easily converted to an LMI condition, namely,

$$I - J < 0. \tag{5.7}$$

The reader is referred to [329] for more details on the proof of inequality (5.7). In the following discussion, at first in Theorem 5.1, sufficient conditions for the $H_\infty$ performance indices (I)–(IV) are formulated as LMI feasibility problems. The sufficient LMI conditions for the generalized $H_2$ performance index (VI) are obtained in Theorem 5.2. In Theorem 5.3, sufficient conditions for the $l_1$ performance indices (VII)–(VIII) are formulated as LMI feasibility problems. Finally, in Algorithm 5.1, a feasible solution to the E-IFDIC problem is obtained by simultaneously considering all the indices (I)–(VIII).

**Theorem 5.1.** *Assume that data is transmitted from the sensor to the E-IFDIC module and from the E-IFDIC module to the actuator subject to violations of the conditions (5.3) and (5.4), respectively. The augmented system (5.5) is stable and the $H_\infty$ performance indices (I)–(IV) are guaranteed if there exist symmetric positive-definite matrices $X$ and $Y$, matrices $H$, $J$, $M$, $N$, $Q$, and positive scalars $\varepsilon_u$, $\varepsilon_y$, $\sigma_u$, $\sigma_y$, $\gamma_1$, $\gamma_2$, $\gamma_3$, $\gamma_4$, such that for $i = 1, \ldots, 4$ the following LMI conditions hold*

$$\begin{bmatrix} E_{11} & 0 & 0 & 0 & E_{12} & E_{13i} & E_{14} & E_{15} \\ * & E_{21i} & 0 & 0 & E_{22i} & E_{23i} & E_{24i} & 0 \\ * & * & E_{31} & 0 & E_{32} & E_{33i} & 0 & 0 \\ * & * & * & E_{41} & E_{42} & E_{43i} & 0 & 0 \\ * & * & * & * & E_{51} & 0 & 0 & 0 \\ * & * & * & * & * & -I & 0 & 0 \\ * & * & * & * & * & * & E_{71} & 0 \\ * & * & * & * & * & * & * & E_{81} \end{bmatrix} < 0, \tag{5.8}$$

*where*

$$E_{11} = E_{51} = -\begin{bmatrix} Y & I \\ I & X \end{bmatrix}, \quad E_{14} = \begin{bmatrix} YC^\mathrm{T} \\ C^\mathrm{T} \end{bmatrix}, \quad E_{15} = \begin{bmatrix} Q^\mathrm{T} \\ 0 \end{bmatrix},$$

$$E_{12} = \begin{bmatrix} YA^\mathrm{T} + Q^\mathrm{T}B_u^\mathrm{T} & M^\mathrm{T} \\ A^\mathrm{T} & A^\mathrm{T}X + C^\mathrm{T}N^\mathrm{T} \end{bmatrix},$$

$$E_{131} = E_{132} = \begin{bmatrix} 0 \\ C^\mathrm{T}H^\mathrm{T} \end{bmatrix}, \quad E_{133} = E_{134} = \begin{bmatrix} YE^\mathrm{T} + Q^\mathrm{T}F_u^\mathrm{T} \\ E^\mathrm{T} \end{bmatrix},$$

$$E_{221} = E_{223} = \begin{bmatrix} B_d^\mathrm{T} & B_d^\mathrm{T}X + D_d^\mathrm{T}N^\mathrm{T} \end{bmatrix}, \quad E_{32} = \begin{bmatrix} B_u^\mathrm{T} & B_u^\mathrm{T}X \end{bmatrix},$$

$$E_{222} = E_{224} = \begin{bmatrix} B_f^\mathrm{T} & B_f^\mathrm{T}X + D_f^\mathrm{T}N^\mathrm{T} \end{bmatrix}, \quad E_{42} = \begin{bmatrix} 0 & N^\mathrm{T} \end{bmatrix},$$

$$E_{231} = D_d^\mathrm{T}H^\mathrm{T}, \quad E_{232} = D_f^\mathrm{T}H^\mathrm{T} - J^\mathrm{T}, \quad E_{233} = F_d^\mathrm{T},$$

$$E_{234} = F_f^T, \quad E_{241} = E_{243} = D_d^T, \quad E_{242} = E_{244} = D_f^T,$$

$$E_{331} = E_{332} = 0, \quad E_{333} = E_{334} = F_u^T, \quad E_{431} = E_{432} = H^T,$$

$$E_{433} = E_{434} = 0, \quad E_{71} = -\sigma_y^{-1}I, \quad E_{81} = -\sigma_u^{-1}I,$$

$$E_{21i} = -\gamma_i^2 I, \quad E_{31} = -\varepsilon_u I, \quad E_{41} = -\varepsilon_y I.$$

*Proof.* Consider the Lyapunov function candidate $V(k) = x_{cl}^T(k)Px_{cl}(k)$, where $P$ is a symmetric positive-definite matrix which is partitioned as follows:

$$P = \begin{bmatrix} X & Y^{-1} - X \\ Y^{-1} - X & X - Y^{-1} \end{bmatrix}, \quad P^{-1} = \begin{bmatrix} Y & Y \\ Y & W \end{bmatrix}, \tag{5.9}$$

where $XY + (Y^{-1} - X)W = 0$ [1]. According to the inequality (5.8) for $i = 1$, by substituting $M$, $N$, and $Q$ with

$$M = (X(A + B_u K - LC - A_c) + Y^{-1}(A_c + LC))Y, \tag{5.10}$$
$$N = (Y^{-1} - X)L, \quad Q = KY,$$

and defining $\hat{C} = [C \quad 0]$, $\hat{C}_c = [0 \quad K]$ we have

$$\text{diag}(T^T, I, I, I, R^T, I, I, I) \; \Psi \; \text{diag}(T, I, I, I, R, I, I, I) < 0, \tag{5.11}$$

where $\Psi$ is the following matrix

$$\begin{bmatrix}
-P & 0 & 0 & 0 & \tilde{A}^T & \hat{C}^T & \hat{C}^T & \hat{C}_c^T \\
* & -\gamma_1^2 I & 0 & 0 & \tilde{B}_d^T & \tilde{D}_d^T & D_d^T & 0 \\
* & * & -\varepsilon_u I & 0 & \tilde{B}_{\delta_u}^T & \tilde{D}_{\delta_u}^T & 0 & 0 \\
* & * & * & -\varepsilon_y I & \tilde{B}_{\delta_y}^T & \tilde{D}_{\delta_y}^T & 0 & 0 \\
* & * & * & * & -P^{-1} & 0 & 0 & 0 \\
* & * & * & * & * & -I & 0 & 0 \\
* & * & * & * & * & * & -\sigma_y^{-1}I & 0 \\
* & * & * & * & * & * & * & -\sigma_u^{-1}I
\end{bmatrix},$$

and

$$T = \begin{bmatrix} Y & I \\ Y & 0 \end{bmatrix}, \quad R = \begin{bmatrix} I & X \\ 0 & Y^{-1} - X \end{bmatrix}.$$

Now, by multiplying both sides of $\Psi$ by $[x_{cl}^T(k) \; d^T(k) \; \delta_u^T(k) \; \delta_y^T(k)]$ from the left and $[x_{cl}^T(k) \; d^T(k) \; \delta_u^T(k) \; \delta_y^T(k)]^T$ from the right, one can conclude that

$$\Delta V + r^T r - \gamma_1^2 d^T d + \sigma_y y^T y - \varepsilon_y \delta_y^T \delta_y + \sigma_u u^T u - \varepsilon_u \delta_u^T \delta_u < 0. \tag{5.12}$$

According to the inequalities (5.3) and (5.4), $\sigma_y y^T(k)y(k) - \varepsilon_y \delta_y^T(k)\delta_y(k) > 0$ and $\sigma_u u^T(k)u(k) - \varepsilon_u \delta_u^T(k)\delta_u(k) > 0$ which leads to

$$\Delta V(k) + r^T(k)r(k) - \gamma_1^2 d^T(k)d(k) < 0. \tag{5.13}$$

For $d(k) = 0$, the inequality (5.13) leads to $\Delta V(k) < 0$, which guarantees the stability of the system. To ensure the $H_\infty$ performance criterion, summing both sides of the inequality (5.13) over $[0, \infty)$, one gets

$$V(\infty) - V(0) + \sum_{k=0}^{\infty} (r^{\mathrm{T}}(k)r(k) - \gamma_1^2 d^{\mathrm{T}}(k)d(k)) < 0, \tag{5.14}$$

where $V(\infty) > 0$ for zero initial condition $V(0) = 0$. Therefore, we have

$$\sum_{k=0}^{\infty} r^{\mathrm{T}}(k)r(k) < \gamma_1^2 \sum_{k=0}^{\infty} d^{\mathrm{T}}(k)d(k), \tag{5.15}$$

which leads to ensuring that the $H_\infty$ performance index (I) is satisfied. The performance indices (II)–(IV) can also be shown by employing the same procedure and following along the same lines as in the derivation of the performance index (I). Therefore, they are omitted here for sake of brevity. This completes the proof of the theorem. ☐

The following theorem gives the LMI constraints for the generalized $H_2$ performance index (VI).

**Theorem 5.2.** *Assume that data is transmitted from the sensor to the E-IFDIC module and from the E-IFDIC module to the actuator subject to violations of the conditions (5.3) and (5.4), respectively, and let $D_d = 0$. The augmented system (5.5) is stable and the generalized $H_2$ performance index (VI) is guaranteed if there exist symmetric positive-definite matrices $X$ and $Y$, matrices $M$, $N$, $V$, $Q$, $J$, and positive scalars $\varepsilon_u$, $\varepsilon_y$, $\sigma_u$, $\sigma_y$, $\gamma_5$, such that the following LMI conditions hold*

$$\begin{bmatrix} E_{11} & 0 & 0 & 0 & E_{12} & E_{14} & E_{15} \\ * & -I & 0 & 0 & E_{221} & 0 & 0 \\ * & * & E_{31} & 0 & E_{32} & 0 & 0 \\ * & * & * & E_{41} & E_{42} & 0 & 0 \\ * & * & * & * & E_{51} & 0 & 0 \\ * & * & * & * & * & E_{71} & 0 \\ * & * & * & * & * & * & E_{81} \end{bmatrix} < 0, \tag{5.16}$$

$$\begin{bmatrix} -E_{11} & 0 & 0 & E_{131} & E_{14} & E_{15} \\ * & -E_{31} & 0 & E_{331} & 0 & 0 \\ * & * & -E_{41} & E_{431} & 0 & 0 \\ * & * & * & \gamma_5^2 I & 0 & 0 \\ * & * & * & * & -E_{71} & 0 \\ * & * & * & * & * & -E_{81} \end{bmatrix} > 0, \tag{5.17}$$

*where the $E_{pq}$ and $E_{pq1}$ matrices for $p = 1, \ldots, 8$, $q = 1, \ldots, 5$ are defined as in Theorem 5.1.*

*Proof.* Consider the Lyapunov function candidate $V(k) = x_{cl}^T(k)Px_{cl}(k)$, where $P$ is a symmetric positive-definite matrix which is partitioned as in (5.9). By defining $\hat{C} = [C \ 0]$, $\hat{C}_c = [0 \ K]$ and by substituting $M$, $N$, and $Q$ with (5.10), and by multiplying both sides of (5.16) by $[x_{cl}^T \ d^T \ \delta_u^T \ \delta_y^T]$ from the left and $[x_{cl}^T \ d^T \ \delta_u^T \ \delta_y^T]^T$ from the right, one arrives at

$$\Delta V - d^T d + \sigma_y y^T y - \varepsilon_y \delta_y^T \delta_y + \sigma_u u^T u - \varepsilon_u \delta_u^T \delta_u < 0. \tag{5.18}$$

According to (5.3) and (5.4), we have

$$\Delta V(k) - d^T(k)d(k) < 0. \tag{5.19}$$

If $d(k) = 0$, then $\Delta V(k) < 0$, which guarantees the stability of the system. For zero initial condition, summing both sides of (5.19) over $[0, k]$ and by using the fact that $V(0) = 0$, we get

$$V(k) = x_{cl}^T(k)Px_{cl}(k) < \sum_{k'=0}^{k} d^T(k')d(k') < \sum_{k'=0}^{\infty} d^T(k')d(k'). \tag{5.20}$$

On the other hand, by substituting $M$, $N$, and $Q$ with (5.10) in (5.17), and by applying the Schur complement [271], and multiplying $[x_{cl}^T \ \delta_u^T \ \delta_y^T]$ from the left and $[x_{cl}^T \ \delta_u^T \ \delta_y^T]^T$ from the right, respectively, the inequality (5.17) becomes

$$x_{cl}^T Px_{cl} - \gamma_5^{-2} r^T r - \sigma_y y^T y + \varepsilon_y \delta_y^T \delta_y - \sigma_u u^T u + \varepsilon_u \delta_u^T \delta_u > 0. \tag{5.21}$$

According to (5.3) and (5.4), $-\sigma_y y^T y + \varepsilon_y \delta_y^T \delta_y < 0$ and $-\sigma_u u^T u + \varepsilon_u \delta_u^T \delta_u < 0$, which leads to

$$x_{cl}^T(k)Px_{cl}(k) - \gamma_5^{-2} r^T(k)r(k) > 0. \tag{5.22}$$

Using the expressions (5.20) and (5.22), we have

$$r^T(k)r(k) < \gamma_5^2 \sum_{k'=0}^{\infty} d^T(k')d(k'), \tag{5.23}$$

and this completes the proof of the theorem. □

The following theorem gives the LMI constraints for the $l_1$ performance indices (VII) and (VIII).

**Theorem 5.3.** *Assume that data is transmitted from the sensor to the E-IFDIC module and from the E-IFDIC module to the actuator subject to violations of the conditions (5.3) and (5.4), respectively. The augmented system (5.5) is stable and the $l_1$ performance indices (VII) and (VIII) are guaranteed if there exist symmetric positive-definite matrices $X$ and $Y$, matrices $M$, $N$, $V$, $Q$, $J$, predefined scalar*

$0 < \lambda < 1$, *and positive scalars* $\mu$, $\varepsilon_u$, $\varepsilon_y$, $\sigma_u$, $\sigma_y$, $\gamma_6$, $\gamma_7$ *such that for* $i = 1, 3$ *the following LMI conditions hold*

$$\begin{bmatrix} (1-\lambda)E_{11} & 0 & 0 & 0 & E_{12} & E_{14} & E_{15} \\ * & -\mu I & 0 & 0 & \Xi_{22i} & E_{24i} & 0 \\ * & * & E_{31} & 0 & E_{32} & 0 & 0 \\ * & * & * & E_{41} & E_{42} & 0 & 0 \\ * & * & * & * & E_{51} & 0 & 0 \\ * & * & * & * & * & E_{71} & 0 \\ * & * & * & * & * & * & E_{81} \end{bmatrix} < 0, \tag{5.24}$$

$$\begin{bmatrix} -\lambda E_{11} & 0 & 0 & 0 & E_{13i} & E_{14} & E_{15} \\ * & \Xi_{21i} & 0 & 0 & \Xi_{23i} & E_{24i} & 0 \\ * & * & -E_{31} & 0 & E_{33i} & 0 & 0 \\ * & * & * & -E_{41} & E_{43i} & 0 & 0 \\ * & * & * & * & \Xi_{61i} & 0 & 0 \\ * & * & * & * & * & -E_{71} & 0 \\ * & * & * & * & * & * & -E_{81} \end{bmatrix} > 0, \tag{5.25}$$

*where* $\Xi_{211} = (\gamma_6 - \mu)I$, $\Xi_{213} = (\gamma_7 - \mu)I$, $\Xi_{611} = \gamma_6 I$, $\Xi_{613} = \gamma_7 I$, $\Xi_{221} = \Xi_{223} = \begin{bmatrix} B_w^T & B_w^T X + D_w^T N^T \end{bmatrix}$, $\Xi_{231} = D_w^T H^T$, $\Xi_{233} = F_w^T$, *and the* $E_{pq}$ *and* $E_{pqi}$ *matrices for* $p = 1, \ldots, 8$, $q = 1, \ldots, 5$, *and* $i = 1, 3$ *are defined as in Theorem 5.1.*

*Proof.* Consider the Lyapunov function candidate $V(k) = x_{cl}^T(k)Px_{cl}(k)$, where $P$ is a symmetric positive-definite matrix which is partitioned as in (5.9). Now, according to (5.24) for $i = 1$, by substituting $M$, $N$, and $Q$ with (5.10), and applying the Schur complement, one can conclude that

$$\begin{bmatrix} \Phi_{11} & \Phi_{12} & \tilde{A}^T P \tilde{B}_{\delta_u} & \tilde{A}^T P \tilde{B}_{\delta_y} \\ * & \Phi_{22} & \tilde{B}_w^T P \tilde{B}_{\delta_u} & \tilde{B}_w^T P \tilde{B}_{\delta_y} \\ * & * & \tilde{B}_{\delta_u}^T P \tilde{B}_{\delta_u} - \varepsilon_u I & \tilde{B}_{\delta_u}^T P \tilde{B}_{\delta_y} \\ * & * & * & \tilde{B}_{\delta_y}^T P \tilde{B}_{\delta_y} - \varepsilon_y I \end{bmatrix} < 0, \tag{5.26}$$

where $\Phi_{11} = \tilde{A}^T P \tilde{A} - P + \lambda P + \sigma_y \hat{C}^T \hat{C} + \sigma_u \hat{C}_c^T \hat{C}_c$, $\Phi_{12} = \tilde{A}^T P \tilde{B}_w + \sigma_y \hat{C}^T D_w$, $\Phi_{22} = \tilde{B}_w^T P \tilde{B}_w + \sigma_y D_w^T D_w - \mu I$. The (1,1)-element of (5.26) (i.e., $\Phi_{11}$) implies that $\tilde{A}^T P \tilde{A} - P < 0$, which guarantees the stability of the system.

By multiplying both sides of (5.26) by $[x_{cl}^T \ w^T \ \delta_u^T \ \delta_y^T]$ from the left and $[x_{cl}^T \ w^T \ \delta_u^T \ \delta_y^T]^T$ from the right, one gets

$$\Delta V + \lambda V - \mu w^T w + \sigma_y y^T y - \varepsilon_y \delta_y^T \delta_y + \sigma_u u^T u - \varepsilon_u \delta_u^T \delta_u < 0,$$

and according to (5.3) and (5.4), we have

$$\Delta V + \lambda V - \mu w^T w < 0. \tag{5.27}$$

Based on the fact that $w(k)^\mathrm{T} w(k) \le 1$, then $\Delta V < 0$ holds whenever $V > \mu/\lambda$. Since $V(0) = 0$, this implies that $V$ cannot exceed the value $\mu/\lambda$, and therefore, $V < \mu/\lambda$, for $k \ge 0$ [271].

By substituting $M$, $N$, and $Q$ with (5.10) in the inequality (5.25), and applying the Schur complement, and multiplying both sides of (5.25) by $[x_\mathrm{cl}^\mathrm{T} \ w^\mathrm{T} \ \delta_\mathrm{u}^\mathrm{T} \ \delta_\mathrm{y}^\mathrm{T}]$ from the left and $[x_\mathrm{cl}^\mathrm{T} \ w^\mathrm{T} \ \delta_\mathrm{u}^\mathrm{T} \ \delta_\mathrm{y}^\mathrm{T}]^\mathrm{T}$ from the right, one can conclude that

$$r^\mathrm{T} r < \gamma_6[\lambda V + (\gamma_6 - \mu)w^\mathrm{T} w - \sigma_\mathrm{y} y^\mathrm{T} y + \varepsilon_\mathrm{y} \delta_\mathrm{y}^\mathrm{T} \delta_\mathrm{y} - \sigma_\mathrm{u} u^\mathrm{T} u + \varepsilon_\mathrm{u} \delta_\mathrm{u}^\mathrm{T} \delta_\mathrm{u}]. \qquad (5.28)$$

According to (5.3) and (5.4), $\varepsilon_\mathrm{y} \delta_\mathrm{y}^\mathrm{T} \delta_\mathrm{y} - \sigma_\mathrm{y} y^\mathrm{T} y < 0$ and $\varepsilon_\mathrm{u} \delta_\mathrm{u}^\mathrm{T} \delta_\mathrm{u} - \sigma_\mathrm{u} u^\mathrm{T} u < 0$, which leads to

$$r^\mathrm{T} r < \gamma_6[\lambda V + (\gamma_6 - \mu)w^\mathrm{T} w]. \qquad (5.29)$$

Since $w(k)^\mathrm{T} w(k) \le 1$ and $V < \mu/\lambda$, therefore, $r^\mathrm{T} r < \gamma_6^2$. Consequently, the $l_1$ norm (VII) is satisfied. The $l_1$ performance index (VIII) can also be shown by employing the same procedure as in the derivation of the performance index (VII). These details are omitted for sake of brevity. This completes the proof of the theorem. □

Note that the inequalities (5.24) and (5.25) are only linear if $\lambda$ is fixed. Determining the best bound guaranteed by (5.24) and (5.25) requires that one performs a line search over $0 < \lambda < 1$.

Based on the above results, the following algorithm is now proposed in order to simultaneously specify and determine the E-IFDIC module parameters and also the event-triggering conditions.

**Algorithm 5.1:** *For a given positive constants $\beta_1$ and $\beta_2$, a feasible solution for the problem of E-IFDIC filter design is obtained by solving the following convex optimization problem*

$$\min_{H,J,M,N,P,Q,X,Y} \beta_1(\zeta_1 \gamma_1 + \zeta_2 \gamma_2 + \zeta_3 \gamma_3 + \zeta_4 \gamma_4 + \zeta_5 \gamma_5 \qquad (5.30)$$
$$+ \zeta_6 \gamma_6 + \zeta_7 \gamma_7) + \beta_2(\varepsilon_u + \varepsilon_y - \sigma_u - \sigma_y)$$

*subject to the inequality (5.7), if $\zeta_2 = 1$; the inequality (5.8) for $i = 1$, 2, 3, 4, if $\zeta_1$, $\zeta_2$, $\zeta_3$, $\zeta_4$ are equal to one, respectively; the inequalities (5.16) and (5.17), if $\zeta_5 = 1$; the inequalities (5.24) and (5.25) for $i = 1$, 3, if $\zeta_6$, $\zeta_7$ are equal to one, respectively, and $\zeta_i \in \{0, 1\}$ for $i = 1, \ldots, 7$. Moreover, the E-IFDIC filter parameters $A_c$, $L$ and $K$ are given by*

$$A_c = (Y^{-1} - X)^{-1}(MY^{-1} - NC - X(A + B_u K)),$$
$$L = (Y^{-1} - X)^{-1}N, \ K = QY^{-1}. \qquad (5.31)$$

The values of the weights $\beta_1$ and $\beta_2$ in Algorithm 5.1 can be chosen by the designer to emphasize certain objectives and priorities that are more important than the others. The value of $\beta_1$ is used to emphasize the fault detection, isolation, and control objectives. The value of $\beta_2$ is used to emphasize and control the communication rates among the sensors, E-IFDIC filter, and the actuators. To achieve larger interevent intervals, larger values of $\beta_2$ should be chosen.

On the other hand, the values of the weights $\zeta_i \in \{0, 1\}$ for $i = 1, \ldots, 7$, in Algorithm 5.1 can be chosen by the designer to consider and design the desired problem. For example, in order to design the E-IFDIC problem, all $\zeta_i$s should be selected as 1. In order to solve the codesign of event detector and $l_1$ controller (**Problem 1** in Remark 5.1), the weights $\zeta_i$ should be selected as $\zeta_7 = 1$ and $\zeta_i = 0$ for $i = 1, \ldots, 6$. Consequently, certain problems in control and fault diagnosis of time-triggered systems (e.g., the IFDC design and the $l_1$ controller design) can be studied for event-triggered systems by using the results and methodologies proposed in this chapter.

Moreover, our proposed results can be easily extended for considering certain time-triggered control and fault diagnosis problems that have not been stated explicitly to be within the framework of this work. For example, the problem of generalized $H_2$ fault detection for two-dimensional Markovian jump and switching systems have been studied in [322,323], respectively. According to the results obtained for the generalized $H_2$ performance index (VI) (in Theorem 5.2), the considered problems in [322,323] can be easily extended to event-triggered framework.

Finally, certain existing methodologies in the field of event-triggered design can also be covered by using our proposed methodology. For example, in [1], an event-triggered stabilizable controller design with reduced communication load among the sensors, controllers, and actuators was considered. The considered problem in [1] is a special case of our results subject to the following setting: $\varepsilon_u = \varepsilon_y = 1$, and $B_d = B_w = B_f = D_d = D_w = D_f = E = F_u = F_d = F_w = F_f = 0$.

Based on the above discussion, it can be concluded that our proposed E-IFDIC approach represents quite a general problem that is capable of handling various EFD and control challenges in a unified framework.

**Remark 5.2.** *In Algorithm 5.1, the solution to the E-IFDIC problem is formulated as strict LMI conditions that permits one to carry out the optimization scheme easily. In most existing results, the controller is predetermined before the event-trigger condition. However, in our method, both the IFDIC module and the event-trigger conditions are designed simultaneously that enables one to determine the E-IFDIC module and the best values of the parameters $\varepsilon_u$, $\varepsilon_y$, $\sigma_u$, $\sigma_y$. Also note that for the parameters ($\sigma_u$, $\sigma_y$) and ($\varepsilon_u$, $\varepsilon_y$) smaller and larger than the obtained feasible solutions are always suitable to be chosen, respectively.*

### Residual evaluation criterion

Following the construction of the residual signal $r(k) \in \mathbb{R}^{n_f}$, the final step in the E-IFDIC strategy is to determine the threshold $J_{\text{th}}$ and the evaluation function $J_r(k)$. In this work, upper and lower threshold values are selected as $J_{\text{th}_i}^u = \sup_{f_i=0, d \in l_2, w \in l_\infty} r_i(k)$ and $J_{\text{th}_i}^l = \inf_{f_i=0, d \in l_2, w \in l_\infty} r_i(k)$, respectively. Based on the selected thresholds and the evaluation function taken as $J_{r_i}(k) = r_i(k)$, the occurrence of a fault can then be detected and isolated by using the following decision logic: $r_i(k) > J_{\text{th}_i}^u$ or $r_i(k) < J_{\text{th}_i}^l \implies f_i(k) \neq 0$.

## 5.4 Two industrial case studies

As stated earlier, our proposed methodology has the capability to solve various event-triggered multiobjective control and fault diagnosis problems. In this section, different aspects and certain capabilities of our methodology in event-triggered multiobjective synthesis of feedback controllers and fault diagnosis filters are demonstrated by providing two industrial case studies under different scenarios (totally nine sets of simulation conditions). In the first case study, the effectiveness of our proposed methodology for solving the problem of "E-IFDIC in presence of bounded energy disturbances" is provided. The main objective of the second case study is to illustrate the effectiveness of our proposed methodology for solving the problem of "event-triggered multiobjective $H_\infty/l_1$ control design in presence of both the bounded energy and bounded peak disturbances."

**Case study 1** *(E-IFDIC design)*

Batch reactor is an essential unit operating in almost all batch processing industries. They are particularly prevalent in the food and beverage industries, polymer, pharmaceutical, and chemicals industries. Operating batch reactors efficiently and economically is quite important as far as the overall profitability is concerned [331].

Hence, these systems should be carefully controlled. On the other hand, to enhance reliability and safety of these systems, the possible equipment failures need to be detected, so that corrective actions can be planned in a timely and effective way. However, these goals (controlling and fault diagnosis) are difficult to achieve, since batch reactors are often subject to large disturbances, modeling uncertainties, and constraints on their computational and communication resources. Therefore, the results of this work are applied for designing an event-triggered IFDIC design for a four-dimensional linearized model of an unstable batch reactor process.

Refer to [332] for more details on the continuous-time model of the system $(A, B_u, C)$. Note that the continuous-time model is discretized with a sampling period of 0.2 s. Moreover, the disturbance matrices $B_d$, $D_d$ are selected as $B_d = 0.1B_u$ and $D_d = 0.1I$, which represent occurrence of actuator and sensor noise in the system. The fault matrices $B_f$ and $D_f$ are selected as $B_f = B_u$ and $D_f = 0$, which represent occurrence of actuator faults in the system. The control output $z(k)$ is considered as $z(k) = 0.1y(k)$. Moreover, it is assumed that there are no bounded peak disturbances in the system.

In the simulations conducted, it is assumed that the disturbance signal $d(k) \in \mathbb{R}^2$ is a band-limited white noise with a power of 0.02. The fault signals $f_1(k)$ (corresponding to the first actuator) and $f_2(k)$ (corresponding to the second actuator) are simulated as rectangular pulsed signals with amplitudes of 1 and 1.5 that have occurred during the time intervals $k = [20\,40]$ and $k = [60\,80]$, respectively. It is desired that by using the event-triggered data transmission scheme, the system detects the occurrence of the fault $f \in \mathbb{R}^2$ in presence of the disturbance $d(k) \in \mathbb{R}^2$ and also attenuates the effects of faults and disturbances on the regulated output $z(k) \in \mathbb{R}^2$.

Note that the considered system in this case study is a MIMO LTI system that has two inputs and two outputs. Therefore, the regulated output $z(k) \in \mathbb{R}^2$ has two elements $z_1(k) \in \mathbb{R}$ and $z_2(k) \in \mathbb{R}$. On the other hand, since there are two possible sources (two actuators) for occurrence of faults $f_1(k), f_2(k) \in \mathbb{R}$ in the system, the residual signal $r(k) \in \mathbb{R}^2$ that is generated by the E-IFDIC module has two elements (viz., $r_1(k)$ and $r_2(k)$). Each residual $(r_i(k), i = 1, 2)$ is responsible for detecting and isolating the occurred fault in the corresponding first and second actuators, respectively.

### A. Effects of $\beta_i$ on the E-IFDIC design

To illustrate the trade-offs between the communication rates and the IFDIC objectives, two scenarios are considered below by having different values of $\beta_i$, $i = 1, 2$. The weights $\zeta_i$ are selected as $\zeta_1 = \zeta_2 = \zeta_3 = \zeta_4 = 1$ and $\zeta_5 = \zeta_6 = \zeta_7 = 0$. In the first scenario, the positive constant weights $\beta_1$ and $\beta_2$ are selected as $\beta_1 = 100$, $\beta_2 = 1$, which imply higher importance to the IFDIC objective [the performance indices (I)–(V)] as compared to the communication rates [the conditions (5.3) and (5.4)]. The optimization problem corresponding to Algorithm 5.1 is solved and the parameters $\gamma_1$–$\gamma_4$ and the event-triggered parameters $\sigma_u$, $\sigma_y$, $\varepsilon_u$, $\varepsilon_y$ are obtained as $\gamma_1 = 0.2158$, $\gamma_2 = 2.8975$, $\gamma_3 = 0.1462$, $\gamma_4 = 1.5321$, $\sigma_u = 0.0266$, $\sigma_y = 0.0476$, $\varepsilon_u = 18.1731$, and $\varepsilon_y = 34.5185$.

In the second scenario, the values of $\beta_i$, $i = 1, 2$ are selected as $\beta_1 = 1$, $\beta_2 = 1,000$, which imply that the design interest and priority is on higher communication rates attenuation than the IFDIC objective. The optimization problem corresponding to Algorithm 5.1 is solved and the parameters $\gamma_1$–$\gamma_4$ and the event-triggered parameters $\sigma_u$, $\sigma_y$, $\varepsilon_u$, $\varepsilon_y$ are obtained as $\gamma_1 = 15.3812$, $\gamma_2 = 126.8486$, $\gamma_3 = 14.4893$, $\gamma_4 = 126.0336$, $\sigma_u = 0.1327$, $\sigma_y = 0.230$, $\varepsilon_u = 5.3991$, and $\varepsilon_y = 10.9513$.

The residual signals $r_1(k)$ and $r_2(k)$ (corresponding to the faults $f_1(k)$ and $f_2(k)$) for the first and second scenarios are shown in Figure 5.2(a), (b) and (c), (d), respectively. In these figures, the solid lines denote the residual signals and the dash-dot lines denote the residual upper and lower thresholds. From Figure 5.2(a) and (b), it follows that the robustness against disturbances and the enhanced fault sensitivity are satisfied and faults are satisfactorily discriminated from the disturbances.

Hence, by using our proposed decision logic, the actuator faults $f_1(k)$ and $f_2(k)$ are effectively detected and isolated in the first scenario. Note that the *fault detection times* for both faults are within 1 time step. However, from Figure 5.2(c) and (d), it can be concluded that since the false alarm rates are increased, the faults $f_1(k)$ and $f_2(k)$ have not been effectively detected and isolated in the second scenario. Therefore, it follows that the scenario 1 has a better diagnosis performance in comparison with the scenario 2. In order to clearly illustrate this observation, the minimum detectable faults for each scenario are provided in Table 5.1.

The regulated output $z(k)$ corresponding to the first and second scenarios are shown in Figure 5.3(a) and (b), respectively, where it can be concluded that the effects of disturbances and faults on the regulated outputs have been attenuated. The energy

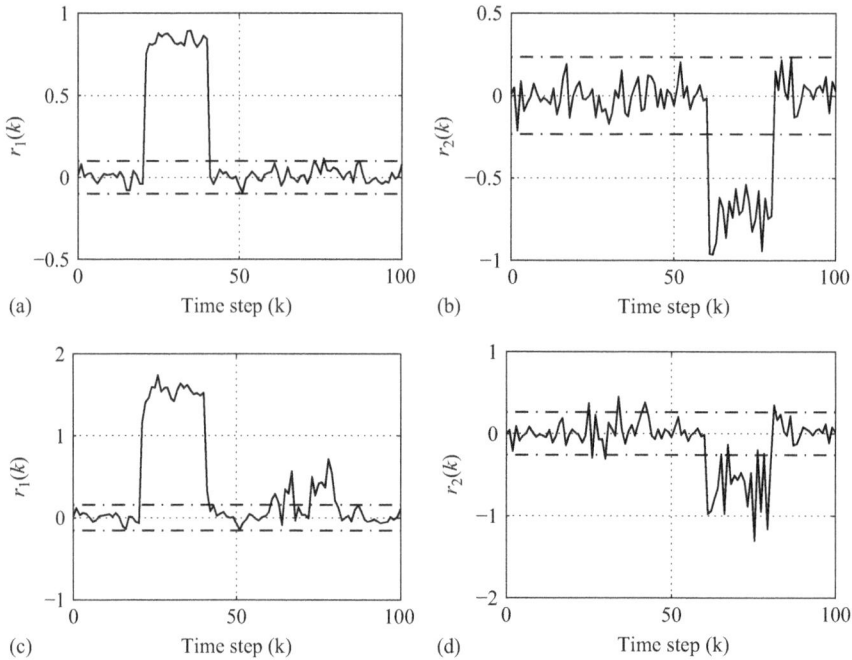

*Figure 5.2    The residual signals: (a) $r_1(k)$ and (b) $r_2(k)$ corresponding to the first
scenario, with (c) $r_1(k)$ and (d) $r_2(k)$ corresponding to the second
scenario*

*Table 5.1    Minimum detectable fault severities*

| Faults | Scenario 1 | Scenario 2 |
|--------|-----------|-----------|
| $f_1$ | 0.17 | 0.19 |
| $f_2$ | 0.8 | 1.35 |

of the regulated output $z(k)$ over the time interval $k = [0\ 100]$ $\left(\text{i.e., } \sum_{k=0}^{100} ||z(k)||^2\right)$
for the first and second scenarios are obtained as 0.2864 and 0.321, respectively.
Hence, from Figure 5.3 and also the above energy values, it can be concluded that the
effects of disturbances and faults on the regulated output have been more attenuated
in the first scenario in comparison with the second scenario.

The interevent intervals of the sensor and the E-IFDIC module event detectors for
both scenarios are depicted in Figures 5.4 and 5.5, respectively, in which each pulse
represents the occurrence of an event that leads to a data transmission. The magnitude
of each pulse specifies the length of time period between that event and the previous
one. Figure 5.4 shows that the total number of sensor events for the first and second
scenarios are 84 and 67, respectively.

(a)

(b)

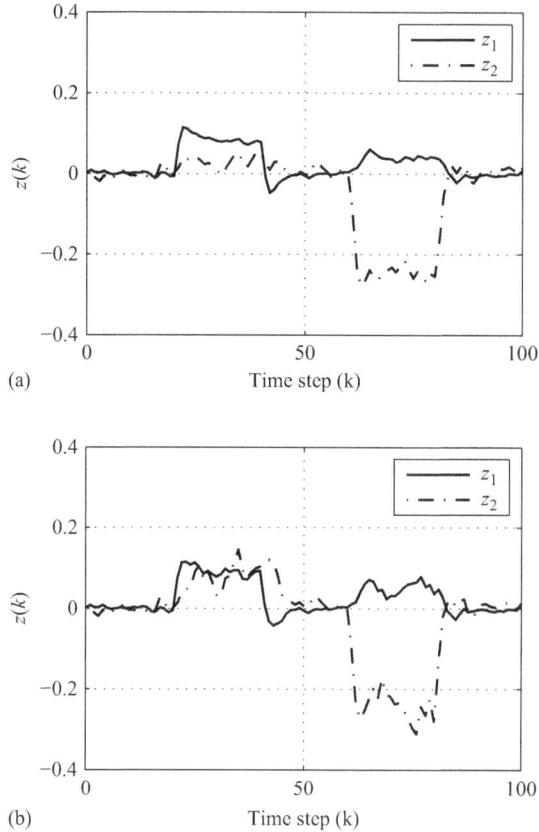

*Figure 5.3   The regulated output z(k): (a) corresponding to the first scenario and (b) corresponding to the second scenario*

Indeed, the total number of events can be easily obtained according to the following equation:

$$TNE = n - \sum_{i=2}^{n} (i-1) \times m_i$$

where $TNE$ denotes the total number of events, $n$ denotes the total number of samples, and $m_i$ denotes the total number of pulses with a magnitude $i$. For example, in Figure 5.4(a), there exist 11 pulses with a magnitude of 2, 1 pulse with a magnitude of 3, and 1 pulse with a magnitude of 4. Therefore, based on the above equation, the total number of sensor events for the first scenario is equal to $TNE = 100 - 11 \times 1 - 2 - 3 = 84$. Moreover, the maximum interevent intervals of the sensor for the first and second scenarios are 4 and 6, respectively.

Hence, the data transmission from the sensor-to-E-IFDIC module channel for the first and second scenarios is reduced by 16% and 33%, respectively. Figure 5.5

(a)

(b)

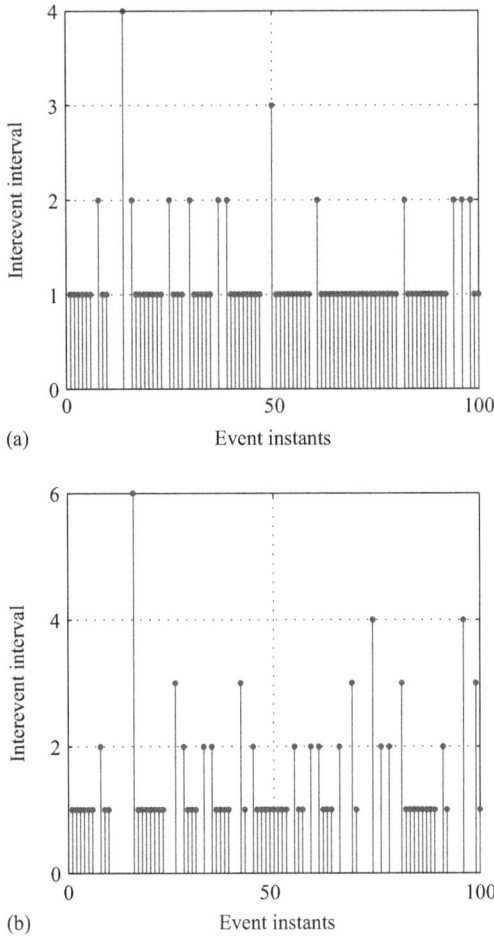

*Figure 5.4    Interevent intervals of the sensor event detector: (a) the first scenario and (b) the second scenario*

shows a reduction in the data transmission from the E-IFDIC module-to-actuator channel by 14% and 25% for the first and second scenarios, respectively. Moreover, by comparing Figures 5.4 and 5.5, it can be concluded that the communication rates have been reduced more in the second scenario in comparison with the first scenario.

From Figures 5.2–5.5, it follows readily that in the second scenario, the IFDIC and communication rates objectives have been degraded and improved, respectively, in comparison with those achieved in the first scenario.

In [329], the problem of time-trigger IFDIC design for continuous-time LTI systems is considered. The setup in [329] is a special case of our results with $\zeta_1 = \zeta_2 = \zeta_3 = \zeta_4 = 1$, $\zeta_5 = \zeta_6 = \zeta_7 = 0$, $\beta_1 = 1$, $\beta_2 = 0$, $\delta_y = \delta_u = 0$, and $\varepsilon_y = \varepsilon_u = \sigma_y = \sigma_u = 0$. The main results presented in this work in Algorithm 5.1 can be reduced

(a)

Event instants

(b)

Event instants

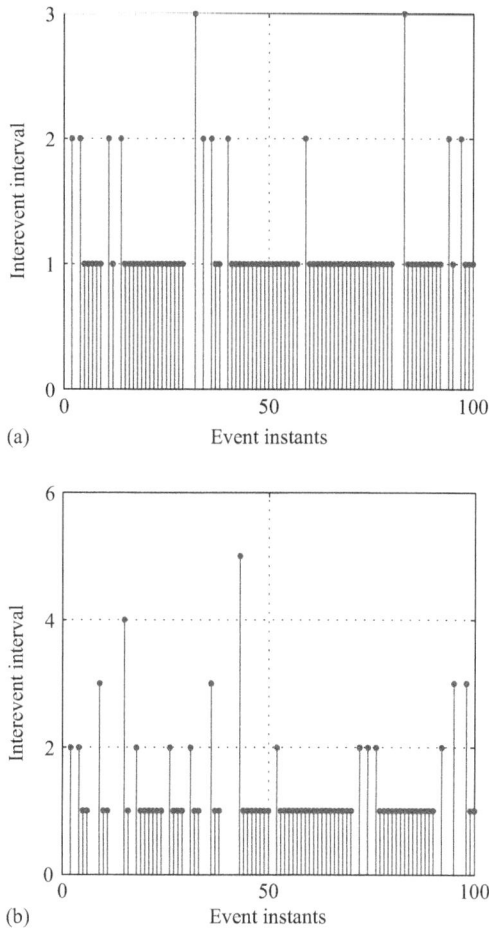

*Figure 5.5    Interevent intervals of the E-IFDIC module event detector: (a) the first scenario and (b) the second scenario*

to this case by eliminating the third, fourth, seventh, and eighth rows and columns of the matrix inequality (5.8). The optimization problem corresponding to this case is now solved and the parameters $\gamma_1-\gamma_4$ are obtained as $\gamma_1 = 0.1634$, $\gamma_2 = 2.4996$, $\gamma_3 = 0.1038$, and $\gamma_4 = 1.1872$. Note that in this case, the percentage of the communication for both sensor-to-E-IFDIC module and E-IFDIC module-to-actuator is 100%.

To summarize, the optimization parameters $\gamma_1-\gamma_4$ and the percentage of communications for both sensor-to-E-IFDIC module and E-IFDIC module-to-actuator networks for time- and event-triggered IFDIC design methodologies are shown in Table 5.2. From this table, it follows that in the time-triggered IFDIC methodology, the IFDIC objective has been improved at the expense of utilizing the entire communication rates.

*Table 5.2    Comparison of time-triggered and event-triggered IFDIC design
methodologies, where S-to-E denotes the "sensor-to-E-IFDIC module"
and E-to-A denotes the "E-IFDIC module-to-actuator"*

| Approach | $\gamma_1$ | $\gamma_2$ | $\gamma_3$ | $\gamma_4$ | The data transmission rate | |
|---|---|---|---|---|---|---|
| | | | | | S-to-E (%) | E-to-A |
| *Scenario 1* | 0.2158 | 2.8975 | 0.1462 | 1.5321 | 84 | 86 |
| *Scenario 2* | 15.3812 | 126.8486 | 14.4893 | 126.0336 | 67 | 75 |
| *Results of [329]* | 0.1634 | 2.4996 | 0.1038 | 1.1872 | 100 | 100 |

Moreover, it is worth noting that although by using our proposed scheme, the performance indices ($\gamma_1, \gamma_2, \gamma_3, \gamma_4$) are degraded in comparison with the time-triggered IFDIC design, they are still sufficient for satisfying the IFDIC objectives (this follows from Figures 5.2 and 5.3). Therefore, our methodology in this work provides a valuable option to the designer to consider a trade-off between the IFDIC and the data transmission reduction objectives of the closed-loop system.

**B. Effectiveness of the E-IFDIC design**

Below another set of simulations (designated as the third scenario) are conducted to demonstrate the effectiveness of our methodology in attenuating the effects of faults and disturbances on the regulated output and also to quantitatively compare our results with those in [1]. At first, by considering Algorithm 5.1 with $\zeta_1 = \beta_2 = 1$, $\beta_1 = \zeta_i = 0$ for $i = 2, \ldots, 7$, and $B_d = B_w = B_f = D_d = D_w = D_f = E = F_u = F_d = F_w = F_f = 0$ (almost similar to Theorem 4.2 of [1]), a stabilizing controller is obtained. Simulations using this stabilizing controller are performed, and the regulated output $z(k)$ is shown in Figure 5.6. Moreover, the energy of the regulated output $z(k)$ over the time interval $k = [0\ 100]$ is obtained as 1.682. The total number of sensor and E-IFDIC module events is 56 and 63, respectively. Hence, the data transmission rates for the sensor and the E-IFDIC module event detectors are reduced by 44% and 37%, respectively.

From Figures 5.3 and 5.6 and the obtained energy values for these three scenarios, it can be concluded that the effects of disturbances and faults have been successfully attenuated by using our methodology in comparison with the results of [1]. Indeed, in comparison with [1], our results not only guarantee stability of the closed-loop system, but also attenuate the effects of external disturbances (with a satisfactory level of $\gamma_3 = 0.1462$ for the scenario 1 and $\gamma_3 = 14.4893$ for the scenario 2) and faults (with a satisfactory level of $\gamma_4 = 1.5321$ for the scenario 1 and $\gamma_4 = 126.0336$ for the scenario 2) that may occur in the system. Moreover, in this scenario, the communication rates objective has been improved in comparison with the previous scenarios. Therefore, although [1] represents as one of the latest work in the field

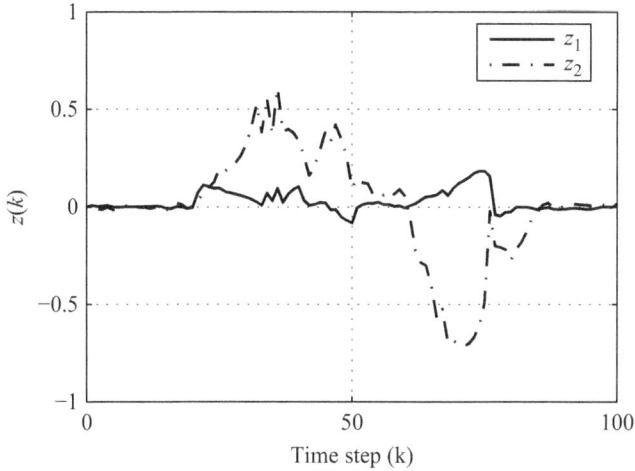

*Figure 5.6   The regulated output $z(k)$ corresponding to the third scenario that is obtained using [1]*

of "codesign of event generator and controller," our results in this paper represent important extensions to the existing work in this field [1,321].

## C. Effects of $\zeta_i$ on the E-IFDIC design

The effects of having different weights $\beta_i$ in the E-IFDIC design were demonstrated in the above simulation results. However, to show the effects of changing the constant weights $\zeta_i$ and also the capability of the generalized $H_2$ performance index on the fault diagnosis problem, the optimization problem of Algorithm 5.1 is solved for two more scenarios. The positive constant weights $\beta_1$ and $\beta_2$ are selected as $\beta_1 = \beta_2 = 1$, which implies the same level of importance or priority in the IFDIC and the communication rates objectives.

Specifically, in the fourth scenario, the weights $\zeta_i$ are selected as $\zeta_1 = \zeta_2 = \zeta_3 = \zeta_4 = \zeta_5 = 1$, and $\zeta_6 = \zeta_7 = 0$, which lead to the same level of importance and priority for all the objectives (I)–(VI). In the fifth scenario, the weights $\zeta_i$ are selected as $\zeta_1 = \zeta_2 = \zeta_3 = \zeta_4 = 1$, $\zeta_5 = 100$ and $\zeta_6 = \zeta_7 = 0$, which imply higher importance for the generalized $H_2$ performance [the performance index (VI) as compared to the performance indices (I)–(V)].

Note that for solving the generalized $H_2$ specification, it should be assumed that the disturbance matrix $D_d = 0$ (refer to Theorem 5.2). However, in the above simulations, it was assumed that $D_d = 0.1I$. Hence, in the following simulations, the disturbance matrices $B_d$, $D_d$ are selected as $B_d = B_u$ and $D_d = 0$. The fault matrices $B_f$ and $D_f$ are selected as $B_f = 0$ and $D_f = I$, which correspond to a sensor fault. Moreover, other matrices and parameters are considered to be selected similar to the previous simulations.

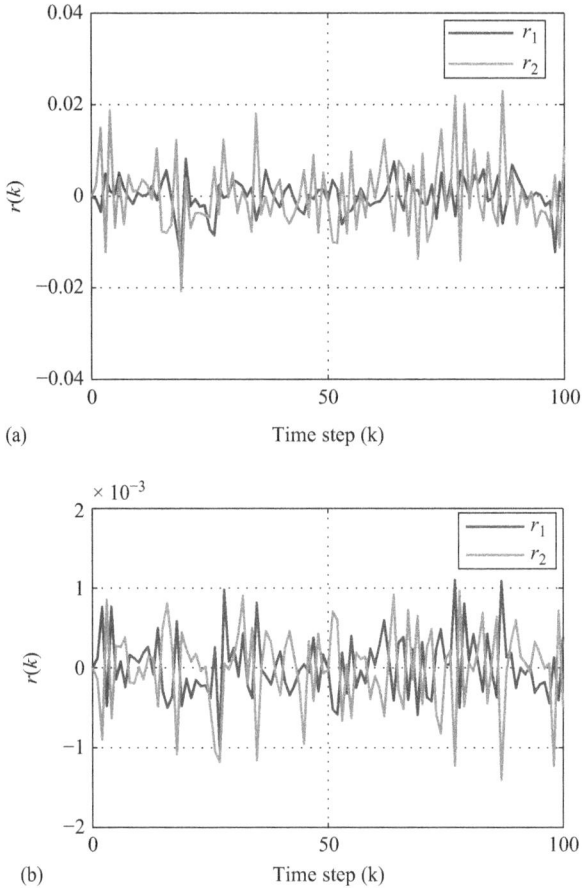

*Figure 5.7　Residual signals corresponding to the healthy system: (a) the fourth scenario and (b) the fifth scenario*

The optimization problem corresponding to Algorithm 5.1 for the scenarios 4 and 5 are solved and the optimization parameters $\gamma_1-\gamma_5$ and the event-triggered parameters $\sigma_u, \sigma_y, \varepsilon_u, \varepsilon_y$ are obtained as follows: for the fourth scenario: $\gamma_1 = 1.0098$, $\gamma_2 = 5.6356$, $\gamma_3 = 1.1501$, $\gamma_4 = 5.0110$, $\gamma_5 = 0.0238$, $\sigma_u = 0.0428$, $\sigma_y = 0.0793$, $\varepsilon_u = 4.3010$, and $\varepsilon_y = 7.9440$, and for the fifth scenario: $\gamma_1 = 1$, $\gamma_2 = 5.8055$, $\gamma_3 = 1.1588$, $\gamma_4 = 4.8802$, $\gamma_5 = 5.7112e - 05$, $\sigma_u = 0.0427$, $\sigma_y = 0.0791$, $\varepsilon_u = 4.2839$, and $\varepsilon_y = 7.9615$. The residual signals corresponding to the healthy system are compared for these two scenarios in Figure 5.7. It can be observed that smaller peak values for the residual signals can be achieved by considering more weights on the generalized $H_2$ performance index.

**Remark 5.3.** *It should be pointed out that from practical and engineering viewpoints, determination of a threshold is to serve as tolerant limits for unknown inputs and* *model uncertainties under fault-free operational conditions. Hence, it is desirable to keep the peak amplitude of the residual signal $r(k)$ below a certain level under the fault-free case. In this chapter, the generalized $H_2$ specification is introduced to tune the peak amplitude of the residual, and this feature becomes very important from a decision making point of view. However, it is clear that although the generalized $H_2$ guarantees a higher robustness to disturbances, and thus it reduces the false alarm rates on the one hand, it increases the missing detection rates on the other. Finally, it should be pointed out that the generalized $H_2$ specification can have other advantages in precomputing the threshold values and also in the integrated design of the residual generator and the residual evaluator in order to improve the fault diagnosis objectives. This extension is beyond the scope of this work and is left as a topic of our future research.*

## Case study 2   *(codesign of event generator and $H_\infty$ ($H_\infty/l_1$) Controller)*

To show the effectiveness of our methodology in attenuation of effects of both the bounded energy and bounded peak disturbances, application of our proposed event-triggered $H_\infty$ and $H_\infty/l_1$ controllers to the heading subsystem of the Subzero III ROV [333] are presented in this case study.

Remotely operated vehicles (ROVs) are underwater robotic platforms that have become increasingly important tools in a wide range of applications including off-shore oil operations, fisheries research, dam inspection, salvage operations, military applications, among others. Since transmission resources are limited under water, using an event-triggered scheme for communication is more practical and efficient. Therefore, the results of this chapter are applied for designing event-triggered $H_\infty$ and $H_\infty/l_1$ controllers for the Subzero III ROV.

Refer to the description provided in Chapter 2 and [333] for more details on the continuous-time mathematical model of the Subzero III ROV. Note that the continuous-time model is discretized with a sampling period of 0.2 s. In simulations conducted, the disturbance matrices $B_d$, $B_w$, $D_d$, $D_w$ are selected as $B_d = B_w = B_u$ and $D_d = D_w = 0.1$, which represent the occurrence of both the bounded energy and bounded peak disturbances in both the actuator and sensor of the system. Moreover, the regulated output $z(k)$ is considered as $z(k) = 0.1y(k)$.

It should be pointed out that waves and ocean currents are the most important sources of environmental disturbances for marine vehicles [262]. The current velocity can be modeled by a first order Gauss–Markov process: $\dot{V}_c + \mu V_c = \nu$, where $\mu \geq 0$ and $\nu$ is a Gaussian white noise. For $\mu = 0$, the model becomes a random walk. Hence, the injected disturbance $d(k)$ to the ROV is a random walk. Moreover, the waves can be modeled by sinusoidal signals or a combination of them. Specifically, the bounded peak disturbance $w(k)$ is taken as $\sin(0.7t)$.

The positive constant weights $\beta_1$ and $\beta_2$ are selected as $\beta_1 = 1$, $\beta_2 = 1$, which implies the same level of importance or priority for the control and communication rates objectives. At first, the weights $\zeta_i$ in Algorithm 5.1 are selected as $\zeta_3 = 1$ and $\zeta_1 = \zeta_2 = \zeta_4 = \zeta_5 = \zeta_6 = \zeta_7 = 0$, which lead to an "event-triggered $H_\infty$ controller" (denoted as the sixth scenario).

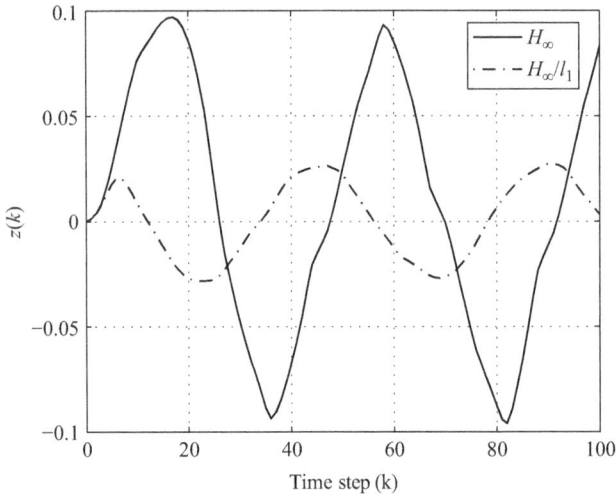

*Figure 5.8    The regulated output z(k) for case study 2*

For $\lambda = 0.5$, the optimization problem corresponding to Algorithm 5.1 is solved and the parameter $\gamma_3$ and the event-triggered parameters $\sigma_u$, $\sigma_y$, $\varepsilon_u$, $\varepsilon_y$ are obtained as $\gamma_3 = 1.6445$, $\sigma_u = 0.4098$, $\sigma_y = 0.4145$, $\varepsilon_u = 1.6555$, and $\varepsilon_y = 1.6113$.

The optimization problem corresponding to Algorithm 5.1 for $\zeta_3 = \zeta_7 = 1$ and $\zeta_1 = \zeta_2 = \zeta_4 = \zeta_5 = \zeta_6 = 0$ is solved which leads to an "event-triggered $H_\infty/l_1$ controller" (denoted as the seventh scenario). The parameters $\gamma_3$, $\gamma_7$ and the event-triggered parameters $\sigma_u$, $\sigma_y$, $\varepsilon_u$, $\varepsilon_y$ in this scenario are obtained as $\gamma_3 = 9.5347$, $\gamma_7 = 17.0437$, $\sigma_u = 0.0298$, $\sigma_y = 0.0265$, $\varepsilon_u = 20.6262$, and $\varepsilon_y = 24.1568$.

For the sixth and seventh scenarios, simulations using the obtained controllers are now conducted. Figure 5.8 compares the regulated output $z(k)$ of the mixed event-triggered $H_\infty/l_1$ feedback controller to the event-triggered $H_\infty$ controller. Note that the mixed-norm $H_\infty/l_1$ controller reduces the maximum peak response by 78.77% over the $H_\infty$ controller. The interevent intervals of the sensor and the E-IFDIC module event detectors for both scenarios are shown in Figures 5.9 and 5.10, respectively.

Figure 5.9 shows a reduction of data transmission from the sensor-to-E-IFDIC module channel by 65% and 26% for the sixth and seventh scenarios, respectively. Figure 5.10 shows a reduction of data transmission from the E-IFDIC module-to-actuator channel by 64% and 23% for the sixth and seventh scenarios, respectively.

It can be seen from Figures 5.8–5.10 that by using our proposed controllers, one can attenuate the effects of external disturbances in the ROV and simultaneously decrease the amount of data that is sent through the sensor-to-E-IFDIC module and E-IFDIC module-to-actuator channels. Moreover, by using the event-triggered multiobjective $H_\infty/l_1$ controller, the effects of external disturbances are more attenuated in comparison with the single-objective event-triggered $H_\infty$ controller.

(a)

(b)

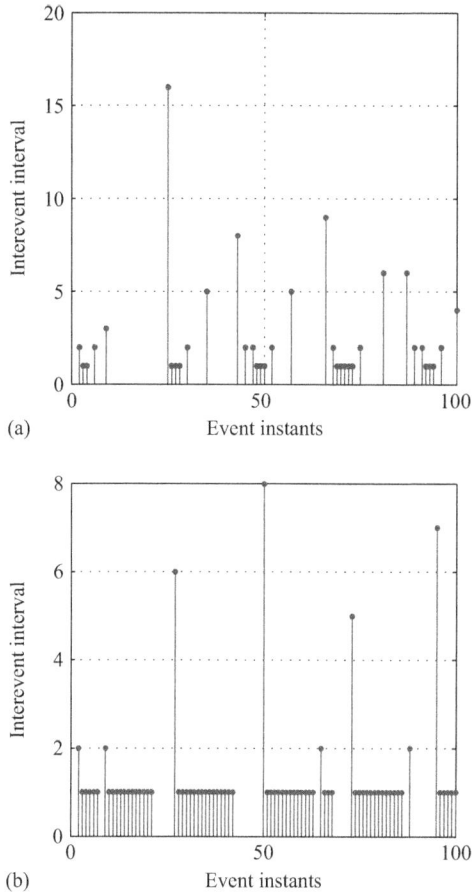

*Figure 5.9   Interevent intervals of the sensor event detector: (a) for the $H_\infty$*
*controller and (b) for the $H_\infty/l_1$ controller corresponding to*
*case study 2*

However, the event-triggered $H_\infty/l_1$ controller yields a slight degradation in the communication rates reduction objective in comparison with the single-objective $H_\infty$ controller. Consequently, our proposed methodology in this chapter provides an option to the designer to evaluate a trade-off between the control and data transmission reduction objectives of the considered closed-loop system.

**Comparison with results in [3]**

In the following, another set of simulations is conducted to compare our proposed methodology in "codesign of event generator and $H_\infty$ controller" with the results of [3] which represents as one of the recent work in this field.

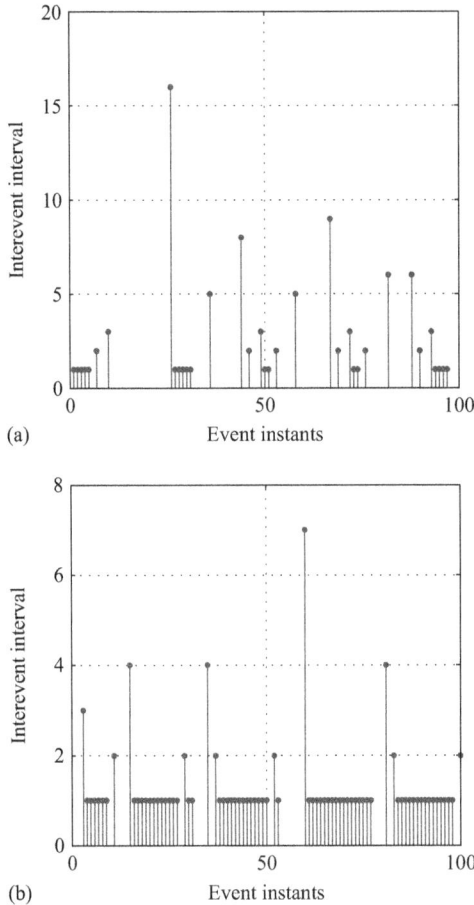

*Figure 5.10    Interevent intervals of the E-IFDIC module event detector: (a) for the*
*$H_\infty$ controller and (b) for the $H_\infty/l_1$ controller corresponding to case*
*study 2*

Let us consider the system that is used in [3] which corresponds to the linearized model of an unstable batch reactor process as in the first case study. For fair
comparison, we take a sampling period of 0.15 s and also let

$$B_d = \begin{bmatrix} 0.2 & 0 & 0.2 & 0 \\ 0 & 0.1 & 0 & 0.1 \end{bmatrix}^T$$

which is identical to [3]. Moreover, in order to have a similar event-trigger structure
the parameter $N$ in [3] is selected as $N = 1$ which implies that there are no predictions
for the future control values. The external disturbance $d(k)$ is considered as $d(k) = $
$\sin(3\pi \cdot k/25)$ for $0 \leqslant k \leqslant 30$ and $d(k) = 0$ for $k > 30$.

*Table 5.3* *Comparison of our proposed methodology with the*
*results of [3], where S-to-E denotes the*
*"Sensor-to-E-IFDIC module" and E-to-A denotes*
*the "E-IFDIC module-to-actuator"*

| Approach | $\gamma_3$ | Number of transmissions | |
|---|---|---|---|
| | | S-to-E | E-to-A |
| **Our results for $\beta_1 = \beta_2 = 1$** | 5.2772 | 172 | 177 |
| **Our results for $\beta_1 = 1, \beta_2 = 100$** | 48.0112 | 132 | 141 |
| **The results of [3]** | 100 | 48 | 167 |

The weights $\zeta_i$ in Algorithm 5.1 are selected as $\zeta_3 = 1$ and $\zeta_1 = \zeta_2 = \zeta_4 = \zeta_5 = \zeta_6 = \zeta_7 = 0$, which lead to an "event-triggered $H_\infty$ controller". Two different scenarios having different values of $\beta_i$, $i = 1, 2$ are considered.

In the eighth scenario, the positive constant weights $\beta_1$ and $\beta_2$ are selected as $\beta_1 = 1, \beta_2 = 1$, which implies the same level of importance or priority for the control and communication rates objectives. In the ninth scenario, the values of $\beta_i$, $i = 1, 2$ are selected as $\beta_1 = 1$, $\beta_2 = 100$, which imply that the design interest and priority is on higher communication rates attenuation than the disturbance attenuation objective. For both scenarios, the optimization problem corresponding to Algorithm 5.1 is solved and the parameter $\gamma_3$ and the event-triggered parameters $\sigma_u$, $\sigma_y$, $\varepsilon_u$, $\varepsilon_y$ are obtained.

Simulations for $k \in [0, 200]$ are conducted and comparison of the results with those in [3] are provided in Table 5.3. From this table, it can be concluded that our methodology has a better performance in attenuating the disturbance signals in comparison with the results of [3], while the proposed methodology in [3] leads to a better communication rate reduction in comparison with our results. Moreover, by using our proposed methodology, it is possible to consider a trade-off analysis between the control and data transmission reduction objectives for the closed-loop system, which is not readily feasible according to the results provided in [3].

**Remark 5.4.** *In simulation results conducted above, different aspects and certain capabilities of our methodology were demonstrated by providing two case studies for two important industrial systems, namely, an unstable batch reactor process and the Subzero III ROV. However, it should be noted that the application of our proposed methodology is not restricted to these systems and can be easily considered for many other classes of engineering systems such as aero engines, chemical processes, manufacturing systems, power networks, wind energy conversion systems, oil and gas systems, and industrial electronic equipment. Indeed, the considered system in (1) is quite general and the linearized model of many industrial and engineering systems in presence of external disturbances and faults can be expressed according to (1).*

*Hence, our methodology which is based on the general model in (1), can be applied to problems of event-triggered multiobjective synthesis of feedback controllers and fault diagnosis filters of different industrial applications.*

Finally, it is also possible to compare our methodology with a *separate* design of the event-triggered controller and the event-triggered fault detection and isolation (FDI) filter.

**Comparison with a separate design:**

In the integrated methodology developed in this chapter, one specifies and designs the E-IFDIC module in a single step that leads to a lower computational complexity in comparison with an alternative separate design of the C-EDC and the EFD. Specifically, in case of the separate design, the dimension of the resulting detector/controller module is $2n$. However, our methodology yields an E-IFDIC module that has the dimension of $n$. The higher dimensionality in the separate design is due to the necessity of including two filters in the controller/detector modules.

Moreover, a fundamental problem in separate design of the controller and FDI units is due to neglecting the rather significant interactions that occur between the two modules in the closed-loop system operation [37]. On the other hand, if quality of the available model is poor, design of the control system and the FDI system has to be undertaken simultaneously in order to improve the overall system functionality.

## 5.5 Conclusion

In this chapter, we have developed a new LMI approach to study the problems of event-triggered multiobjective synthesis of feedback controllers and fault diagnosis filters in a unified framework. Toward this end, a mixed generalized $H_2/H_\infty/H_-/l_1$ formulation of an E-IFDIC problem for LTI systems with reduced communication in the sensor and the actuator networks has been developed and designed. An LMI approach for codesign of the IFDIC problem and event-triggered conditions was also introduced, that in addition to stabilizing the closed-loop system, simultaneously guarantees that one accomplishes the fault detection, isolation, and control objectives. Applications of our methodology to two case studies were presented to demonstrate and illustrate the effectiveness of our proposed approach and also to compare our results with the recent literature in [1,3,329].

*Chapter 6*

# Event-triggered fault estimation and accommodation design for linear systems

The problem of event-triggered active fault-tolerant control (E-AFTC) of discrete-time linear systems is addressed in this chapter by using an integrated design of event-triggered fault/state estimator with a fault-tolerant controller. An event-triggered observer is proposed which can simultaneously provide an estimate of the system states and faults. Through an event-triggered transmission mechanism, it is shown that the amount of data sent to the observer module is significantly reduced. Moreover, an observer-based fault-tolerant controller based on fault and state estimates is designed.

A robust $H_\infty$ formulation of the problem is given that guarantees stability of the resulting closed-loop system, as well as it attenuates the effects of external disturbances and compensates the effects of the faults. Linear matrix inequality (LMI) sufficient conditions are derived to simultaneously obtain the observer and controller parameters and the event-triggered condition. Simulation results for an autonomous remotely operated vehicle (ROV) is given to illustrate the effectiveness of the proposed design methodology. The work presented in this chapter has partially appeared in [334].

This chapter is organized as follows. We begin with a brief literature review. In Section 6.2, the event-triggered fault/state observer and the fault-tolerant controller are described, and the definition of E-AFTC problem is presented. Moreover, the E-AFTC problem is formulated based on the $H_\infty$ criterion. The main results are given in Section 6.3, where the LMI-based solution to the E-AFTC problem is obtained. E-AFTC design for the Subzero III ROV is considered in Section 6.4, to demonstrate the validity of the proposed approach.

## 6.1 Introduction

As opposed to conventional periodic digital control systems, event-triggered feedback control design that drives the control action aperiodically has received considerable attention in recent years [335]. The basic idea of event-triggered control is that the control action is only driven when required and it is determined based on a certain condition on plant measurements, known as the event-triggering mechanism [336, 337]. This new approach reduces the usage of computational and communication resources without compromising on the overall desired system performance [338]. It is therefore a highly crucial methodology for practical applications especially when communications are costly or limited [328,339].

When, for example, a sensor or an actuator fault occurs in the system that produces undesirable effects, it is known that classical control schemes cannot work properly and could lead to degraded performance or even to system instability. To overcome this concern, fault-tolerant control (FTC) strategies are necessary to achieve the system objectives in spite of presence of faults and to avoid catastrophic behavior [340]. Most of the existing FTC methodologies are based on time-triggered framework, and there exist very limited study in event-triggered FTC design. The problem of event-triggered simultaneous fault diagnosis and control design for linear systems is recently considered in [341,342].

However, in [341,342], only integration of the control module with the fault detection and isolation (FDI) module is investigated, and the integrated design of FDI within an active FTC (AFTC) system is not considered. Indeed in [341,342], similar to all passive FTC (PFTC) methodologies, the faults are considered as special kinds of uncertainties, and a fixed structure controller is designed to be robust against a class of considered faults [343].

These types of passive fault-tolerant controllers are usually easy to design and implement; however, the obtained results are conservative and may result in limited fault-tolerant capability. Moreover, in practical application, faults can be critical and PFTC methodologies may not be capable of ensuring the stability and performance of the system in presence of faults. Therefore, the AFTC strategy, where the control law is reconfigured according to the occurring fault, is required for compensating the fault effects and to guarantee stability and desired performance of the closed-loop system.

AFTC systems can usually lead to a better performance since, based on characterization of the occurred fault, they modify the controller or synthesize a new control law online. This can be accomplished by simultaneous estimation of the fault and the system state. This information is then used within the FTC system with the purpose of compensating for the faults [344]. During the past decade, several methodologies in integrated design of fault estimation (FE) and FTC have been introduced, e.g., by using adaptive observer [340], sliding mode observer (SMO) [345], unknown input observer [346], a combination of SMO, and extended state observer (ESO) [347], among others.

Inspired by the above discussion, this chapter develops a novel E-AFTC design for discrete-time linear systems subject to presence of faults and external disturbances. To reduce the communication rates, an event-triggered condition is designed to determine whether the newly measured data should be transmitted or not. The proposed approach is achieved by designing the FE and FTC modules and the event-triggered condition collectively via $H_\infty$ optimization using a single-step LMI formulation.

## 6.2   Problem statement

In this section, we first describe a discrete-time linear time-invariant (LTI) system subject to faults and external disturbances. Then, the event-triggered fault/state observer, the fault-tolerant controller, and definition of the E-AFTC problem are presented. Finally, the E-AFTC problem is formulated based on the $H_\infty$ criterion.

## 6.2.1 System discerption

Consider the discrete-time LTI system $\mathscr{G}$ of the following form:

$$\begin{aligned}
x(k+1) &= Ax(k) + B_u u(k) + B_d d(k) + B_f f(k), \\
y(k) &= Cx(k) + D_d d(k), \\
z(k) &= Ex(k) + F_d d(k),
\end{aligned} \tag{6.1}$$

with the state $x(k) \in \mathbb{R}^n$, the control input $u(k) \in \mathbb{R}^{n_u}$, the measured output $y(k) \in \mathbb{R}^{n_y}$, the regulated output $z(k) \in \mathbb{R}^{n_z}$, the external disturbance $d(k) \in \mathbb{R}^{n_d}$, and the possible fault $f(k) \in \mathbb{R}^{n_f}$. It is further assumed that the disturbance and fault signals are $l_2$-norm bounded. The matrices $A$, $B_u$, $B_d$, $B_f$, $C$, $D_d$, $E$, and $F_d$ are constant with appropriate dimensions. It is assumed that $\text{rank}(B_u, B_f) = \text{rank}(B_u)$, which indicates that the space spanned by the columns of $B_f$ is a subset of the space spanned by the columns of $B_u$. Using this assumption and based on the results of [340], there exists a matrix $B_u^* \in \mathbb{R}^{n_u \times n}$ such that:

$$(I - B_u B_u^*)B_f = 0. \tag{6.2}$$

## 6.2.2 Problem definition

To simultaneously guarantee the stability of the system and control performance and avoid catastrophic failures, the following AFTC structure is utilized:

$$u(k) = u_1(k) + u_2(k), \tag{6.3}$$

where $u_1(k)$ denotes the nominal controller and $u_2(k)$ is the additional controller term that is used to compensate for the effects of the fault on the system.

Note that full-state measurements are not available in many real physical systems, and therefore state feedback controller cannot be used, and instead, an output feedback controller (or observer-based state feedback controller) should be used. Therefore, in this work the nominal controller $u_1(k)$ is selected as $u_1(k) = K_1 \hat{x}(k)$, where $\hat{x}(k)$ denotes the estimate of the state $x(k)$. On the other hand, in order to have an ideal AFTC system, $u_2(k)$ should be selected such that $B_u u_2(k) + B_f f(k) = 0$, i.e., $u_2(k) = -K_2 f(k)$ with $B_u K_2 = B_f$ [348]. However, in practical systems, it is usually impossible to obtain the fault information exactly.

Based on the above discussion, the following AFTC is now considered in this work:

$$u(k) = K_1 \hat{x}(k) - K_2 \hat{f}(k) \tag{6.4}$$

where $\hat{f}(k)$ denotes the estimate of the fault $f(k)$, $K_2 = B_u^* B_f$ and $K_1$ will be designed subsequently.

Based on the proposed controller above, it is necessary to estimate both the fault and the states of the system. Therefore, the following event-triggered integrated fault/state observer is now proposed to simultaneously estimate the fault and the states of the system:

$$\hat{x}(k+1) = Ax(k) + B_u u(k) + B_f \hat{f}(k) + L(\bar{y}(k) - \hat{y}(k)),$$
$$\hat{y}(k) = C\hat{x}(k), \tag{6.5}$$
$$\hat{f}(k+1) = \hat{f}(k) + F(\bar{y}(k) - \hat{y}(k)),$$

where the matrices $L \in \mathbb{R}^{n \times n_y}$ and $F \in \mathbb{R}^{n_f \times n_y}$ denote the observer gains to be designed subsequently, $\hat{y}(k) \in \mathbb{R}^{n_y}$ denotes the observer output, and $\bar{y}(k) \in \mathbb{R}^{n_y}$ denotes the last measurement that is sent from the sensor to the fault/state observer.

In an event-triggered strategy, the information of the sensor ($y(k)$) is transmitted to the observer only at the transmission times $k_i^y$, $i \in \mathbb{N}$. Therefore, the input to the observer module is described as follows:

$$\bar{y}(k) = y(k_i^y), \quad \text{for } k \in [k_i^y, k_{i+1}^y), \; i, k \in \mathbb{N}. \tag{6.6}$$

It is assumed that the system output is sent through the network only when the following event condition is violated:

$$\sigma y^{\mathrm{T}}(k)y(k) - \delta^{\mathrm{T}}(k)\delta(k) > 0, \tag{6.7}$$

where $\sigma > 0$ is a weighting constant which controls the rate of the communication, and $\delta(k) = \bar{y}(k) - y(k)$.

By integrating the AFTC and the observer modules (6.4) and (6.5) with the system equation (6.1), the augmented system dynamics $\mathcal{G}_{\mathscr{C}}$ can be written as follows:

$$\tilde{x}_{\mathrm{cl}}(k+1) = \tilde{A}_{\mathrm{cl}}\tilde{x}_{\mathrm{cl}}(k) + \tilde{B}_{\mathrm{dcl}}d(k) + \tilde{B}_{\mathrm{fcl}}\Delta f(k) + \tilde{B}_{\delta \mathrm{cl}}\delta(k),$$
$$y(k) = \tilde{C}_{\mathrm{cl}}\tilde{x}_{\mathrm{cl}}(k) + D_d d(k), \tag{6.8}$$
$$z(k) = \tilde{E}_{\mathrm{cl}}\tilde{x}_{\mathrm{cl}}(k) + F_d d(k),$$

where $\tilde{x}_{\mathrm{cl}}(k) = [x(k)^{\mathrm{T}} \; \tilde{e}(k)^{\mathrm{T}}]^{\mathrm{T}}$, $\tilde{e}(k) = [e_x(k)^{\mathrm{T}} \; e_f(k)^{\mathrm{T}}]^{\mathrm{T}}$, $e_x(k) = x(k) - \hat{x}(k)$, $e_f(k) = f(k) - \hat{f}(k)$, $\Delta f(k) = f(k+1) - f(k)$ and

$$\tilde{A}_{\mathrm{cl}} = \begin{bmatrix} A + B_u K_1 & -B_u K_1 \tilde{I} + \tilde{B}_f \\ 0 & \tilde{A} + \tilde{L}C\tilde{I} \end{bmatrix}, \quad \hat{I} = \begin{bmatrix} 0 \\ I \end{bmatrix},$$

$$\tilde{B}_{\mathrm{dcl}} = \begin{bmatrix} B_d \\ \tilde{B}_d + \tilde{L}D_d \end{bmatrix}, \quad \tilde{B}_{\mathrm{fcl}} = \begin{bmatrix} 0 \\ \hat{I} \end{bmatrix}, \quad \tilde{B}_{\delta \mathrm{cl}} = \begin{bmatrix} 0 \\ \tilde{L} \end{bmatrix},$$

$$\tilde{A} = \begin{bmatrix} A & B_f \\ 0 & I \end{bmatrix}, \quad \tilde{B}_d = \begin{bmatrix} B_d \\ 0 \end{bmatrix}, \quad \tilde{L} = \begin{bmatrix} -L \\ -F \end{bmatrix},$$

$$\tilde{B}_f = \begin{bmatrix} 0 & B_f \end{bmatrix}, \quad \tilde{C}_{\mathrm{cl}} = \begin{bmatrix} C & 0 \end{bmatrix}, \quad \tilde{E}_{\mathrm{cl}} = \begin{bmatrix} E & 0 \end{bmatrix}, \quad \tilde{I} = \begin{bmatrix} I & 0 \end{bmatrix}.$$

It should be noted that due to the structure of the AFTC (6.4), the fault difference term $\Delta f$ appears in the dynamics of the augmented system (6.8). In some work

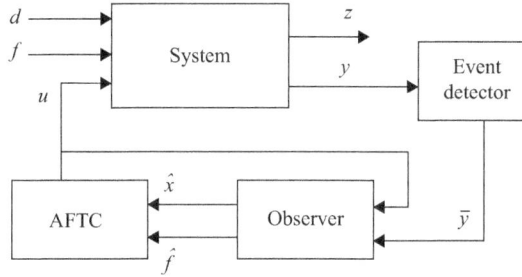

*Figure 6.1 Schematic of the E-AFTC methodology*

(e.g., [349]), it is assumed that this term is very small and can be neglected. However, in many practical fault scenarios (such as in abrupt faults), faults can produce a large value of $\Delta f$ at a certain time, and therefore, it is necessary to consider its effects on the augmented system (6.8).

The main concept for an event-triggered AFTC problem is depicted in Figure 6.1. The event generator is employed to reduce the transmissions from the sensor to the fault/state observer. Therefore, it is possible to significantly reduce the amount of communication for fault/state estimation and control tasks in comparison with a time-triggered AFTC methodology.

## 6.2.3 Problem formulation

In order to solve the integrated fault/state estimation and fault-tolerant controller design problem (AFTC problem), the AFTC system (6.4) and the fault/state observer (6.5) should be designed simultaneously. However, based on (6.8), it is obvious that the design of the observer parameters $L$ and $F$, and the controller parameter $K_1$ are independent from each other.

Consequently, in most of the existing time-triggered AFTC design methodologies [340,348,350–352], the AFTC module is implemented in two steps; *first*, the fault/state estimator is designed, and *second*, the AFTC is designed. Indeed, both estimator and controller modules are designed separately and then their performances are considered together, which is convenient to calculate the design parameters.

However, in presence of event-trigger strategy, it is not possible to break down the augmented system (6.8) into two separate systems to design the observer parameters independent from the controller gain. Indeed, in order to implement a two-step procedure, the augmented system (6.8) should be separated into two subsystems with the states of $x(k)$ and $\tilde{e}(k)$, respectively, where the first is used in designing the controller gain and the second is used in designing the observer parameters.

On the other hand, since our proposed observer in this work is an event-triggered observer and based on the event condition (6.7), the state of the system $(x(k))$ is required for formulation of the observer. Therefore, in E-AFTC problem, the design of the AFTC system (6.4) and the fault/state observer (6.5) should be considered simultaneously, which motivates us to study the following problem.

***The E-AFTC problem***:

Given the system (6.8) with external disturbances and fault, it is required to collectively design the fault/state observer (6.5), the AFTC system (6.4), and the event-triggered condition (6.7) such that the following performance indices are satisfied, namely, (i) the augmented system (6.8) is stable and (ii) the effects of disturbances and fault difference term $\Delta f(k)$ on the regulated output $z(k)$ are minimized. In other words, the fault/state observer (6.5), the AFTC system (6.4) and the event-triggered condition (6.7) are designed simultaneously such that the augmented system (6.8) is stable and the following conditions are satisfied:

$$(\text{I}) \ \sup_{f=0} \frac{||z||_2}{||d||_2} < \gamma_1,$$

$$(\text{II}) \ \sup_{d=0} \frac{||z||_2}{||\Delta f||_2} < \gamma_2. \tag{6.9}$$

The $H_\infty$ norm performances (I) and (II) are used to minimize the effects of disturbances and fault difference term $\Delta f$ on the regulated output.

## 6.3 Main results

In this section, sufficient LMI conditions for designing the E-AFTC problem are provided. First, the following theorem provides sufficient LMI conditions for the $H_\infty$ performance index (I) for the augmented system (6.8).

**Theorem 6.1.** *Consider the augmented system (6.8) with $f(k) = 0$ and suppose that the data is transmitted from the sensor to the fault/state observer when the condition (6.7) is violated. The augmented system (6.8) is stable and the performance index (I) is guaranteed if there exist symmetric positive-definite matrix $P_1$, matrices $X$, $\hat{X}_1$, $M$, $Q$, and positive scalars $\sigma$, $\gamma_1$, such that the following LMI holds*

$$\begin{bmatrix} P_1 - E_{11} & E_{12} & E_{13} & E_{14} \\ * & -P_1 + E_{22} & E_{23} & 0 \\ * & * & E_{33} & 0 \\ * & * & * & -I \end{bmatrix} < 0, \tag{6.10}$$

$$X_1^{\mathrm{T}} B_u = B_u \hat{X}_1^{\mathrm{T}},$$

*where* $E_{11} = X + X^{\mathrm{T}}$, $E_{22} = \tilde{E}_{cl}^{\mathrm{T}} \tilde{E}_{cl} + \sigma \tilde{C}_{cl}^{\mathrm{T}} \tilde{C}_{cl}$, $E_{23} = \tilde{E}_{cl}^{\mathrm{T}} F_d + \sigma \tilde{C}_{cl}^{\mathrm{T}} D_d$, $E_{33} = F_d^{\mathrm{T}} F_d - \gamma_1^2 I + \sigma D_d^{\mathrm{T}} D_d$, $X = \mathrm{diag}(X_1, X_2)$,

$$E_{12} = \begin{bmatrix} X_1^{\mathrm{T}}A + B_u M^{\mathrm{T}} & -B_u M^{\mathrm{T}}\tilde{I} + X_1^{\mathrm{T}}\tilde{B}_f \\ 0 & X_2^{\mathrm{T}}\tilde{A} + Q^{\mathrm{T}}C\tilde{I} \end{bmatrix},$$

$$E_{13} = \begin{bmatrix} X_1^{\mathrm{T}}B_d \\ X_2^{\mathrm{T}}\tilde{B}_d + Q^{\mathrm{T}}D_d \end{bmatrix}, \quad E_{14} = \begin{bmatrix} 0 \\ Q^{\mathrm{T}} \end{bmatrix}.$$

*Moreover, the parameters of the fault/state observer and the fault-tolerant controller are obtained as follows:*

$$\begin{bmatrix} L^{\mathrm{T}} & F^{\mathrm{T}} \end{bmatrix}^{\mathrm{T}} = -X_2^{-\mathrm{T}}Q^{\mathrm{T}} \ and \ K_1 = \hat{X}_1^{-\mathrm{T}}M^{\mathrm{T}}. \tag{6.11}$$

*Proof.* In order to ensure the system stability and the performance index (I) for the augmented system (6.8), we need to satisfy the next condition, namely,

$$\Delta V + z^{\mathrm{T}}z - \gamma_1^2 d^{\mathrm{T}}d + \sigma y^{\mathrm{T}}y - \delta^{\mathrm{T}}\delta < 0. \tag{6.12}$$

where $\Delta V = V(k+1) - V(k)$, and $V(k)$ is a positive definite Lyapunov function candidate. Indeed, according to the inequality (6.7), $\sigma y^{\mathrm{T}}(k)y(k) - \delta^{\mathrm{T}}(k)\delta(k) > 0$, and hence,

$$\Delta V(k) + z^{\mathrm{T}}(k)z(k) - \gamma_1^2 d^{\mathrm{T}}(k)d(k) < 0. \tag{6.13}$$

For $d(k) = 0$, the inequality (6.13) leads to $\Delta V(k) < 0$, that ensures the system stability.

To guarantee the $H_\infty$ performance measure, summing both sides of the inequality (6.13) over $[0, \infty)$, one gets

$$V(\infty) - V(0) + \sum_{k=0}^{\infty} (z^{\mathrm{T}}(k)z(k) - \gamma_1^2 d^{\mathrm{T}}(k)d(k)) < 0, \tag{6.14}$$

where $V(\infty) > 0$ with zero initial condition $V(0) = 0$. Therefore, we have

$$\sum_{k=0}^{\infty} z^{\mathrm{T}}(k)z(k) < \gamma_1^2 \sum_{k=0}^{\infty} d^{\mathrm{T}}(k)d(k), \tag{6.15}$$

which guarantees the $H_\infty$ performance index (I).

In the following, sufficient conditions for satisfying the condition (6.12) are obtained in terms of the LMI feasibility conditions. Let the Lyapunov function candidate $V(k)$ be selected as follows:

$$V(k) = \tilde{x}_{\mathrm{cl}}^{\mathrm{T}}(k)P_1\tilde{x}_{\mathrm{cl}}(k), \tag{6.16}$$

where $P_1$ is a positive definite symmetric matrix. From (6.8), (6.12), and (6.16), the following inequality is obtained:

$$\begin{bmatrix} \tilde{x}_{\mathrm{cl}}(k) \\ d(k) \\ \delta(k) \end{bmatrix}^{\mathrm{T}} \begin{bmatrix} \varphi_{11} & \varphi_{12} & \varphi_{13} \\ * & \varphi_{22} & \varphi_{23} \\ * & * & \varphi_{33} \end{bmatrix} \begin{bmatrix} \tilde{x}_{\mathrm{cl}}(k) \\ d(k) \\ \delta(k) \end{bmatrix} < 0, \tag{6.17}$$

where $\varphi_{11} = \tilde{A}_{cl}^T P_1 \tilde{A}_{cl} - P_1 + \tilde{E}_{cl}^T \tilde{E}_{cl} + \sigma \tilde{C}_{cl}^T \tilde{C}_{cl}$, $\varphi_{12} = \tilde{A}_{cl}^T P_1 \tilde{B}_{dcl} + \tilde{E}_{cl}^T F_d + \sigma \tilde{C}_{cl}^T D_d$,
$\varphi_{13} = \tilde{A}_{cl}^T P_1 \tilde{B}_{\delta cl}$, $\varphi_{22} = \tilde{B}_{dcl}^T P_1 \tilde{B}_{dcl} + F_d^T F_d + \sigma D_d^T D_d - \gamma_1^2 I$, $\varphi_{23} = \tilde{B}_{dcl}^T P_1 \tilde{B}_{\delta cl}$, and
$\varphi_{33} = \tilde{B}_{\delta cl}^T P_1 \tilde{B}_{\delta cl} - I$.

The following inequality is sufficient to imply (6.17), namely,

$$N_U^T Z N_U < 0, \tag{6.18}$$

where $N_U$ and $Z$ are given by

$$Z = \begin{bmatrix} P_1 & 0 & 0 & 0 \\ * & \Phi & \tilde{E}_{cl}^T F_d + \sigma \tilde{C}_{cl}^T D_d & 0 \\ * & * & F_d^T F_d - \gamma_1^2 I + \sigma D_d^T D_d & 0 \\ * & * & * & -I \end{bmatrix},$$

$$N_U = \begin{bmatrix} \tilde{A}_{cl} & \tilde{B}_{dcl} & \tilde{B}_{\delta cl} \\ I & 0 & 0 \\ 0 & I & 0 \\ 0 & 0 & I \end{bmatrix} \rightarrow U = \begin{bmatrix} -I & \tilde{A}_{cl} & \tilde{B}_{dcl} & \tilde{B}_{\delta cl} \end{bmatrix},$$

and $\Phi = -P_1 + \tilde{E}_{cl}^T \tilde{E}_{cl} + \sigma \tilde{C}_{cl}^T \tilde{C}_{cl}$. Note that in the inequality (6.18), the Lyapunov matrix $P_1$ and the system matrices are coupled with one another.

Now, the Projection Lemma [238] is employed to reduce the conservativeness in the solution to the E-AFTC problem. Specially, by choosing the matrix $N_V$ in (1.9b) as

$$N_V = \begin{bmatrix} 0 & 0 & 0 \\ I & 0 & 0 \\ 0 & I & 0 \\ 0 & 0 & I \end{bmatrix}^T \rightarrow V = \begin{bmatrix} I & 0 & 0 & 0 \end{bmatrix}, \tag{6.19}$$

and using Lemma 1.1, it is then possible to show that the inequality (6.18) is equivalent to

$$Z + \text{Herm} \left\{ \begin{bmatrix} -I \\ \tilde{A}_{cl}^T \\ \tilde{B}_{dcl}^T \\ \tilde{B}_{\delta cl}^T \end{bmatrix} \begin{bmatrix} X & 0 & 0 & 0 \end{bmatrix} \right\} < 0. \tag{6.20}$$

By partitioning $X$ into $X = \text{diag}(X_{1(n \times n)}, X_{2((n+n_f) \times (n+n_f)})$, and using (6.8), (6.20), and also the Schur complement, the following inequality is obtained:

$$\begin{bmatrix} P_1 - \mathcal{E}_{11} & \mathcal{E}_{12} & \mathcal{E}_{13} & \mathcal{E}_{14} \\ * & -P_1 + \mathcal{E}_{22} & E_{23} & 0 \\ * & * & E_{33} & 0 \\ * & * & * & -I \end{bmatrix} < 0, \tag{6.21}$$

where $E_{11} = X + X^\mathrm{T}$, $E_{23} = \breve{E}_{\mathrm{cl}}^\mathrm{T} F_\mathrm{d} + \sigma \tilde{C}_{\mathrm{cl}}^\mathrm{T} D_\mathrm{d}$, $E_{33} = F_\mathrm{d}^\mathrm{T} F_\mathrm{d} - \gamma_1^2 I + \sigma D_\mathrm{d}^\mathrm{T} D_\mathrm{d}$, $E_{22} - \tilde{E}_{\mathrm{cl}}^\mathrm{T} \tilde{E}_{\mathrm{cl}} + \sigma \tilde{C}_{\mathrm{cl}}^\mathrm{T} \tilde{C}_{\mathrm{cl}}$, and

$$\mathscr{E}_{12} = \begin{bmatrix} X_1^\mathrm{T} A + X_1^\mathrm{T} B_\mathrm{u} K_1 & -X_1^\mathrm{T} B_\mathrm{u} K_1 \tilde{I} + X_1^\mathrm{T} \tilde{B}_\mathrm{f} \\ 0 & X_2^\mathrm{T} \tilde{A} + X_2^\mathrm{T} \tilde{L} C \tilde{I} \end{bmatrix},$$

$$\mathscr{E}_{13} = \begin{bmatrix} X_1^\mathrm{T} B_\mathrm{d} \\ X_2^\mathrm{T} \tilde{B}_\mathrm{d} + X_2^\mathrm{T} \tilde{L} D_\mathrm{d} \end{bmatrix}, \mathscr{E}_{14} = \begin{bmatrix} 0 \\ X_2^\mathrm{T} \tilde{L} \end{bmatrix}.$$

Since the inequality (6.21) is not in the form of an LMI, by using the equality constraint [252] $X_1^\mathrm{T} B_\mathrm{u} = B_\mathrm{u} \hat{X}_1^\mathrm{T}$ and substituting $Q^\mathrm{T} = X_2^\mathrm{T} \tilde{L}$ and $M^\mathrm{T} = \hat{X}_1^\mathrm{T} K_1$ into the inequality (6.21), the LMI condition in (6.10) is obtained. □

The following theorem gives the LMI condition for the performance index (II).

**Theorem 6.2.** *Consider the augmented system* (6.8) *with* $d(k) = 0$ *and suppose that the data is transmitted from the sensor to the fault/state observer when the condition* (6.7) *is violated. The augmented system* (6.8) *is stable and the performance index* (II) *is guaranteed if there exist symmetric positive-definite matrix* $P_2$, *matrices* $X$, $\hat{X}_1$, $M$, $Q$, *and positive scalars* $\sigma$, $\gamma_2$, *such that the following LMI holds*

$$\begin{bmatrix} P_2 - E_{11} & E_{12} & \Xi_{13} & E_{14} \\ * & -P_2 + E_{22} & 0 & 0 \\ * & * & -\gamma_2^2 I & 0 \\ * & * & * & -I \end{bmatrix} < 0,$$
(6.22)

$$X_1^\mathrm{T} B_\mathrm{u} = B_\mathrm{u} \hat{X}_1^\mathrm{T},$$

*where* $\Xi_{13} = \begin{bmatrix} 0 & \hat{I}^\mathrm{T} X_2 \end{bmatrix}^\mathrm{T}$, *and* $E_{ij}$ *matrices for* $i = 1, 2$ *and* $j = 1, 2, 4$ *are defined in Theorem 6.1.*

*Proof.* Similar to the proof of Theorem 6.1, the following condition should be satisfied in order to ensure that the system stability and the performance index (II) for the augmented system (6.8) can be ensured, namely,

$$\Delta V + z^\mathrm{T} z - \gamma_2^2 \Delta f^\mathrm{T} \Delta f + \sigma y^\mathrm{T} y - \delta^\mathrm{T} \delta < 0.$$
(6.23)

where $\Delta V = V(k+1) - V(k)$, and $V(k)$ is a positive definite Lyapunov function candidate. According to the inequality (6.7), $\sigma y^\mathrm{T}(k) y(k) - \delta^\mathrm{T}(k) \delta(k) > 0$, and therefore from (6.23) it follows that

$$\Delta V(k) + z^\mathrm{T}(k) z(k) - \gamma_2^2 \Delta f^\mathrm{T}(k) \Delta f(k) < 0.$$
(6.24)

For $\Delta f(k) = 0$, the inequality (6.24) leads to $\Delta V(k) < 0$, that ensures the system stability.

To guarantee the $H_\infty$ performance measure, summing both sides of the inequality (6.24) over $[0, \infty)$, leads to

$$V(\infty) - V(0) + \sum_{k=0}^{\infty} (z^T(k)z(k) - \gamma_2^2 \Delta f^T(k)\Delta f(k)) < 0, \qquad (6.25)$$

where $V(\infty) > 0$ with zero initial condition $V(0) = 0$. Therefore, we have

$$\sum_{k=0}^{\infty} z^T(k)z(k) < \gamma_2^2 \sum_{k=0}^{\infty} \Delta f^T(k)\Delta f(k), \qquad (6.26)$$

which guarantees the $H_\infty$ performance index (II).

In the following, sufficient LMI conditions for satisfying the condition (6.23) are obtained. Let the Lyapunov function candidate $V(k)$ be selected as follows:

$$V(k) = \tilde{x}_{cl}^T(k)P_2\tilde{x}_{cl}(k), \qquad (6.27)$$

where $P_2$ is a positive definite symmetric matrix. From (6.8), (6.23), and (6.27), the following inequality is obtained:

$$\begin{bmatrix} \tilde{x}_{cl}(k) \\ \Delta f(k) \\ \delta(k) \end{bmatrix}^T \begin{bmatrix} \varphi_{11} & \varphi_{12} & \varphi_{13} \\ * & \varphi_{22} & \varphi_{23} \\ * & * & \varphi_{33} \end{bmatrix} \begin{bmatrix} \tilde{x}_{cl}(k) \\ \Delta f(k) \\ \delta(k) \end{bmatrix} < 0, \qquad (6.28)$$

where $\varphi_{11} = \tilde{A}_{cl}^T P_2 \tilde{A}_{cl} - P_2 + \tilde{E}_{cl}^T \tilde{E}_{cl} + \sigma \tilde{C}_{cl}^T \tilde{C}_{cl}$, $\varphi_{12} = \tilde{A}_{cl}^T P_2 \tilde{B}_{fcl}$, $\varphi_{13} = \tilde{A}_{cl}^T P_2 \tilde{B}_{\delta cl}$, $\varphi_{22} = \tilde{B}_{fcl}^T P_2 \tilde{B}_{fcl} - \gamma_2^2 I$, $\varphi_{23} = \tilde{B}_{fcl}^T P_2 \tilde{B}_{\delta cl}$, and $\varphi_{33} = \tilde{B}_{\delta cl}^T P_2 \tilde{B}_{\delta cl} - I$.

The following inequality is sufficient to imply (6.28), namely,

$$N_U^T Z N_U < 0, \qquad (6.29)$$

where $N_U$ and $Z$ are given by

$$Z = \begin{bmatrix} P_2 & 0 & 0 & 0 \\ * & \Phi & 0 & 0 \\ * & * & -\gamma_2^2 I & 0 \\ * & * & * & -I \end{bmatrix},$$

$$N_U = \begin{bmatrix} \tilde{A}_{cl} & \tilde{B}_{fcl} & \tilde{B}_{\delta cl} \\ I & 0 & 0 \\ 0 & I & 0 \\ 0 & 0 & I \end{bmatrix} \rightarrow U = \begin{bmatrix} -I & \tilde{A}_{cl} & \tilde{B}_{fcl} & \tilde{B}_{\delta cl} \end{bmatrix},$$

and $\Phi = -P_2 + \tilde{E}_{cl}^T \tilde{E}_{cl} + \sigma \tilde{C}_{cl}^T \tilde{C}_{cl}$. Note that in the inequality (6.29), the Lyapunov matrix $P_2$ and the system matrices are coupled with one another.

Now, the Projection Lemma 1.1 is employed to reduce the conservativeness in the solution to the E-AFTC problem. Especially, by choosing the matrix $N_V$ in equation (1.9b) as

$$N_V = \begin{bmatrix} 0 & 0 & 0 \\ I & 0 & 0 \\ 0 & I & 0 \\ 0 & 0 & I \end{bmatrix}^{\mathrm{T}} \rightarrow V = \begin{bmatrix} I & 0 & 0 & 0 \end{bmatrix}, \tag{6.30}$$

and using Lemma 1.1, it is then possible to show that the inequality (6.29) is equivalent to

$$Z + \mathrm{Herm}\left\{ \begin{bmatrix} -I \\ \tilde{A}_{\mathrm{cl}}^{\mathrm{T}} \\ \tilde{B}_{\mathrm{fcl}}^{\mathrm{T}} \\ \tilde{B}_{\delta\mathrm{cl}}^{\mathrm{T}} \end{bmatrix} \begin{bmatrix} X & 0 & 0 & 0 \end{bmatrix} \right\} < 0. \tag{6.31}$$

By partitioning $X$ into $X = \mathrm{diag}(X_{1(n\times n)}, X_{2((n+n_f)\times(n+n_f))})$, and using (6.8), (6.31) and also the Schur complement, the following inequality is obtained:

$$\begin{bmatrix} P_2 - E_{11} & \mathcal{E}_{12} & \mathcal{E}_{13} & \mathcal{E}_{14} \\ * & -P_2 + E_{22} & 0 & 0 \\ * & * & -\gamma_2^2 I & 0 \\ * & * & * & -I \end{bmatrix} < 0, \tag{6.32}$$

where $E_{11} = X + X^{\mathrm{T}}$, $E_{22} = \tilde{E}_{\mathrm{cl}}^{\mathrm{T}}\tilde{E}_{\mathrm{cl}} + \sigma\tilde{C}_{\mathrm{cl}}^{\mathrm{T}}\tilde{C}_{\mathrm{cl}}$, and

$$\mathcal{E}_{12} = \begin{bmatrix} X_1^{\mathrm{T}}A + X_1^{\mathrm{T}}B_u K_1 & -X_1^{\mathrm{T}}B_u K_1 \tilde{I} + X_1^{\mathrm{T}}\tilde{B}_f \\ 0 & X_2^{\mathrm{T}}\tilde{A} + X_2^{\mathrm{T}}\tilde{L}C\tilde{I} \end{bmatrix},$$

$$\mathcal{E}_{13} = \begin{bmatrix} 0 \\ X_2^{\mathrm{T}}\hat{I} \end{bmatrix}, \quad \mathcal{E}_{14} = \begin{bmatrix} 0 \\ X_2^{\mathrm{T}}\tilde{L} \end{bmatrix}.$$

Since the inequality (6.32) is not in the form of an LMI, by using the equality constraint [252] $X_1^{\mathrm{T}}B_u = B_u\hat{X}_1^{\mathrm{T}}$ and substituting $Q^{\mathrm{T}} = X_2^{\mathrm{T}}\tilde{L}$ and $M^{\mathrm{T}} = \hat{X}_1^{\mathrm{T}}K_1$ into the inequality (6.21), the LMI condition in (6.22) is now obtained. This completes the proof of the theorem. □

**Remark 6.1.** *It should be noted that the equality constraint $X_1^{\mathrm{T}}B_u = B_u\hat{X}_1^{\mathrm{T}}$ is difficult to solve by using the LMI toolbox. However, for a positive scalar $\alpha$, it is possible to* *convert it into the following optimization problem which is solvable by using the LMI toolbox [353]:*

minimize   $\alpha$

subject to:

$$\begin{bmatrix} \alpha I & X_1^T B_u - B_u \hat{X}_1^T \\ * & \alpha I \end{bmatrix} > 0. \tag{6.33}$$

To design an E-AFTC that has the optimal robustness to external disturbances and compensates for the effects of fault signals while reduces the communication rate, Theorems 6.1 and 6.2 are combined to find the observer, the controller, and the event-triggered condition such that both $H_\infty$ performance indices in (6.9) are satisfied. The next corollary summarizes the main result of this chapter.

**Corollary 6.1.** *If there exist symmetric positive-definite matrices $P_1$, $P_2$, matrices $X$, $\hat{X}_1$, M, Q, and positive scalars $\sigma$, $\alpha$, $\gamma_1$, and $\gamma_2$ such that there is a solution to the following optimization problem:*

minimize   $\beta_1(\gamma_1 + \gamma_2) - \beta_2\sigma + \alpha$

subject to: (6.10), (6.22), (6.33)

$$\tag{6.34}$$

*then under the event-triggered condition (6.7), the augmented system (6.8) is stable and the performance indices (I)–(II) are satisfied.*

**Remark 6.2.** *The positive constant weights $\beta_1$ and $\beta_2$ in (6.34) are used to emphasize the importance of the AFTC and the communication objectives relative to one another. The value of $\beta_1$ is used to emphasize the estimation and control objectives. The value of $\beta_2$ is used to control the data transmission between the sensor event detector and the fault/state observer module.*

## 6.4   Case study

In this section, we present numerical simulation results to demonstrate the effectiveness of our proposed results in this chapter. For this purpose, a speed subsystem of the Subzero III ROV [261] with the following parameters is considered:

$$A = \begin{bmatrix} 0.2148 & -0.2905 \\ 0.1578 & 0.9459 \end{bmatrix}, \quad B_u = B_f = \begin{bmatrix} 0.1578 \\ 0.0294 \end{bmatrix},$$

$$C = E = \begin{bmatrix} 0 & 0.0029 \end{bmatrix}, \quad B_d = 0.1B_u, \quad D_d = F_d = 0.$$

The disturbance signal $d(k)$ is considered as a band-limited white noise with a power of 0.0001. The fault signal $f(k)$ is simulated as

$$f(k) = \begin{cases} 1 & 50 < k < 110, \\ \sin(0.1k) & 160 < k < 250, \\ 0 & \text{otherwise.} \end{cases}$$

### A. Effects of $\beta_i$ on the E-AFTC design

To illustrate the trade-offs between the communication rates and the AFTC objectives, two scenarios are considered below by selecting different values for $\beta_i$, $i = 1, 2$.

In the first scenario, the positive constant weights $\beta_1$ and $\beta_2$ are assumed to be equal to one, which implies the same level of importance between the E-AFTC objectives. The optimization problem corresponding to Corollary 6.1 is solved and the optimization parameters $\gamma_1$–$\gamma_2$ and the event-triggered parameter $\sigma$ are obtained as $\gamma_1 = 3.0641e - 06$, $\gamma_2 = 3.6588e - 06$, and $\sigma = 0.2$, respectively.

Figure 6.2 shows the FE simulation results. It can be seen that the fault is estimated efficiently. Figure 6.3 shows the regulated output $z(k)$ with AFTC (6.4) and the nominal controller $u_1(k) = -K_1\hat{x}(k)$, respectively. From this figure, it can be concluded that the proposed AFTC system has a much better control performance than the simple nominal controller in which the fault is not compensated by using the estimation of the fault.

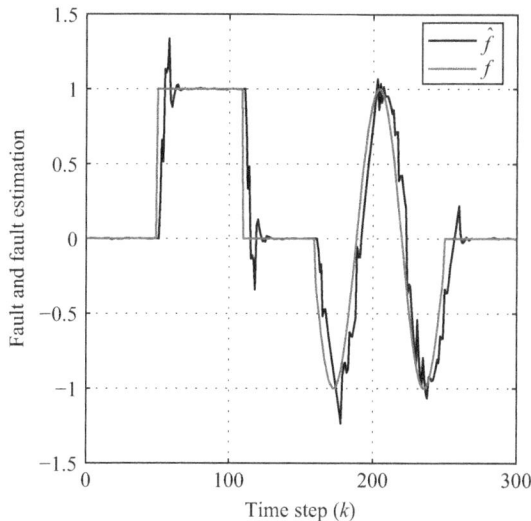

*Figure 6.2   Fault $f$ and its estimate $\hat{f}$*

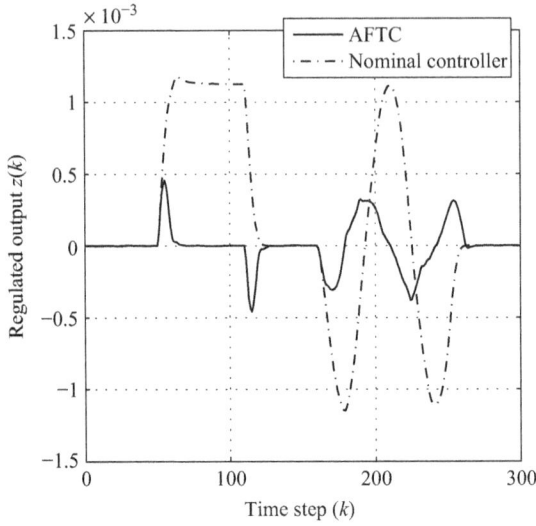

*Figure 6.3    Regulated output z(k) corresponding to AFTC and the nominal controller*

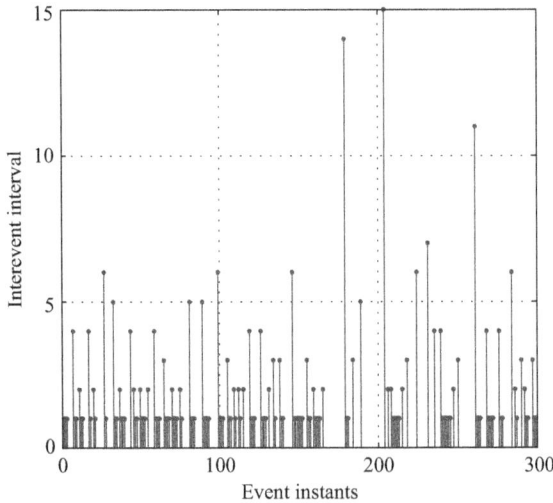

*Figure 6.4    Interevent intervals*

The interevent interval is shown in Figure 6.4. The value of each stem shows the length of the time period between that event and the previous one, which shows a reduction of data transmission to 48.66% for sending the sensor data to the fault/state observer.

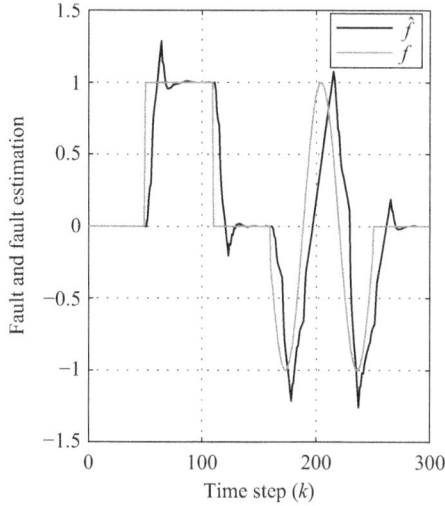

*Figure 6.5   Fault f and its estimate $\hat{f}$*

The above simulation results show that our proposed E-AFTC methodology does satisfactorily retain the FE and tolerant-control requirements with lower resource usage as compared to a time-triggered AFTC methodology.

In the second scenario, the values of $\beta_i$, $i = 1, 2$ are selected as $\beta_1 = 0.001$ and $\beta_2 = 1$, which imply that the design interest and priority is on higher communication rates attenuation than the AFTC objective. The optimization problem corresponding to Corollary 6.1 is solved and the optimization parameters $\gamma_1$–$\gamma_2$ and the event-triggered parameter $\sigma$ are obtained as $\gamma_1 = 1.0524e - 04$, $\gamma_2 = 8.9936e - 05$, and $\sigma = 0.35$, respectively.

The FE simulation results are depicted in Figure 6.5. Moreover, Figure 6.6 shows the regulated output $z(k)$ with AFTC. From these figures, it can be seen that the fault is not adequately estimated and consequently the fault is not effectively compensated by using the designed AFTC. Therefore, the designed AFTC system in this scenario has less performance than the one in the first scenario.

The interevent interval is depicted in Figure 6.7. This figure shows a reduction of data transmission to 37% for sending the sensor data to the fault/state observer. Consequently, by comparing Figures 6.4 and 6.7, it can be concluded that the communication rates have been reduced more in this scenario in comparison with the results obtained in the previous simulations.

## B. Robustness with respect to parametric uncertainties

The proposed E-AFTC module has an intrinsic robustness with respect to parametric uncertainties. To examine the robustness of the proposed scheme, it is assumed that the dynamic parameters of the considered ROV are uncertain but bounded. Indeed, the E-AFTC module is designed by using the nominal parameters of the Subzero III

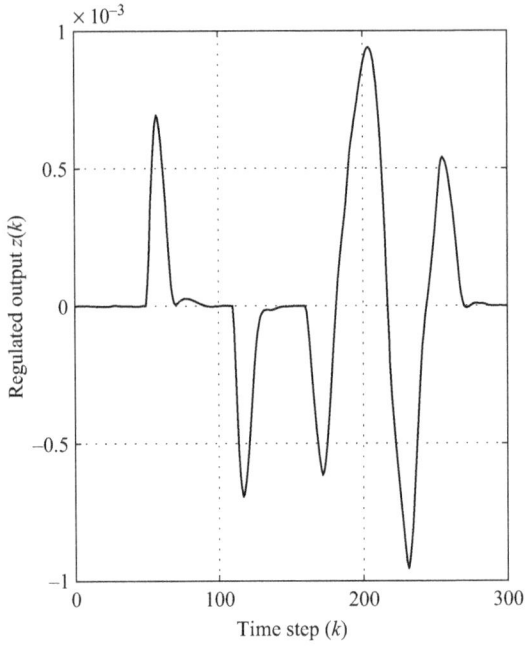

*Figure 6.6    Regulated output z(k)*

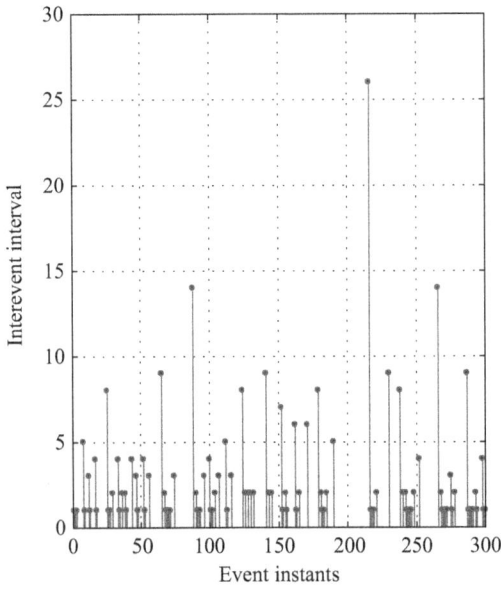

*Figure 6.7    Interevent intervals*

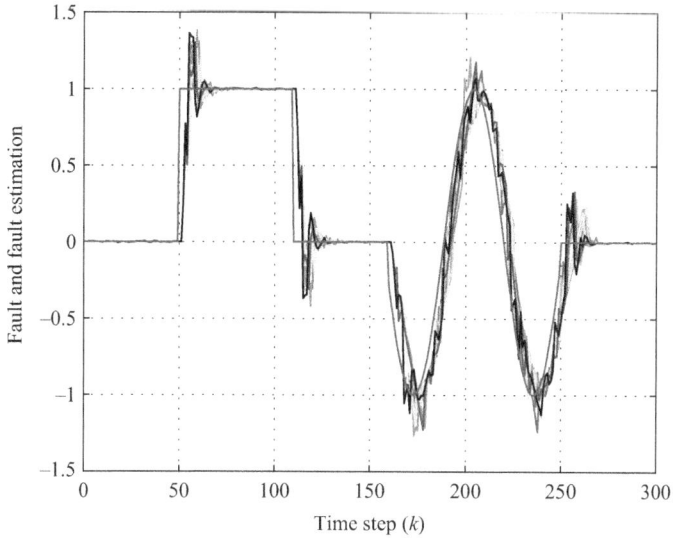

*Figure 6.8   Fault f and its estimate f̂*

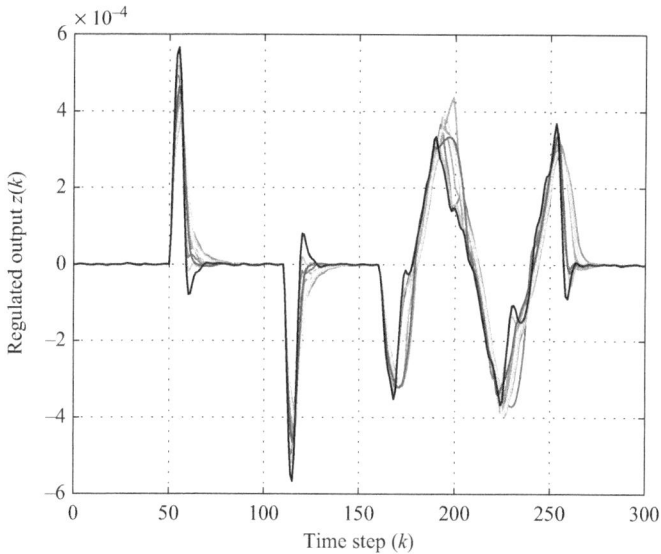

*Figure 6.9   Regulated output z(k)*

ROV $(p^*)$ while the actual values of the parameters are randomly selected in the bound $[p^* - \varepsilon p^*, p^* + \varepsilon p^*]$. To be more specific, it is assumed that the entries $A[1, 1]$ and $A[2, 1]$ are uncertain with $\varepsilon = 0.5$.

The simulation results for nine sets of parameters are shown in Figures 6.8, 6.9, and Table 6.1. From Figure 6.8, it follows that the fault is estimated efficiently.

Table 6.1  *Data transmission rates for different values*
*of the parameters*

|  |  | $A[1,1]$ | | |
|---|---|---|---|---|
|  |  | **0.1074** | **0.2** | **0.3222** |
| $A[2,1]$ | 0.0789 | 46.67% | 47% | 48.67% |
|  | 0.14 | 49% | 49.67% | 49% |
|  | 0.2367 | 52.67% | 53.67% | 53.33% |

Moreover, from Figure 6.9, it is clear that the effects of disturbance and fault on the regulated output have been effectively attenuated. Furthermore, from Table 6.1, it can be concluded that the data transmission rates for sending the sensor data to the fault/state observer have been significantly reduced. Consequently, it can be concluded that the proposed E-AFTC methodology is quite robust against parametric uncertainties.

## 6.5  Conclusion

In this chapter, we considered an $H_\infty$ formulation of the E-AFTC problem for LTI systems. An LMI approach for codesign of the AFTC problem and the event-triggered condition was presented, that in addition to reducing the communication exchanges, simultaneously stabilizes the closed-loop system, attenuates the effects of external disturbances, and compensates for the effects of a fault. Application of our methodology to the speed subsystem of Subzero III ROV was also presented which shows the capability of our proposed approach in AFTC and communication rate objectives.

*Chapter 7*

# Integrated fault detection and consensus control design for a network of multiagent systems

In this chapter, we address the problem of integrated fault detection and consensus control (IFDCC) of linear continuous-time multiagent systems. A mixed $H_\infty/H_-$ formulation of the IFDCC problem is presented and distributed detection filters are designed using only relative output information among the agents. With our proposed methodology, all agents reach either a state consensus (SC) or a model reference consensus (MRC), while the agents simultaneously collaborate with one another to detect the occurrence of faults in the team. Indeed, each agent not only can detect its own fault but also is also capable of detecting its neighbor's faults.

It is shown that through a decomposition approach, the computational complexity of solving the distributed problem is significantly reduced as compared to an optimal centralized solution. The extended linear matrix inequalities (LMIs) are used to reduce the conservativeness of the IFDCC results by introducing additional matrix variables to eliminate the couplings of Lyapunov matrices with the system matrices. It is shown that under a special condition on the network topology, the faulty agent can be isolated in the team. Simulation results corresponding to two teams of autonomous underwater vehicles (AUVs) demonstrate and illustrate the effectiveness and capabilities of our proposed design methodology. The work presented in this chapter has partially appeared in [354,355],

This chapter is organized as follows. After a brief overview of the relevant literature in Section 7.1, we begin in Section 7.2 by presenting the statement of distributed IFDCC problem for a team of multiagent systems. In the next section, we employ a model transformation methodology in Section 7.3.1 and a model decomposition technique in Section 7.3.2, to reduce the computational complexity of the design as well as to specify IFDCC modules with a distributed architecture. An LMI formulation and design solution to the IFDCC problem is obtained in Section 7.4. Moreover, the residual evaluation criterion and the proposed distributed fault isolation scheme are studied in Sections 7.4.1 and 7.4.2, respectively. The effectiveness and capabilities of the proposed results are shown in Section 7.5.

## 7.1 Introduction

Multiagent systems have recently received considerable attention in the control system community due to their potential applications in many areas, such as formation

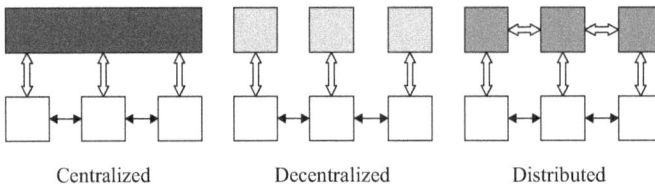

*Figure 7.1* *Difference between the centralized, the decentralized and the distributed architectures for a system, where narrow arrows denote physical interaction between subsystems and thick arrows denote communication and measuring channels [2]*

control of unmanned aerial and underwater vehicles, flocking of mobile vehicles, distributed optimization of multiple mobile robotic systems, and scheduling of automated highway systems [88].

Multiagent systems represent complex distributed systems. The term distributed describes systems whose structure can be analyzed as being constituted by multiple subsystems (agents or nodes) that communicate with other neighboring subsystems. This is different from the notion of decentralized systems that describes systems whose structure can be made of multiple subsystems that have quite limited interactions with each other, as well as the notion of centralized, where every part of the system interacts with every other one [2]. The difference between the concepts of centralized, decentralized, and distributed systems is depicted in Figure 7.1.

It is worth noting that the terms centralized, decentralized, and distributed can be used both for referring to physical systems and also architectures. The notion of system or physical system is considered for the object that is being controlled and monitored against presence of disturbances, faults, and uncertainties, while the notion of architecture is considered for combination of hardware and software systems that are used to implement and execute the control and fault diagnosis tasks.

On the other hand, automatic detection of system faults is of growing importance as the size and complexity of systems rapidly increase. A great deal of attention has recently been devoted to the problem of detection filters design to accomplish fault detection and isolation (FDI) tasks [30]. Based on the above discussion, there are three different architectures for FDI design of multiagent systems.

In the centralized architecture, only one central FDI block is employed and all the information is sent to this central unit through the entire communication channels. Most of the literature on FDI of multiagent systems are based on centralized fault diagnosis architectures [270]. In practice, it is quite challenging to address the problem of FDI in a network of multiagent systems using a centralized architecture due to the stringent computational resources and communication bandwidth limitations [356].

The simplest approach to overcome these drawbacks is to use a decentralized architecture. In the decentralized scheme, the problem of designing FDI algorithm that uses only local information is investigated. In this architecture, as many local

FDI blocks as the numbers of agents are employed. Each block needs to receive the measurements from its corresponding agent and will execute only the computations that are needed to solve the part of the FDI problem corresponding to the specific agent. It is intuitive that this approach will reduce the computation power and the communication capacity needed by each node, with respect to a centralized architecture. However, in a decentralized implementation since the FDI modules only receive local information, they will inevitably be unable to take into account in their solution the information corresponding to the neighboring agents [2].

In some engineering problems, the coupling between subsystems is so weak that it can be ignored or considered as simply a disturbance; however, this clearly does apply to only a very limited subset of cases of interest. Hence, for most practical cases, it is unrealistic and infeasible to solve the FDI problem for a network of multiagent systems by using a decentralized architecture.

Consequently, the only viable and feasible architecture is the distributed one. Similar to the decentralized case, in the distributed architecture, as many FDI modules as the number of physical agents are employed. However, they are more effective than decentralized solutions due to the fact that they take into account information exchanges among the neighboring agents. Moreover, they are more suitable than a centralized architecture due to their lower complexity and use of fewer network resources.

On the other hand, one of the most frequent tasks to be accomplished by autonomous agents is to agree and reach a consensus upon some parameters. Agreement variables represent quantities of interest such as the work load in a network of parallel computers, the clock speed for wireless sensor networks, the velocity, or the formation pattern for a team of autonomous vehicles. From the fault diagnosis point of view, the agreement of certain agents on the occurrence of a fault in any part of the multiagent system can increase the reliability and applicability of the system. Hence, reaching unanimity in a faulty system is an important problem that is studied by computer scientists interested in distributed computing.

In the decentralized architecture, faulty agents are unaware of the structure and state of the network and ignore the presence of faults in other parts of the system. However, in the distributed architecture, the faulty agents have knowledge of the structure and state of the network and may cooperate with others to decide on the occurrence of a fault in the agents. Indeed, by using the distributed FDI architectures, each agent can detect and isolate not only its own faults but also faults of its neighbors. Therefore, the distributed architectures increase the reliability and applicability of the proposed methodologies.

However, despite the critical necessity of distributed FDI solutions for a network of multiagent systems, there exist only a few contributions in the literature on design of distributed FDI filters [31,269,270,357]. The problem of distributed FDI in a network of multiagent systems having double integrator dynamics is considered in [270]. In [269], development, design, and analysis of actuator FDI filters for a network of unmanned vehicles are investigated. Recently, the problem of distributed FDI in a network of heterogeneous multiagent systems having different dynamics and order from one another is also considered in [273]. In [358], a robust semidecentralized fault detection (FD) strategy for a network of homogeneous multiagent systems is presented and applied to a team of microair vehicles.

On the other hand, cooperative control of multiagent systems has recently received significant attention from various scientific communities. One critical issue arising from cooperative control of multiagent systems is determining how to develop distributed control policies that are based on local information and enable all agents to reach an agreement on certain quantities of interest. This problem is known as the consensus problem [359].

Consensus problem has a long-standing tradition in computer science. In the context of multiagent systems, recent years have witnessed dramatic advances of various distributed strategies that achieve agreements. MRC [360] is another basic concept in cooperative control of multiagent systems. Here, the objective is not only to achieve consensus but also to make the agents follow a reference system or a virtual leader.

According to the above literature review, many of the robust model-based approaches in multiagent systems research can be classified into as belonging to one of the following two categories. First, there are robust model-based approaches for achieving cooperative control, and second, there are robust model-based approaches with the objective of FDI capabilities. Notwithstanding the above, there are scenarios where it is possible and, more importantly, desirable to consider an integrated design of feedback controllers and FD filters. This simultaneous design unifies both the control and the detection modules into a single framework.

Hence, it is conceivable and expected that an integrated fault detection and control (IFDC) design should lead to a far less overall complexity as compared to an approach where these two modules are designed separately [361]. It should be pointed out that the IFDC design has other advantages in comparison with the separate design. The reader can refer to [46,361] for more details on these advantages.

The contributions of this work can therefore be summarized as follows:

1. An LMI-based approach for the mixed $H_\infty/H_-$ distributed IFDCC design of continuous-time linear multiagent systems using relative output information is proposed. A set of distributed modules are designed that generate two signals, namely, the residual and the control signals. Using the residual signal, not only each agent's faults but also faults of its neighbors can be detected. Using the control signal, all the agents can either achieve (i) an SC or (ii) an MRC. The IFDCC modules are designed such that the effects of disturbances and faults on the residual signals are minimized and maximized, respectively (for accomplishing the FD task), while the effects of disturbances and faults on the specified control outputs are minimized (for accomplishing the SC problem or the MRC problem).

2. Motivated by [360,362], a decomposition technique is employed in the IFDCC design procedure that allows the system to be partitioned into a set of lower order subsystems for reducing the computational complexity of the design as well as for specifying the IFDCC modules with a distributed architecture. The decomposition approach that is used in [360,363] leads to conservativeness in

the solution since one needs to equate the Lyapunov variables between all the conditions. To overcome this difficulty, the authors in [362] applied the extended LMI formulations for a class of discrete-time systems to avoid common Lyapunov variables and solved the problem of designing distributed controllers with $H_\infty$ and $H_2$ performance indices. However, continuous-time systems were not addressed and solved in [362]. It should be pointed out that it is not possible to relate the results in the continuous-time and the discrete-time domains by simple transformations since the underlying control objective involves $H_\infty$ and $H_2$ performances that are distinctly and separately formulated for their respective controller designs.

In this work, motivated by [277], the extended LMI formulation for solving our proposed $H_\infty/H_-$ distributed IFDCC problem for a network of continuous-time multiagent systems is used to avoid the use of common Lyapunov variables.

3. A required sufficient condition on the network topology and residual signals for guaranteeing isolation of the faulty agent in the team is also obtained. This distributed strategy is based on flags that are generated corresponding to the residual signals of the team agents.

## 7.2 System description and problem statement

### 7.2.1 Multiagent team representation

Consider a team of $N$ homogeneous agents, where the $i$th agent has a linear dynamic representation as governed by the following model:

$$\begin{aligned}
\dot{x}_i(t) &= Ax_i(t) + B_u u_i(t) + B_d d_i(t) + B_f f_i(t), \\
y_i(t) &= Cx_i(t) + D_d d_i(t) + D_f f_i(t), \quad i = 1, 2, \ldots, N
\end{aligned} \tag{7.1}$$

where $x_i(t) \in \mathbb{R}^n$ denotes the state vector, $d_i(t) \in \mathbb{R}^{n_d}$ denotes an $\mathbb{L}_2$-norm bounded external disturbance, $f_i(t) \in \mathbb{R}^{n_f}$ denotes the fault signal, $u_i(t) \in \mathbb{R}^{n_u}$ denotes the control input, and $y_i(t) \in \mathbb{R}^{n_y}$ with $n_y \geq n_f$ denotes the measured output of the agent $i$.

It is assumed that the system $(A, B_f, C, D_f)$ does not have any transmission zeros on the extended imaginary axis. This guarantees that the occurred faults in the system (7.1) are structurally detectable. This assumption also implies that $T_{y_i f_i}(\infty) = D_f$ has full column rank. There are two ways to relax this assumption. In the first, the definition of the $H_-$ index is restricted to a frequency range for which $T_{y_i f_i}$ has a full rank using the generalized Kalman-Yakubovich-Popov (KYP) lemma [41]; see Lemma 1.10 for more details regarding the finite frequency $H_-$ index design. In the second, frequency weights are introduced to cancel out the zeros on the imaginary axis [364]. These extensions are beyond the scope of this work and are left as topics of our future research.

For simplicity, the model uncertainties are assumed to be recast as disturbances. It is assumed that $(A, B_u, C)$ is stabilizable and detectable. Moreover, multiple faults cannot occur simultaneously in the team.

**Problem statement**

The objective of this work is to design a set of distributed modules designated as the "detector/controller units," where the detectors are state observers and the controllers are observer-based controllers. The modules are tasked to generate both the residual signals to detect faults as well as the control signals to ensure that the desired control specifications are satisfied. Two cooperative control goals (corresponding to two different requirements) are considered as follows:

**Requirement 1.** *SC.* In this problem, the objective is to have all the agents' states driven to a common value. In order to reach consensus, a common approach is to define a controlled output function $z_{sc_i}(t) = x_i(t) - (1/N)\sum_{j=1}^{N} x_j(t), i = 1, \ldots, N$, for measuring the disagreement of $x_i(t)$ with the average state of all the agents [365]. Note that if $z_{sc_i}(t) = 0, \forall i \in \mathcal{V}$, then $x_i(t) = x_j(t)$ for $\forall i,j \in \mathcal{V}$, which implies that the consensus is achieved.

**Requirement 2.** *MRC.* In this problem, the objective is not only to achieve SC but also to ensure that the states follow a reference model/virtual leader that is governed by

$$\begin{aligned}
\dot{x}_r(t) &= Ax_r(t) + B_u u_r(t), \\
y_r(t) &= Cx_r(t),
\end{aligned} \tag{7.2}$$

where $x_r(t) \in \mathbb{R}^n$, $u_r(t) \in \mathbb{R}^{n_u}$, and $y_r(t) \in \mathbb{R}^{n_y}$ are the state vector, the input, and the output of the reference system, respectively. Given that the agent's states or combination of them are required to converge to the reference state $x_r$ satisfying (7.2), it is convenient to define performance outputs $z_{mrc_i} \in \mathbb{R}^{n_z}$ as $z_{mrc_i}(t) = C_2(x_i(t) - x_r(t)), i = 1, \ldots, N$, where $C_2$ is a constant matrix with appropriate dimension that is defined according to the designer's specifications.

## 7.2.2 Design of the detector/controller units

Below, we provide the details regarding the methodology that we will be pursuing for Requirements 1 and 2 as stated formally in the previous subsection.

**Requirement 1.** It is assumed that each agent generates its *relative output information* $\zeta_i(t) = \sum_{j \in \mathcal{N}_i} (y_i(t) - y_j(t))$. Note that although agents operate with the $\zeta_i$s as the available information, however we will also use $z_{sc_i}$ as a measure of the agent's performance.

To achieve consensus, a detector and a controller filter for the $i$th agent is proposed as follows:

$$\begin{aligned}
\dot{\hat{x}}_i(t) &= A\hat{x}_i(t) + B_u u_i(t) + Fr_i(t), \\
u_i(t) &= K\hat{x}_i(t), \\
r_i(t) &= \zeta_i(t) - \hat{\zeta}_i(t),
\end{aligned} \tag{7.3}$$

where $\hat{x}_i(t) \in \mathbb{R}^n$ denotes the detection filter or the observer state vector, $r_i(t) \in \mathbb{R}^{n_y}$ denotes the residual signal, and the constant matrices $F \in \mathbb{R}^{n \times n_y}$ and $K \in \mathbb{R}^{n_u \times n}$ denote the detector/controller filter gains to be selected and specified subsequently, and $\hat{\zeta}_i(t) = \sum_{j \in \mathcal{N}_i} C(\hat{x}_i(t) - \hat{x}_j(t))$.

**Requirement 2.** In the MRC problem, it is assumed that a nonempty subset of the agents (or at least one agent) has access to the output of the reference model that is given by (7.2).

Therefore, it is assumed that each agent has access to the *relative output information* $\xi_i(t) = \sum_{j \in \mathcal{N}_i} (y_i(t) - y_j(t)) + q_i(y_i(t) - y_r(t))$, where $q_i > 0$ if the $i$th agent has access to the reference model (7.2) and zero otherwise. The following detection and controller filter is proposed for the $i$th agent:

$$\dot{\hat{x}}_i(t) = A\hat{x}_i(t) + B_u u_i(t) + Fr_i(t),$$

$$\dot{\hat{x}}_r(t) = A\hat{x}_r(t) + B_u u_r(t),$$

$$u_i(t) = K(\hat{x}_i(t) - \hat{x}_r(t)) + u_r(t),$$

$$r_i(t) = \xi_i(t) - \hat{\xi}_i(t),$$

(7.4)

where $\hat{\xi}_i(t) = C \sum_{j \in \mathcal{N}_i} ((\hat{x}_i - \hat{x}_j) + q_i(\hat{x}_i - \hat{x}_r))$.

## 7.2.3 Closed-loop team dynamics

Let us now define the following variables, namely, $e_{sc_i} = x_i - \hat{x}_i$, $x_{e_i} = x_i - x_r$, $\hat{x}_{e_i} = \hat{x}_i - \hat{x}_r$, $e_{mrc_i} = x_{e_i} - \hat{x}_{e_i}$, $x_e = [x_{e_1}^T, \ldots, x_{e_N}^T]^T$, $e_{sc} = [e_{sc_1}^T, \ldots, e_{sc_N}^T]^T$, $e_{mrc} = [e_{mrc_1}^T, \ldots, e_{mrc_N}^T]^T$, $x = [x_1^T, \ldots, x_N^T]^T$, $\hat{x} = [\hat{x}_1^T, \ldots, \hat{x}_N^T]^T$, $d = [d_1^T, \ldots, d_N^T]^T$, $f = [f_1^T, \ldots, f_N^T]^T$, $r = [r_1^T, \ldots, r_N^T]^T$, $z_{mrc} = [z_{mrc_1}^T, \ldots, z_{mrc_N}^T]^T$, and $z_{sc} = [z_{sc_1}^T, \ldots, z_{sc_N}^T]^T$. By concatenating the governing equation of the detector/controller module (7.3) [or (7.4)] with that of the agent governing equation (7.1), the following closed-loop system dynamics is obtained:

$$\dot{x}_{cl}(t) = A_{cl} x_{cl}(t) + B_{dcl} d(t) + B_{fcl} f(t),$$

$$r(t) = C_{cl} x_{cl}(t) + D_{dcl} d(t) + D_{fcl} f(t),$$

$$z_{sc}(t) = C_{1cl} x_{cl}(t) \quad \text{(to measure the SC goal)},$$

$$z_{mrc}(t) = C_{2cl} x_{cl}(t) \quad \text{(to measure the MRC goal)}.$$

(7.5)

Corresponding to Requirement 1, $x_{cl} = [e_{sc}^T \ x^T]^T$ and $\tilde{L} = L$. Corresponding to Requirement 2, $\tilde{L} = L_2 = L + Q$, $Q = \text{diag}\{q_1, q_2, \ldots, q_N\}$, and $x_{cl} = [e_{mrc}^T \ x_e^T]^T$. Moreover, $D_{dcl} = \tilde{L} \otimes D_d$, $D_{fcl} = \tilde{L} \otimes D_f$,

$$A_{cl} = \begin{bmatrix} I \otimes A - \tilde{L} \otimes FC & 0 \\ -I \otimes B_u K & I \otimes (A + B_u K) \end{bmatrix},$$

$$B_{fcl} = \begin{bmatrix} I \otimes B_f - \tilde{L} \otimes FD_f \\ I \otimes B_f \end{bmatrix}, \quad B_{dcl} = \begin{bmatrix} I \otimes B_d - \tilde{L} \otimes FD_d \\ I \otimes B_d \end{bmatrix},$$

$$C_{cl} = [\tilde{L} \otimes C \ \ 0], \quad C_{1cl} = [0 \ \ L_C \otimes I], \quad C_{2cl} = [0 \ \ I \otimes C_2],$$

and $L_C = [L_{C_{ij}}]$ is a symmetric matrix with $L_{C_{ii}} = ((N-1)/N)$ and $L_{C_{ij}} = (-1/N)$ for $i \neq j$. Note that since the graph $\mathfrak{G}$ is assumed to be undirected and at least one agent has access to the reference model, it follows from Lemma 1 in [359] that the matrix $L_2$ is positive definite.

## 7.2.4  The distributed IFDCC problem definition

The distributed IFDCC problem is concerned with designing and developing a control strategy as to how agents can reach the SC (for satisfying Requirement 1) or the MRC (for satisfying Requirement 2) and simultaneously each agent is capable of detecting not only its own faults but also faults of its neighbors, by using only the available information accessible to each agent. The distributed IFDCC problem that is addressed in this work is now stated explicitly as follows.

**The distributed IFDCC problem definition:**

For the network model (7.1), a detector/controller (7.3) for Requirement 1 or a detector/controller (7.4) for Requirement 2 should be designed such that the closed-loop system (7.5) is

- stable,
- the effects of disturbances on the control outputs and residual outputs are minimized, and
- the effects of faults on the control outputs are minimized while their effects on the residual outputs are maximized.

The $H_\infty$ norm is used to attenuate the disturbance effects on the residuals and the control outputs, as well as to attenuate the fault effects on the control outputs. Moreover, the $H_-^+$ index is used for Requirement 1 and $H_-$ index is used for Requirement 2 to guarantee the sensitivity of the residuals to the faults. Note that, due to singularity of $L$, the performance index $H_-$ is always zero for Requirement 1. Hence, instead of $H_-$, the performance index $H_-^+$ is considered for this requirement.

Consequently, our specific problem can indeed be cast into the following multiobjective $H_\infty/H_-^+$ IFDCC (detector/controller) design for Requirement 1,

$$\text{minimize}\quad \rho_1\gamma_1 + \rho_2\gamma_2 + \rho_3\gamma_3 - \rho_4\beta$$

subject to:

(i) $\|T_{rd}\|_\infty < \gamma_1$,    (ii) $\|T_{rf}\|_-^+ > \beta$,

(iii) $\|T_{z_{sc}d}\|_\infty < \gamma_2$,    (iv) $\|T_{z_{sc}f}\|_\infty < \gamma_3$,

$$\text{(7.6)}$$

and the following multiobjective $H_\infty/H_-$ IFDCC design for Requirement 2,

$$\text{minimize}\quad \rho_1\gamma_1 + \rho_2\gamma_2 + \rho_3\gamma_3 - \rho_4\beta$$

subject to:

(i) $\|T_{rd}\|_\infty < \gamma_1$,    (ii) $\|T_{rf}\|_- > \beta$,

(iii) $\|T_{z_{mrc}d}\|_\infty < \gamma_2$,    (iv) $\|T_{z_{mrc}f}\|_\infty < \gamma_3$.

$$\text{(7.7)}$$

The positive constant weights $\rho_1$, $\rho_2$, $\rho_3$, and $\rho_4$ can be used by the designer as a trade-off analysis among the objectives (i)–(iv).

It should be pointed out that the optimization problems (7.6) and (7.7) correspond to a distributed IFDCC architecture due to the fact that relative information between each agent and its neighbors ($\sum_{j \in \mathcal{N}_i}(y_i - y_j)$) for Requirement 1 and $\sum_{j \in \mathcal{N}_i}(y_i - y_j) + q_i(y_i - y_r)$ for Requirement 2) are used in the problem formulation and its solution (corresponding to each of the IFDCC modules).

Based on the performance index (iv), the effects of faults on the control outputs are minimized that indeed satisfy the fault-tolerant control (FTC) objective. Moreover, the FD information is not explicitly used in the controller design. Hence, our proposed IFDCC methodology represents a passive FTC methodology. However, the diagnosis information can be used subsequently in the framework of design of an active FTC system. This issue is left as a topic of our future research.

In the next section, in order to reduce the computational complexity of the design as well as to specify the IFDCC modules with a distributed architecture, using a unified approach both problems (7.6) and (7.7) are formulated as a mixed $H_\infty / H_-$ problem for $N - 1$ decoupled systems for Requirement 1 and $N$ decoupled systems for Requirement 2.

## 7.3 Distributed IFDCC decomposition

### 7.3.1 Model transformation

The matrix $\tilde{L}$ in Requirement 1 is singular; hence, it can be shown that the state matrix of the closed-loop system (7.5) in Requirement 1 is unstable if the matrix $A$ is unstable. This indicates that the closed-loop system is not completely stabilizable. To address this problem, Lemma 1.4 is used to transform the overall closed-loop system (7.5) in Requirement 1 (with the dimension of $2n \times N$ states) into a reduced order system (with the dimension of $2n \times (N - 1)$ states).

Let us define $\overline{e}(t) = (L_C \otimes I_n)e_{sc}(t)$ and $\overline{x}(t) = (L_C \otimes I_n)x(t)$. By invoking Lemma 1.4, there exists an orthogonal matrix $U = [U_1, U_2] \in \mathbb{R}^{N \times N}$ with $U_2 = \mathbf{1}_N / \sqrt{N}$ such that $U^T L_C U = \overline{L}_C = \text{diag}(I_{N-1}, 0)$ and $U^T L U = \overline{L} = \text{diag}(L_1, 0)$, where $L_1 \in \mathbb{R}^{(N-1) \times (N-1)}$ is a positive definite matrix if and only if the graph is connected.

By applying the orthogonal transformation $\check{e} = [\check{e}_1^T, \ldots, \check{e}_N^T]^T = (U^T \otimes I_n)\overline{e}$, $\check{x} = [\check{x}_1^T, \ldots, \check{x}_N^T]^T = (U^T \otimes I_n)\overline{x}$, $\check{r} = [\check{r}_1^T, \ldots, \check{r}_N^T]^T = (U^T \otimes I_{n_y})r$, $\check{z} = [\check{z}_1^T, \ldots, \check{z}_N^T]^T = (U^T \otimes I_n)z_{sc}$, $\check{d} = [\check{d}_1^T, \ldots, \check{d}_N^T]^T = (U^T \otimes I_{n_d})d$, $\check{f} = [\check{f}_1^T, \ldots, \check{f}_N^T]^T = (U^T \otimes I_{n_f})f$, and after some algebraic manipulations the closed-loop system (7.5) under Requirement 1 is transformed into the following closed-loop model:

$$
\begin{bmatrix} \dot{\check{e}}(t) \\ \dot{\check{x}}(t) \end{bmatrix} = \begin{bmatrix} \overline{L}_C \otimes A - \overline{L}_C \overline{L} \otimes FC & 0 \\ -\overline{L}_C \otimes B_u K & \overline{L}_C \otimes (A + B_u K) \end{bmatrix} \begin{bmatrix} \check{e}(t) \\ \check{x}(t) \end{bmatrix}
$$
$$
+ \begin{bmatrix} \overline{L}_C \otimes B_d - \overline{L}_C \overline{L} \otimes FD_d \\ \overline{L}_C \otimes B_d \end{bmatrix} \check{d}(t) + \begin{bmatrix} \overline{L}_C \otimes B_f - \overline{L}_C \overline{L} \otimes FD_f \\ \overline{L}_C \otimes B_f \end{bmatrix} \check{f}(t)
$$

$$
\check{r}(t) = \begin{bmatrix} \overline{L} \otimes C & 0 \end{bmatrix} \begin{bmatrix} \check{e}(t) \\ \check{x}(t) \end{bmatrix} + \overline{L} \otimes D_d \check{d}(t) + \overline{L} \otimes D_f \check{f}(t)
$$

$$
\check{z}(t) = \begin{bmatrix} 0 & \overline{L}_C \otimes I \end{bmatrix} \begin{bmatrix} \check{e}(t) \\ \check{x}(t) \end{bmatrix}
$$

$$(7.8)$$

Note that the last rows of the symmetric matrices $\bar{L}_C$ and $\bar{L}_C\bar{L}$ are both zeros. Therefore, by rearranging elements of the state vector in the system (7.8), one can obtain an equivalent system:

$$
\begin{bmatrix} \dot{\hat{x}}_{cl}(t) \\ \dot{\check{x}}_{clN}(t) \end{bmatrix} = \begin{bmatrix} \hat{A}_{cl} & 0 \\ 0 & 0 \end{bmatrix} \begin{bmatrix} \hat{x}_{cl}(t) \\ \check{x}_{clN}(t) \end{bmatrix} + \begin{bmatrix} \hat{B}_{dcl} & 0 \\ 0 & 0 \end{bmatrix} \begin{bmatrix} \hat{d}(t) \\ \check{d}_N(t) \end{bmatrix} + \begin{bmatrix} \hat{B}_{fcl} & 0 \\ 0 & 0 \end{bmatrix} \begin{bmatrix} \hat{f}(t) \\ \check{f}_N(t) \end{bmatrix}
$$

$$
\begin{bmatrix} \hat{r}(t) \\ \check{r}_N(t) \end{bmatrix} = \begin{bmatrix} \hat{C}_{1cl} & 0 \\ 0 & 0 \end{bmatrix} \begin{bmatrix} \hat{x}_{cl}(t) \\ \check{x}_{clN}(t) \end{bmatrix} + \begin{bmatrix} \hat{D}_{dcl} & 0 \\ 0 & 0 \end{bmatrix} \begin{bmatrix} \hat{d}(t) \\ \check{d}_N(t) \end{bmatrix} + \begin{bmatrix} \hat{D}_{fcl} & 0 \\ 0 & 0 \end{bmatrix} \begin{bmatrix} \hat{f}(t) \\ \check{f}_N(t) \end{bmatrix} \qquad (7.9)
$$

$$
\begin{bmatrix} \hat{z}_{sc}(t) \\ \check{z}_N(t) \end{bmatrix} = \begin{bmatrix} \hat{C}_{2cl} & 0 \\ 0 & 0 \end{bmatrix} \begin{bmatrix} \hat{x}_{cl}(t) \\ \check{x}_{clN}(t) \end{bmatrix}
$$

where $\hat{x}_{cl} = [\hat{e}^T \; \hat{x}^T]^T$, $\check{x}_{clN} = [\check{e}_N^T \; \check{x}_N^T]^T$, $\hat{x} = [\hat{x}_1^T, \ldots, \hat{x}_{N-1}^T]^T$, $\hat{e} = [\hat{e}_1^T, \ldots, \hat{e}_{N-1}^T]^T$, $\hat{z}_{sc} = [\hat{z}_1^T, \ldots, \hat{z}_{N-1}^T]^T$, $\hat{f} = [\hat{f}_1^T, \ldots, \hat{f}_{N-1}^T]^T$, $\hat{d} = [\hat{d}_1^T, \ldots, \hat{d}_{N-1}^T]^T$, $\hat{r} = [\hat{r}_1^T, \ldots, \hat{r}_{N-1}^T]^T$, and

$$
\hat{A}_{cl} = \begin{bmatrix} I_{N-1} \otimes A - L_1 \otimes FC & 0 \\ -I_{N-1} \otimes B_u K & I_{N-1} \otimes (A + B_u K) \end{bmatrix},
$$

$$
\hat{B}_{dcl} = \begin{bmatrix} I_{N-1} \otimes B_d - L_1 \otimes FD_d \\ I_{N-1} \otimes B_d \end{bmatrix}, \quad \hat{B}_{fcl} = \begin{bmatrix} I_{N-1} \otimes B_f - L_1 \otimes FD_f \\ I_{N-1} \otimes B_f \end{bmatrix},
$$

$$
\hat{C}_{1cl} = \begin{bmatrix} L_1 \otimes C & 0 \end{bmatrix}, \quad \hat{C}_{2cl} = \begin{bmatrix} 0 & I_{N-1} \otimes I \end{bmatrix}, \quad \hat{D}_{dcl} = L_1 \otimes D_d, \quad \hat{D}_{fcl} = L_1 \otimes D_f.
$$

From (7.8) and (7.9), it can be concluded that

$$
T_{\check{r}\hat{f}} = \mathrm{diag}(T_{\hat{r}\hat{f}}, 0), \quad T_{\check{r}\hat{d}} = \mathrm{diag}(T_{\hat{r}\hat{d}}, 0),
$$

$$
T_{\hat{z}\check{f}} = \mathrm{diag}(T_{\hat{z}_{sc}\hat{f}}, 0), \quad T_{\hat{z}\check{d}} = \mathrm{diag}(T_{\hat{z}_{sc}\hat{d}}, 0).
$$

Consequently, the closed-loop system (7.5) under Requirement 1 is transformed into the following reduced-order model representation:

$$
\dot{\hat{x}}_{cl}(t) = \hat{A}_{cl}\hat{x}_{cl}(t) + \hat{B}_{dcl}\hat{d}(t) + \hat{B}_{fcl}\hat{f}(t),
$$

$$
\hat{r}(t) = \hat{C}_{1cl}\hat{x}_{cl}(t) + \hat{D}_{dcl}\hat{d}(t) + \hat{D}_{fcl}\hat{f}(t), \qquad (7.10)
$$

$$
\hat{z}_{sc}(t) = \hat{C}_{2cl}\hat{x}_{cl}(t),
$$

Note that the orthogonal transformation $U$ does not change the various norms of the system. Consequently, it can be shown that after some algebraic manipulations the following relationships hold based on the above definitions, namely,

$$
||T_{z_{sc}d}||_\infty = ||T_{\hat{z}_{sc}\hat{d}}||_\infty, \; ||T_{rd}||_\infty = ||T_{\hat{r}\hat{d}}||_\infty,
$$

$$
||T_{z_{sc}f}||_\infty = ||T_{\hat{z}_{sc}\hat{f}}||_\infty, \; ||T_{rf}||_-^+ = ||T_{\hat{r}\hat{f}}||_-^+ = ||T_{\hat{r}\hat{f}}||_-.
$$

In the next subsection, at first, the reduced-order model (7.10) for Requirement 1 and the closed-loop model (7.5) for Requirement 2 are decomposed into a set of decoupled models. Subsequently, the distributed IFDCC problem for the decomposed system is formally formulated.

## 7.3.2 Model decomposition

The matrices $L_1$ and $L_2$ are symmetric; hence, there exist orthogonal matrices $J \in \mathbb{R}^{(N-1) \times (N-1)}$ and $\tilde{J} \in \mathbb{R}^{N \times N}$ such that $J^\mathsf{T} L_1 J = \mathrm{diag}\{\lambda_1, \ldots, \lambda_{N-1}\}$ and $\tilde{J}^\mathsf{T} L_2 \tilde{J} = \mathrm{diag}\{\lambda_1, \ldots, \lambda_N\}$, respectively [360,362].

In order to decompose the system (7.10) for Requirement 1 and system (7.5) for Requirement 2, the error signals $\hat{x}$ (or $x_e$) and $\hat{e}$ (or $e_{mrc}$), the disturbance signal $\hat{d}$ (or $d$), the fault signal $\hat{f}$ (or $f$), the output signal $\hat{z}_{sc}$ (or $z_{mrc}$), and the residual signal $\hat{r}$ (or $r$) should be reformulated into new variables as follows.

Corresponding to Requirement 1:
$[\tilde{x}_1^\mathsf{T}, \ldots, \tilde{x}_{N-1}^\mathsf{T}]^\mathsf{T} = (J^\mathsf{T} \otimes I_n)\hat{x}$, $[\tilde{e}_1^\mathsf{T}, \ldots, \tilde{e}_{N-1}^\mathsf{T}]^\mathsf{T} = (J^\mathsf{T} \otimes I_n)\hat{e}$, $[\tilde{d}_1^\mathsf{T}, \ldots, \tilde{d}_{N-1}^\mathsf{T}]^\mathsf{T} = (J^\mathsf{T} \otimes I_{n_d})\hat{d}$, $[\tilde{f}_1^\mathsf{T}, \ldots, \tilde{f}_{N-1}^\mathsf{T}]^\mathsf{T} = (J^\mathsf{T} \otimes I_{n_f})\hat{f}$, $[\tilde{z}_1^\mathsf{T}, \ldots, \tilde{z}_{N-1}^\mathsf{T}]^\mathsf{T} = (J^\mathsf{T} \otimes I_n)\hat{z}_{sc}$, $[\tilde{r}_1^\mathsf{T}, \ldots, \tilde{r}_{N-1}^\mathsf{T}]^\mathsf{T} = (J^\mathsf{T} \otimes I_{n_y})\hat{r}$.

Corresponding to Requirement 2:
$[\tilde{x}_1^\mathsf{T}, \ldots, \tilde{x}_N^\mathsf{T}]^\mathsf{T} = (\tilde{J}^\mathsf{T} \otimes I_n)x_e$, $[\tilde{e}_1^\mathsf{T}, \ldots, \tilde{e}_N^\mathsf{T}]^\mathsf{T} = (\tilde{J}^\mathsf{T} \otimes I_n)e_{mrc}$, $[\tilde{d}_1^\mathsf{T}, \ldots, \tilde{d}_N^\mathsf{T}]^\mathsf{T} = (\tilde{J}^\mathsf{T} \otimes I_{n_d})d$, $[\tilde{f}_1^\mathsf{T}, \ldots, \tilde{f}_N^\mathsf{T}]^\mathsf{T} = (\tilde{J}^\mathsf{T} \otimes I_{n_f})f$, $[\tilde{z}_1^\mathsf{T}, \ldots, \tilde{z}_N^\mathsf{T}]^\mathsf{T} = (\tilde{J}^\mathsf{T} \otimes I_{n_z})z_{mrc}$, $[\tilde{r}_1^\mathsf{T}, \ldots, \tilde{r}_N^\mathsf{T}]^\mathsf{T} = (\tilde{J}^\mathsf{T} \otimes I_{n_y})r$.

After some algebraic manipulations, the system (7.10) for Requirement 1 and the system (7.5) for Requirement 2 can be rewritten into the following unified decomposed system, namely,

$$
\begin{aligned}
\dot{\tilde{x}}_{cl_i} &= A_{cl_i}\tilde{x}_{cl_i} + B_{dcl_i}\tilde{d}_i(t) + B_{fcl_i}\tilde{f}_i(t), \\
\tilde{r}_i(t) &= C_{cl_i}\tilde{x}_{cl_i}(t) + D_{dcl_i}\tilde{d}_i(t) + D_{fcl_i}\tilde{f}_i(t), \\
\tilde{z}_i(t) &= H_{cl_i}\tilde{x}_{cl_i},
\end{aligned}
\tag{7.11}
$$

where $A_{cl_i} = \begin{bmatrix} A - \lambda_i FC & 0 \\ -B_u K & A + B_u K \end{bmatrix}$, $\tilde{x}_{cl_i} = \begin{bmatrix} \tilde{e}_i(t) \\ \tilde{x}_i(t) \end{bmatrix}$, $B_{dcl_i} = \begin{bmatrix} B_d - \lambda_i FD_d \\ B_d \end{bmatrix}$, $B_{fcl_i} = \begin{bmatrix} B_f - \lambda_i FD_f \\ B_f \end{bmatrix}$, $C_{cl_i} = \begin{bmatrix} \lambda_i C & 0 \end{bmatrix}$, $D_{dcl_i} = \lambda_i D_d$, $D_{fcl_i} = \lambda_i D_f$, $H_{cl_i} = \begin{bmatrix} 0 & H \end{bmatrix}$. Also, $H = I$ and $i = 1, \ldots, N - 1$ for Requirement 1 and $H = C_2$ and $i = 1, \ldots, N$ for Requirement 2.

Since $(J \otimes I)$ and $(\tilde{J} \otimes I)$ are unitary matrices, one can easily obtain $\|T_{z_{sc}d}\|_\infty = \max_i \|T_{\tilde{z}_i \tilde{d}_i}\|_\infty$, $\|T_{z_{sc}f}\|_\infty = \max_i \|T_{\tilde{z}_i \tilde{f}_i}\|_\infty$, $\|T_{rf}\|_-^+ = \min_i \|T_{\tilde{r}_i \tilde{f}_i}\|_-$, $\|T_{z_{mrc}d}\|_\infty = \max_i \|T_{\tilde{z}_i \tilde{d}_i}\|_\infty$, $\|T_{rd}\|_\infty = \max_i \|T_{\tilde{r}_i \tilde{d}_i}\|_\infty$, $\|T_{z_{mrc}f}\|_\infty = \max_i \|T_{\tilde{z}_i \tilde{f}_i}\|_\infty$, and $\|T_{rf}\|_- = \min_i \|T_{\tilde{r}_i \tilde{f}_i}\|_-$, where $i = 1, \ldots, N - 1$ in Requirement 1 and $i = 1, \ldots, N$ in Requirement 2.

We are now in a position to reformulate the distributed IFDCC problem for the system (7.11).

### Distributed IFDCC problem formulation

For a network of multiagent systems governed by (7.1), there exists a distributed detection and controller filter (7.3) for Requirement 1 and a distributed detection and controller filter (7.4) for Requirement 2 that detects the occurred faults in the team and satisfies the consensus problem for Requirement 1 and the MRC for Requirement 2, if the decomposed system in (7.11) is simultaneously stable and the following optimization problem is satisfied

$$\text{minimize} \quad \rho_1\gamma_1 + \rho_2\gamma_2 + \rho_3\gamma_3 - \rho_4\beta$$

subject to:

(i) $\|T_{\tilde{r}_i\tilde{d}_i}\|_\infty < \gamma_1,$     (ii) $\|T_{\tilde{r}_i\tilde{f}_i}\|_- > \beta,$

(iii) $\|T_{\tilde{z}_i\tilde{d}_i}\|_\infty < \gamma_2,$     (iv) $\|T_{\tilde{z}_i\tilde{f}_i}\|_\infty < \gamma_3,$

(7.12)

where $i = 1,\dots,N-1$ for Requirement 1 and $i = 1,\dots,N$ for Requirement 2.

**Remark 7.1.** *As far as the computational complexity of the IFDCC problem is concerned, note that when the number of agents $N$ is large, the problems (7.6) and (7.7) for the closed-loop dynamics (5.5) are exceedingly challenging to implement. However, through our proposed "model transformation" and the "distributed IFDCC problem" formulation, we have indeed relaxed this concern significantly by converting the IFDCC problem of the team (7.5) into a set of independent and decoupled $N-1$ subsystems for Requirement 1 and $N$ subsystems for Requirement 2, where each subsystem dimension is the same as that of a single agent. Motivated by [354,362], it can be shown that by using our proposed methodology the order of the computational complexity is also reduced by a factor of $N^2/(N-1)$ for Requirement 1 and by a factor $N$ for Requirement 2 in comparison to the problems (7.6) and (7.7).*

The set of transmission zeros of the systems $(A, B_f, C, D_f)$ and $(A_{cl_i}, B_{fcl_i}, C_{cl_i}, D_{fcl_i})$ are denoted by $\mathscr{Z}_1$ and $\mathscr{Z}_2$, respectively. It is worth noting that if any element of $\mathscr{Z}_2$ is on the extended imaginary axis, then there exists a class of faults that independent of their severity are structurally not detectable [364,366]. Therefore, determining the transmission zeros of the system $(A_{cl_i}, B_{fcl_i}, C_{cl_i}, D_{fcl_i})$ that are on the extended imaginary axis is of great importance.

In the following, at first in Theorem 7.1 the relationship between $\mathscr{Z}_1$ and $\mathscr{Z}_2$ will be given. It will then be shown that none of the elements of $\mathscr{Z}_2$ are on the extended imaginary axis, which implies that there does not exist any structurally nondetectable faults in the system.

**Theorem 7.1.** *The transmission zeros $\mathscr{Z}_2$ are the same as the transmission zeros $\mathscr{Z}_1$ together with the modes of $A + B_uK$.*

*Proof.* The transmission zeros $\mathscr{Z}_2$ coincide with those complex values $s$ at which the equation $\mathscr{Z}_2 = \{s | \text{Rank}P(s)\} < 2n + n_f$ is satisfied [366], where $P(s)$ is defined as

$$\begin{bmatrix} A - \lambda_iFC - sI & 0 & B_f - \lambda_iFD_f \\ -B_uK & A + B_uK - sI & B_f \\ \lambda_iC & 0 & \lambda_iD_f \end{bmatrix}.$$

(7.13)

By interchanging the rows and columns of $P(s)$, the rank will be the same as the rank of the following matrix

$$\begin{bmatrix} A - \lambda_iFC - sI & B_f - \lambda_iFD_f & 0 \\ \lambda_iC & \lambda_iD_f & 0 \\ -B_uK & B_f & A + B_uK - sI \end{bmatrix}.$$

It follows that the transmission zeros in $\mathscr{Z}_2$ are those values of $s$ for which

$$\text{Rank}\begin{bmatrix} A - \lambda_i FC - sI & B_f - \lambda_i FD_f \\ \lambda_i C & \lambda_i D_f \end{bmatrix} = \text{Rank}\begin{bmatrix} A - sI & B_f \\ C & D_f \end{bmatrix} < n + n_f, \qquad (7.14)$$

and/or

$$\text{Rank}[-sI + A + B_u K] < n. \qquad (7.15)$$

Equation (7.14) introduces the transmission zeros $\mathscr{Z}_1$ and (7.15) specifies that the eigenvalues of $A + B_u K$ are a subset of the set $\mathscr{Z}_2$. This completes the proof of the theorem. $\qquad\square$

Using Theorem 7.1, it can be concluded that the closed-loop faulty system $(A_{cl_i}, B_{fcl_i}, C_{cl_i}, D_{fcl_i})$ does not have any transmission zeros on the extended imaginary axis. This leads to detectability of all faults from the residuals.

Indeed, in this work, it is assumed that the open-loop system $(A, B_f, C, D_f)$ does not have any transmission zeros on the extended imaginary axis. On the other hand, since the controller gain $K$ that is designed in Theorem 7.2 is such that the closed-loop system (7.11) is stable, then all the eigenvalues of $A + B_u K$ are on the left-half imaginary plane. Hence, $A + B_u K$ and consequently $(A_{cl_i}, B_{fcl_i}, C_{cl_i}, D_{fcl_i})$ do not have any transmission zeros on the extended imaginary axis.

Since it is assumed that multiple faults cannot occur simultaneously in the team (this implies that faults do not "hide" effects of each other) and also the system $(A_{cl_i}, B_{fcl_i}, C_{cl_i}, D_{fcl_i})$ does not have any transmission zeros on the extended imaginary axis, the residual signal corresponding to each agent is affected by each fault that occurs in each agent or its neighbors. Moreover, since $(A, B_u, C)$ is stabilizable and detectable, the detector/controller module (7.3) for Requirement 1 [or the detector/controller module (7.4) for Requirement 2] can be obtained from the solution of the optimization problem (7.12) to detect not only each agent's faults but also faults of its nearest neighbors.

Stated more explicitly, considering the closed-loop system (7.5), it follows based on the rows of the network topology matrix $\tilde{L}$, each residual $r_i(t)$ is a function of the $i$th agent's fault and also the faults of its neighbors. On the other hand, the detector/controller parameters are designed such that the effects of faults on the residual outputs are maximized. Therefore, the $i$th residual (corresponding to the $i$th agent) is sensitive to the $i$th fault and also the possible faults of the neighbors of the $i$th agent. In view of the above, it can be concluded that our proposed scheme enables each agent to detect its own faults and faults of its neighboring agents.

In the next section, the LMI solution of the proposed IFDCC problem is obtained corresponding to both requirements.

## 7.4  LMI formulation and design for the IFDCC problem

There are four performance indices (i)–(iv) (for both requirements) that must be satisfied simultaneously for solving the distributed IFDCC problem of the decomposed system (7.11). The main result of this section is now provided in Theorem 7.2 where a feasible solution to the IFDCC problem is obtained by considering all the indices simultaneously.

Since the LMI conditions for solving the optimization problem (7.12) involve product of Lyapunov matrices with the system state-space matrices, by selecting identical Lyapunov matrices between all the conditions one would obtain a conservative solution. Consequently, to avoid this problem, we will employ the projection lemma (Lemma 1.1) to reduce the conservativeness of the IFDCC design solution by introducing additional matrix variables $(X_1, X_2)$, to avoid the coupling of the Lyapunov matrices with the system state-space matrices. The main result of this work is now stated below.

**Theorem 7.2.** *Consider the decomposed system (7.11) for both requirements and let $\theta$, $\rho_1$, $\rho_2$, $\rho_3$, and $\rho_4$ be given positive constants. The governing system (7.11) is stable and the performance indices (i)–(iv) are guaranteed, if there exist positive scalars $\gamma_1$, $\gamma_2$, $\gamma_3$, $\beta$, positive definite symmetric matrices $P_{1i}$, $P_{2i}$, $P_{3i}$, $P_{4i}$, and matrices $X_1$, $X_2$, $\hat{X}_2$, $M_1$, $M_2$, such that for $i = 1, \ldots, N - 1$ in Requirement 1 and for $i = 1, \ldots, N$ in Requirement 2 the following optimization problem is solved for $k = 1, 2, 3, 4$:*

$$\underset{P_{ki}, X_1, X_2, \hat{X}_2, M_1, M_2}{\text{minimize}} \quad \rho_1\gamma_1 + \rho_2\gamma_2 + \rho_3\gamma_3 - \rho_4\beta$$

subject to:

$$\begin{bmatrix} Herm(\Pi_i) + \Xi_{1i} & \Xi_{21i} & \Psi_i + \Xi_{3i}D_d \\ * & -\theta(X + X^T) & \theta\Psi_i \\ * & * & \lambda_i^2 D_d^T D_d - \gamma_1^2 I \end{bmatrix} < 0,$$

$$\begin{bmatrix} Herm(\Pi_i) - \Xi_{1i} & \Xi_{22i} & \Phi_i - \Xi_{3i}D_f \\ * & -\theta(X + X^T) & \theta\Phi_i \\ * & * & \beta^2 I - \lambda_i^2 D_f^T D_f \end{bmatrix} < 0, \qquad (7.16)$$

$$\begin{bmatrix} Herm(\Pi_i) + \Xi_4 & \Xi_{23i} & \Psi_i \\ * & -\theta(X + X^T) & \theta\Psi_i \\ * & * & -\gamma_2^2 I \end{bmatrix} < 0,$$

$$\begin{bmatrix} Herm(\Pi_i) + \Xi_4 & \Xi_{24i} & \Phi_i \\ * & -\theta(X + X^T) & \theta\Phi_i \\ * & * & -\gamma_3^2 I \end{bmatrix} < 0,$$

$$B_u^T X_2 = \hat{X}_2 B_u^T,$$

*where* $\Xi_{1i} = \lambda_i^2 \begin{bmatrix} C & 0 \end{bmatrix}^T \begin{bmatrix} C & 0 \end{bmatrix}$, $\Xi_4 = \begin{bmatrix} 0 & \Upsilon \end{bmatrix}^T \begin{bmatrix} 0 & \Upsilon \end{bmatrix}$, $\Xi_{3i} = \lambda_i^2 \begin{bmatrix} C & 0 \end{bmatrix}^T$, $\Xi_{2ki} = P_{ki} + \theta\Pi_i - X^T$, $X = diag(X_1, X_2)$, *and*

$$\Pi_i = \begin{bmatrix} A^T X_1 - \lambda_i C^T M_2^T & -M_1^T B_u^T \\ 0 & A^T X_2 + M_1^T B_u^T \end{bmatrix},$$

$$\Psi_i = \begin{bmatrix} X_1^T B_d - \lambda_i M_2 D_d \\ X_2^T B_d \end{bmatrix},$$

$$\Phi_i = \begin{bmatrix} X_1^T B_f - \lambda_i M_2 D_f \\ X_2^T B_f \end{bmatrix},$$

and $\Upsilon = I$ for Requirement 1 and $\Upsilon = C_2$ for Requirement 2. The IFDCC module gains $F$ and $K$ are now specified by $K = (M_1^T \hat{X}_2^{-1})^T$ and $F = (M_2^T X_1^{-1})^T$, respectively.

*Proof.* To guarantee the stability and the performance index (i) for the system (7.11), the condition $\dot{V}_i(t) < -\tilde{r}_i^T(t)\tilde{r}_i(t) + \gamma_1^2 \tilde{d}_i^T(t)\tilde{d}_i(t)$ must be satisfied where $V_i(t) = \tilde{x}_{cli}^T(t)P_{1i}\tilde{x}_{cli}(t)$ is a positive definite Lyapunov function candidate. From the stated condition and (7.11), the following inequality is obtained:

$$
\begin{bmatrix} I & 0 \\ A_{cl_i} & B_{dcl_i} \end{bmatrix}^T \begin{bmatrix} 0 & P_{1i} \\ P_{1i} & 0 \end{bmatrix} \begin{bmatrix} I & 0 \\ A_{cl_i} & B_{dcl_i} \end{bmatrix}
$$
$$
+ \begin{bmatrix} 0 & I \\ C_{cl_i} & D_{dcl_i} \end{bmatrix}^T \underbrace{\begin{bmatrix} -\gamma_1^2 I & 0 \\ 0 & I \end{bmatrix}}_{\Theta} \begin{bmatrix} 0 & I \\ C_{cl_i} & D_{dcl_i} \end{bmatrix} < 0. \tag{7.17}
$$

Now the inequality (7.17) can be reformulated as $N_U^T Z N_U < 0$, where $N_U$ and $Z$ are given by

$$
Z = \begin{bmatrix} C_{cl_i}^T C_{cl_i} & P_{1i} & C_{cl_i}^T D_{dcl_i} \\ * & 0 & 0 \\ * & * & \Gamma_{1i} \end{bmatrix}, \quad N_U = \begin{bmatrix} I & 0 \\ A_{cl_i} & B_{dcl_i} \\ 0 & I \end{bmatrix}, \tag{7.18}
$$

where $\Gamma_{1i} = D_{dcl_i}^T D_{dcl_i} - \gamma_1^2 I$. By defining the matrices $N_V$ and $V$ as follows:

$$
N_V = \begin{bmatrix} \theta I & -I & 0 \\ 0 & 0 & I \end{bmatrix}^T, \quad V = \begin{bmatrix} I & \theta I & 0 \end{bmatrix}, \tag{7.19}
$$

and by invoking the result in Lemma 1.1, it can be concluded that the inequality $N_U^T Z N_U < 0$ is equivalent to

$$
Z + \mathrm{Herm}\left([A_{cl_i} \quad -I \quad B_{dcl_i}]^T [X \quad \theta X \quad 0]\right) < 0. \tag{7.20}
$$

By partitioning $X$ into $X = \mathrm{diag}(X_1, X_2)$, $X_i \in \mathbb{R}^{n \times n}$, $i = 1, 2$ and by using (7.18) and (7.20), the following inequality is obtained:

$$
\begin{bmatrix} \mathrm{Herm}(\varphi_1) + \Xi_{1i} & P_{1i} + \theta\varphi_1 - X^T & \varphi_2 + \Theta_i D_d \\ * & -\theta(X + X^T) & \theta\varphi_2 \\ * & * & \lambda_i^2 D_d^T D_d - \gamma_1^2 I \end{bmatrix} < 0, \tag{7.21}
$$

where

$$
\varphi_1 = \begin{bmatrix} A^T X_1 - \lambda_i C^T F^T X_1 & -K^T B_u^T X_2 \\ 0 & A^T X_2 + K^T B_u^T X_2 \end{bmatrix},
$$

$$
\varphi_2 = \begin{bmatrix} X_1^T B_d - \lambda_i X_1^T F D_d \\ X_2^T B_d \end{bmatrix}.
$$

Since the inequality (7.21) is not in the form of an LMI, therefore we employ the equality constraint [252] $B_u^T X_2 = \hat{X}_2 B_u^T$ and substitute $M_1^T = K \hat{X}_2$ and $M_2^T = F^T X_1$ into the inequality (7.21), so that the first inequality in (7.16) is obtained.

The proof of the performance index (ii) for the system (7.11) is similar to that of the performance index (i). Indeed by changing $\Theta$ and $Z$ in the proof above as follows:

$$Z = \begin{bmatrix} -C_{cl_i}^{\mathrm{T}} C_{cl_i} & P_{2i} & -C_{cl_i}^{\mathrm{T}} D_{fcl_i} \\ * & 0 & 0 \\ * & * & \Gamma_{2i} \end{bmatrix},$$

$$\Theta = \begin{bmatrix} \beta^2 I & 0 \\ 0 & -I \end{bmatrix},$$

(7.22)

where $\Gamma_{2i} = \beta^2 I - D_{fcl_i}^{\mathrm{T}} D_{fcl_i}$, the second inequality in (7.16) is satisfied.

The performance indices (iii) and (iv) can also be shown by similarly employing the same techniques as in the derivations of the performance indices (i) and (ii). Therefore, these details are omitted for sake of brevity. This completes the proof of the theorem.  □

Note that the IFDCC module design gains, for both requirements are obtained from Theorem 7.2. Indeed, according to the designer specifications, Theorem 7.2 for each of the requirements should be solved separately. In other words, by selecting, for example, Requirement 1, one ends up with achieving the FD and consensus objectives simultaneously. However, by selecting Requirement 2, one ends up with achieving the FD and MRC objectives simultaneously. Therefore, Theorem 7.2 for each of these requirements should be solved independently, which leads to different design gains (different $K$ and $F$ gains for the two requirements).

**Remark 7.2.** *The multiagent network topology has a direct impact on the performance of the IFDCC design problem. Specifically, according to the derivations in the proof of Theorem 7.2, we have*

$$\lambda_i^2 D_d^{\mathrm{T}} D_d - \gamma_1^2 I < 0, \quad \beta^2 I - \lambda_i^2 D_f^{\mathrm{T}} D_f < 0.$$

*On the other hand, to obtain residuals that have high sensitivities to the faults, $\beta^2$ must be sufficiently large, and to obtain residuals that have high robustness to disturbances, $\gamma_1^2$ must be sufficiently small. Since $\lambda_1$ and $\lambda_{N-1}$ ($\lambda_N$) for Requirement 1 (Requirement 2) are the minimum and the maximum eigenvalues of $\tilde{L}$, respectively, it follows clearly that the multiagent network topology has a direct impact on the performance of the IFDCC filter design solution.*

## 7.4.1  Residual evaluation criterion

Following the construction of the residuals $r_i(t) \in \mathbb{R}^{n_y}, \forall i \in \{1, \ldots, N\}$, the final step in the IFDCC strategy is to determine the thresholds $J_{\mathrm{th}_i}$ and the evaluation functions $J_{r_i}(t)$ for developing a formal FD decision logic. Various evaluation functions can be considered as described in [30].

The upper and lower threshold values are selected as $J_{\mathrm{th}_i}^u = \sup_{f_i=0, d_i \in \mathfrak{D}} r_i(t)$ and $J_{\mathrm{th}_i}^l = \inf_{f_i=0, d_i \in \mathfrak{D}} r_i(t)$, respectively, where $d_i$ represents the disturbance and $\mathfrak{D}$ denotes the set of all allowable disturbances. For instance, $\mathfrak{D} = \mathcal{L}_2$ or $\mathfrak{D}$ can be selected as a set of Gaussian white noise. Based on the selected thresholds and the evaluation function taken as $J_{r_i}(t) = r_i(t)$, the occurrence of a fault can then be detected by using the following decision logic: if $r_i(t) > J_{\mathrm{th}_i}^u$ or $r_i(t) < J_{\mathrm{th}_i}^l \implies f_i \neq 0$ or $f_j \neq 0, j \in \mathcal{N}_i$, that is the $i$th agent or one of its neighbors is faulty.

## 7.4.2 *Distributed fault isolation*

In a network of multiagent systems where agents communicate with one another through the network topology $\mathfrak{G}$ with the structure of $\tilde{\mathcal{N}}_i \neq \tilde{\mathcal{N}}_j$, for all $i, j \in \mathcal{V}$, where $\tilde{\mathcal{N}}_i = \{i, \mathcal{N}_i\}$, our methodology is capable of isolating the faulty agent in the team by using the so-called flags $\varepsilon_i$, $i \in \mathcal{V}$ (where the $i$th flag is 1 if $r_i(t) > J^u_{\text{th}_i}$ or $r_i(t) < J^l_{\text{th}_i}$, and 0 otherwise).

Indeed, it is assumed that the neighboring agents share their flags with each other and each agent constructs the defined fault pattern $\Upsilon_i = \{\varepsilon_j : j \in \tilde{\mathcal{N}}_i\}$, $i \in \mathcal{V}$. Finally, the faulty agent can be isolated according to the following decision logic: if all the elements of $\Upsilon_i$ are equal to one then the $i$th agent is faulty.

## 7.5 Case study

To illustrate the effectiveness and capabilities of our proposed methodology and solution for Requirements 1 and 2, two case studies corresponding to two teams of AUVs are given in this section.

### 7.5.1 *Case study 1 (corresponding to Requirement 1)*

Consider a team of four agents that are communicating with one another according to the graph $\mathfrak{G}$ with $\mathcal{N}_1 = \{2, 3\}$, $\mathcal{N}_2 = \{1, 4\}$, $\mathcal{N}_3 = \{1, 4\}$, and $\mathcal{N}_4 = \{2, 3\}$ as shown in Figure 7.2. Each agent is considered as a speed subsystem of the Subzero III AUV [261] with the following transfer function:

$$T(s) = \frac{0.002864}{(s + 0.439)(s + 4.195)}.$$

The system matrices $A$, $B_u$, and $C$ are obtained from the above transfer function. The fault matrices $B_f$ and $D_f$ are considered as $B_f = [0 \ 0]^T$ and $D_f = 1$ for the case of sensor faults. Moreover, the disturbance matrices $B_d$ and $D_d$ are selected as $B_d = [0.2 \ 0.5]^T$ and $D_d = 0.1$. In the simulations conducted, and without loss of generality, the disturbance vector $d(t)$ is set to $d(t) = [0.05w(t) \, 0.01w(t) \, 0.05w(t) \, 0.03w(t)]^T$, where $w(t)$ is an energy-limited white noise.

The controlled output function $z_{sc_i}(t)$ is defined as $z_{sc_i}(t) = x_i(t) - (1/4)\sum_{j=1}^{4} x_j(t), i = 1, 2, 3, 4$. The positive constant weights $\rho_1$, $\rho_2$, $\rho_3$, and $\rho_4$ are selected to be equal to one, which implies the same level of importance or priority in the diagnosis and control objectives. For a given $\theta = 0.05$, the optimization problem of Theorem 7.2 is solved, and the parameters $\gamma_1$, $\gamma_2$, $\gamma_3$, and $\beta$ are numerically obtained as $\gamma_1 = 0.267$, $\gamma_2 = 0.2281$, $\gamma_3 = 1.8e - 04$, and

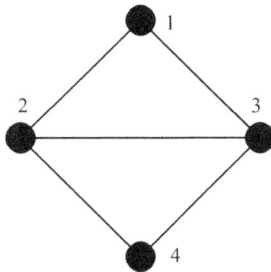

*Figure 7.2   The information graph $\mathfrak{G}$*

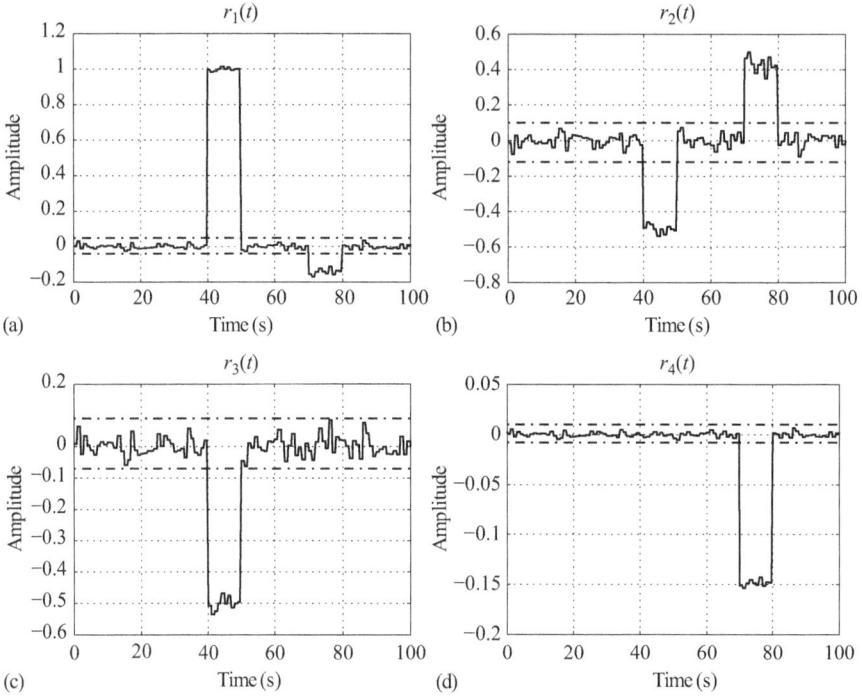

*Figure 7.3    The residuals $r_i(t)$, $i = 1, \ldots, 4$ generated for (a) agent 1, (b) agent 2, (c) agent 3, and (d) agent 4, where the solid lines denote the residual signals and the dash-dot lines denote the residual upper and lower thresholds*

$\beta = 3.99$, respectively. Moreover, the IFDCC module gains $F$ and $K$ are obtained as $F = [0.0153 \ -0.002]^T$ and $K = [-0.0061 \ 0.0059]$, respectively.

In the following, we demonstrate the performance capabilities of our proposed distributed IFDCC methodology through considering different faults with various severities. Indeed, it is assumed that multiple abrupt sensor faults $f_1(t)$ and $f_2(t)$ occur in agents 1 and 2, respectively. The fault $f_1(t)$ is simulated as a rectangular pulsed signal with an amplitude of 0.5 that has occurred during the time interval 40–50 s. The fault $f_2(t)$ is simulated as a rectangular pulsed signal with an amplitude of 0.15 that has occurred during the time interval 70–80 s. Moreover, it is assumed that there are no faults in agents 3 and 4.

The initial states of the agents are selected as $x_1(0) = [2.5 \ 6]^T$, $x_2(0) = [0.5 \ 2]^T$, $x_3(0) = [0.2 \ 1]^T$, and $x_4(0) = [0 \ 0]^T$.

It should be noted that by considering the worst case analysis of the residuals corresponding to the healthy operation of the team being subjected to various noise and disturbances, the threshold values $J_{th}^u = [-0.04 \ -0.12 \ -0.07 \ -0.008]^T$ and $J_{th}^l = [0.05 \ 0.1 \ 0.09 \ 0.01]^T$ are numerically selected.

The residual signals that are generated for all the agents are shown in Figure 7.3. From Figure 7.3, it can be concluded that the IFDCC scheme is robust against the disturbances, and moreover, the fault sensitivities are enhanced and faults are well discriminated from the external disturbances. This illustrates that our proposed IFDCC method can successfully be

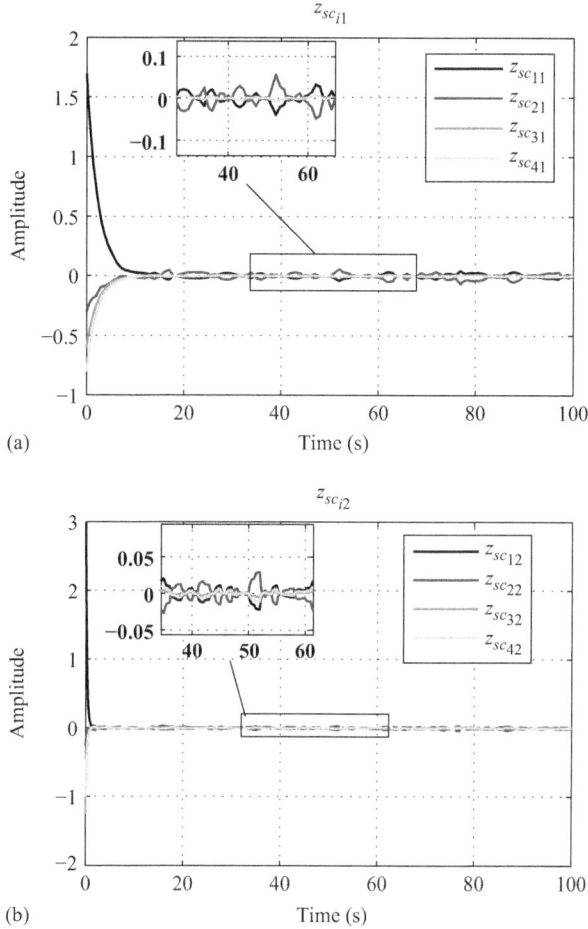

*Figure 7.4* *The state error trajectories of the four AUV agents closed-loop system with the same level of importance in the diagnosis and control objectives: (a) the first states of $z_{sc_i}$ and (b) the second states of $z_{sc_i}$*

used for FD in a network of multiagent systems, where each agent not only detects its own faults but also faults of its nearest neighboring agents.

This is confirmed by observing the residual $r_1(t)$ which confirms the presence of the fault in agent 1 as well the residuals $r_2(t)$ and $r_3(t)$ which are the neighbors of agent 1 confirming the presence of a fault in agent 1. Also by observing the residual $r_2(t)$, the presence of the fault in agent 2 is confirmed. This fault is also confirmed by inspecting the residuals $r_1(t)$ and $r_4(t)$ associated with agents 1 and 4 that are the nearest neighbors to agent 2.

Given that the required isolation condition ($\mathcal{N}_i \neq \mathcal{N}_j$) for the network topology is satisfied, our methodology is capable of isolating the faulty agent in the team. For the fault signal $f_1(t)$, the fault patterns $\Upsilon_i$, $i = 1, 2, 3, 4$, are obtained as $\Upsilon_1 = \{1, 1, 1\}$, $\Upsilon_2 = \{1, 1, 0\}$, $\Upsilon_3 = \{1, 1, 0\}$, and $\Upsilon_4 = \{0, 1, 1\}$. It is obvious that all elements of $\Upsilon_1$ are equal to one, and

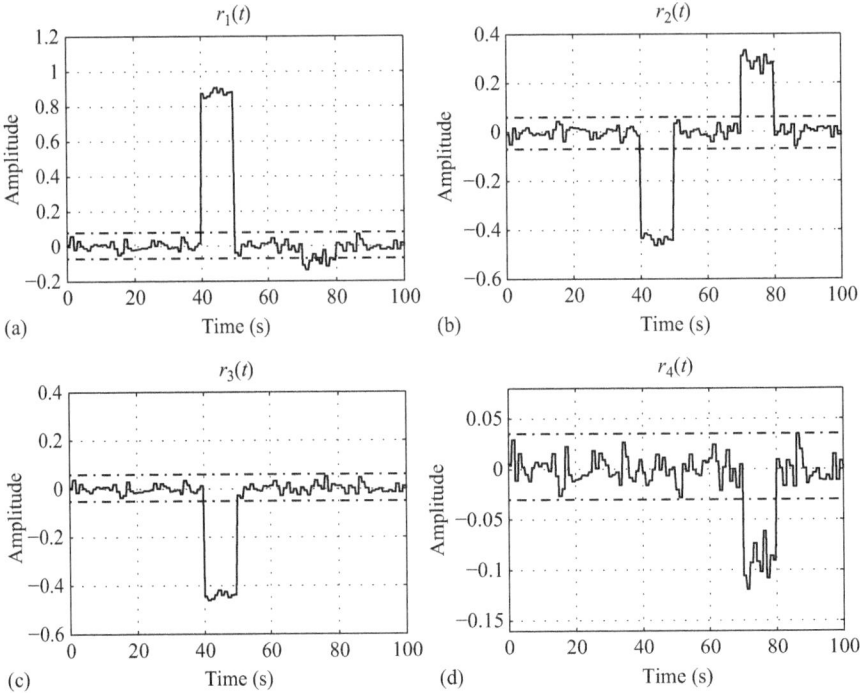

*Figure 7.5*   *The residuals $r_i(t)$, $i = 1, \ldots, 4$ generated for: (a) agent 1, (b) agent 2, (c) agent 3, and (d) agent 4, where the solid lines denote the residual signals and the dash-dot lines denote the residual upper and lower thresholds*

therefore, the first agent is faulty. Moreover, the fault patterns for the fault signal $f_2(t)$ are obtained as $\Upsilon_1 = \{1, 1, 0\}$, $\Upsilon_2 = \{1, 1, 1\}$, $\Upsilon_3 = \{0, 1, 1\}$, and $\Upsilon_4 = \{1, 1, 0\}$. This confirms that the second agent is faulty.

Figure 7.4 depicts the state error trajectories ($z_{sc_i}(t)$). Clearly, it can be observed that the SC is achieved, and this validates the effectiveness and capabilities of our proposed IFDCC methodology.

To illustrate the trade-offs between the control and the FD objectives another set of simulations are conducted. In these simulations, the values of $\rho_k$, $k = 1, 2, 3, 4$, are selected as $\rho_1 = 0.1$, $\rho_2 = 0.1$, $\rho_3 = 10$, and $\rho_4 = 10$, which indicate that the design interest and priority is in more control performance than in the diagnosis performance. The performance indices for this case are obtained as $\gamma_1 = 0.539$, $\gamma_2 = 0.179$, $\gamma_3 = 1.1e - 4$, and $\beta = 3.017$.

The residual signals that are generated for all the agents and the state error trajectories ($z_{sc_i}(t)$) are shown in Figures 7.5 and 7.6, respectively. From Figures 7.5 and 7.6 and by comparing them with Figures 7.3 and 7.4, it follows readily that in these simulations the diagnosis and control objectives have been degraded and improved, respectively. Indeed, by comparing Figures 7.3 and 7.5, it follows that higher sensitivity of the residuals to the fault signals is present as compared to the previous simulation results.

Moreover, from Figure 7.5(a), it can be concluded that the fault $f_2(t)$ cannot be effectively detected by using the first agent in these simulation results. On the other hand, by comparing

(a)

(b)

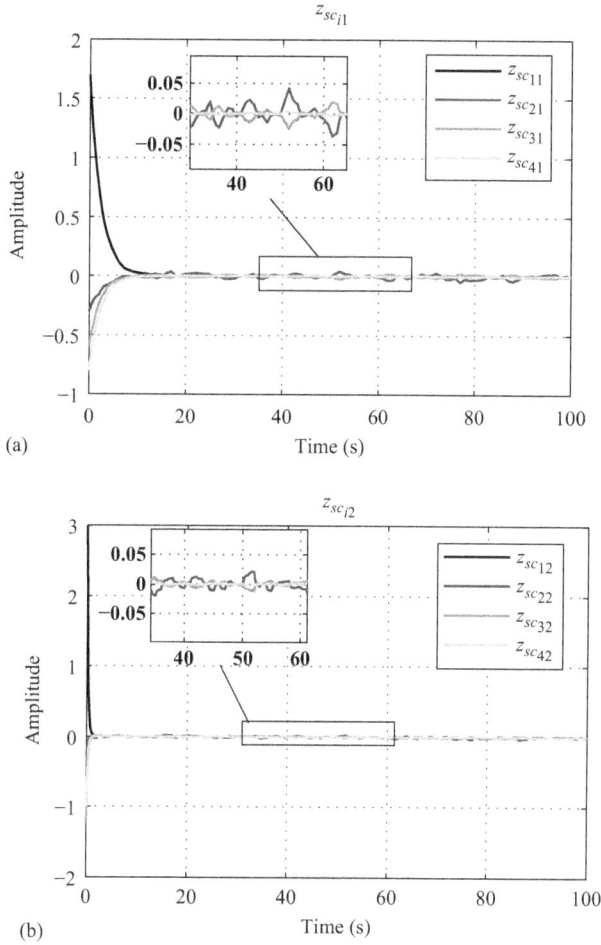

*Figure 7.6* *The state error trajectories of the four AUV agents closed-loop system with more level of importance in the control objective than in the diagnosis objective: (a) the first states of $z_{sc_i}$ and (b) the second states of $z_{sc_i}$*

Figures 7.4 and 7.6, it can be concluded that the effects of disturbances have been more attenuated in these simulation results in comparison with the previous results.

Below we provide comparisons with other schemes in order to illustrate the effectiveness and advantages of our proposed design methodology.

**Comparison with a methodology with a common Lyapunov function**

If we instead apply a common Lyapunov approach (e.g., as in [43]) which uses the same Lyapunov matrices that are coupled with the system matrices, then $\beta$ is obtained as 1.501 for a given $\gamma_1 = 0.267$, $\gamma_2 = 0.2281$, and $\gamma_3 = 1.8e - 04$. Note that by using our approach a larger value of $\beta$ is obtained, namely, $\beta = 3.99$, implying a higher sensitivity of the residual generator to the fault signal. Consequently, by considering the same Lyapunov functions for all the performance indices (i)–(iv), one does indeed impose a more conservativeness in the design solution.

**Comparison with a centralized design**

It is also possible to solve this problem by using the centralized methodology. However, as stated in Remark 7.1 the order of the complexity of the solution is 16/3 times higher than that of our distributed methodology.

**Comparison with a separate design**

Note that our integrated methodology leads to a lower computational complexity in comparison with an alternative separate design. Indeed, in case of a separate design, the order of the resulting detector/controller module is $2n$ for each agent. However, our methodology yields detector/controller modules of dimension $n$ for each agent. The higher dimensionality in the separate design is due to the necessity of including two observers in the controller/detector modules.

## 7.5.2  Case study 2 (corresponding to Requirement 2)

To illustrate the effectiveness of our proposed methodology for Requirement 2, which is the MRC problem, our IFDCC strategy is applied to the problem of longitudinal motion of a team of AUVs. For simulation studies, the relative motion of four AUVs (corresponding to the information graph of Figure 7.2) are considered.

The governing equation corresponding to the dynamics of the longitudinal speed $u$ for the $i$th AUV is given as follows [367]:

$$\dot{u} = -0.1148u^2 + 0.9603\mathcal{Q}_n - 0.0519u\sqrt{\mathcal{Q}_n} + 3.209 \times 10^{-5}\frac{u^3}{\sqrt{\mathcal{Q}_n}}, \tag{7.23}$$

where $\mathcal{Q}_n$ denotes the propeller control. In the above equation, it should be noted that the main dynamics is due to the term $-0.1148u^2$, and the main control is given by the term $0.9603\mathcal{Q}_n$. Furthermore, the other nonlinear terms can be neglected. Refer to [367] for more information on the AUV dynamics.

Therefore, the control-based model of the longitudinal speed can be considered as

$$\dot{u} = \Gamma u^2 + \Lambda\mathcal{Q}_n, \tag{7.24}$$

where $\Gamma = -0.1148$ and $\Lambda = 0.9603$. Equation (7.24) is nonlinear due to the $u^2$ term. Since our IFDCC design methodology is applicable to a team of linear multiagent systems, therefore a nonlinear feedback linearization compensation is applied to the system (7.24) in order to obtain an equivalent linear model. Note that the goal of this compensation is to cancel out the nonlinearity in (7.24) by acting on the control $\mathcal{Q}_n$.

A nonlinear compensation can be selected of the form [367]:

$$\mathcal{Q}_n = \Lambda^{-1}(-\Gamma u^2 + Au + BW), \tag{7.25}$$

where $W$ is now to be considered as the new control input. By substituting the nonlinear compensation (7.25) into (7.24), the feedback linearized model is obtained as follows:

$$\dot{u} = Au + BW,$$

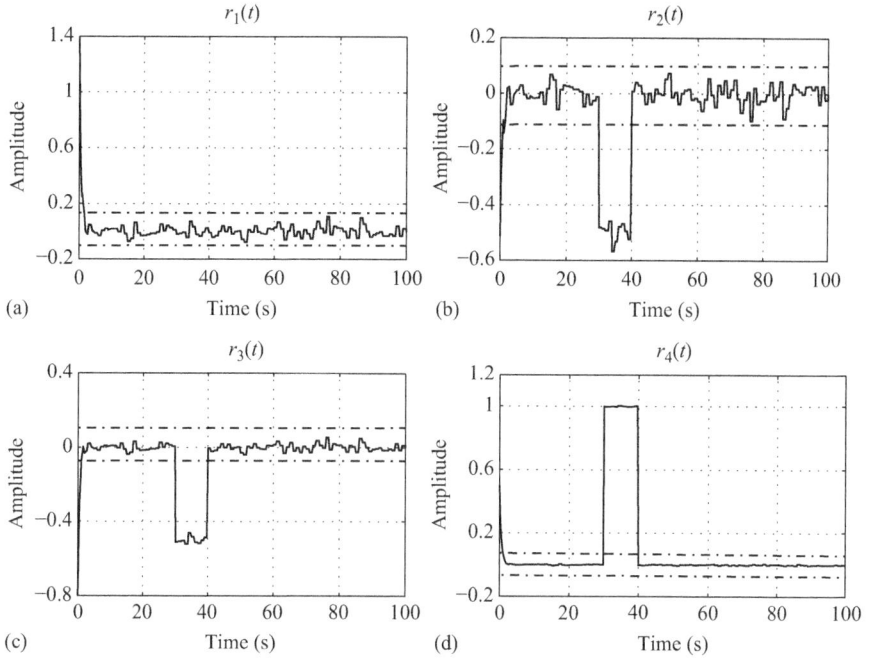

*Figure 7.7   The residuals $r_i(t)$, $i = 1, \ldots, 4$ generated for (a) agent 1, (b) agent 2,*
*(c) agent 3, and (d) agent 4, where the solid lines denote the residual*
*signals and the dash-dot lines denote the residual upper and lower*
*thresholds*

where $A$ is the new desired pole and $B = -A$ in order to have a normalized system. In the simulations conducted below, and without loss of generality, $A$ is selected as $A = -2$.

The disturbance coefficients $B_d, D_d$ are selected as $B_d = 0$ and $D_d = 0.1$, which represent the occurrence of sensor noise in the AUV (as common in industrial applications). The fault coefficients $B_f$ and $D_f$ are selected as $B_f = 0$ and $D_f = 1$, which represent the occurrence of a sensor fault in the system.

In simulations conducted, the disturbances $d_i(t)$ are considered to be the same as in case study 1. The fault signal $f_4(t)$ is simulated as a rectangular pulsed signal with an amplitude of 0.5 that is present during the time interval 30–40 s. Moreover, it is assumed that there are no faults in agents 1, 2, and 3. The initial state of agents 1, 2, 3, and 4 are selected as $u_{1_0} = 0.5$, $u_{2_0} = 0.1$, $u_{3_0} = 0$, and $u_{4_0} = 0.3$, respectively. Moreover, the reference input signal $u_r(t)$ is selected as $u_r(t) = 0.5\sin(0.05t)$.

It is assumed that only the first agent has access to the model reference output $y_r$ ($q_1 = 1$, $q_2 = q_3 = q_4 = 0$). The eigenvalues of the matrix $L_2$ are obtained as $\lambda_1 = 0.1864$, $\lambda_2 = 2.4707$, $\lambda_3 = 4$, and $\lambda_4 = 4.3429$, respectively. For a given $\theta = 0.05$, and $\rho_1 = \rho_2 = \rho_3 = \rho_4 = 1$, the optimization problem of Theorem 7.2 is solved and the optimization parameters $\gamma_1$, $\gamma_2$, $\gamma_3$, and $\beta$ are obtained as $\gamma_1 = 0.15$, $\gamma_2 = 0.15$, $\gamma_3 = 0.1887$, and $\beta = 0.0346$, respectively. The IFDCC module gains $K$ and $F$ are obtained as $K = -0.2043$ and $F = 0.003$, respectively.

The residual signals for agents 1, 2, 3, and 4 are shown in Figure 7.7. From Figure 7.7, it can be observed that the fault signal can be readily distinguished from the disturbances in $r_4(t)$. Moreover, it follows that agent 4 and its neighbors (agents 2 and 3) all detect the fault $f_4(t)$,

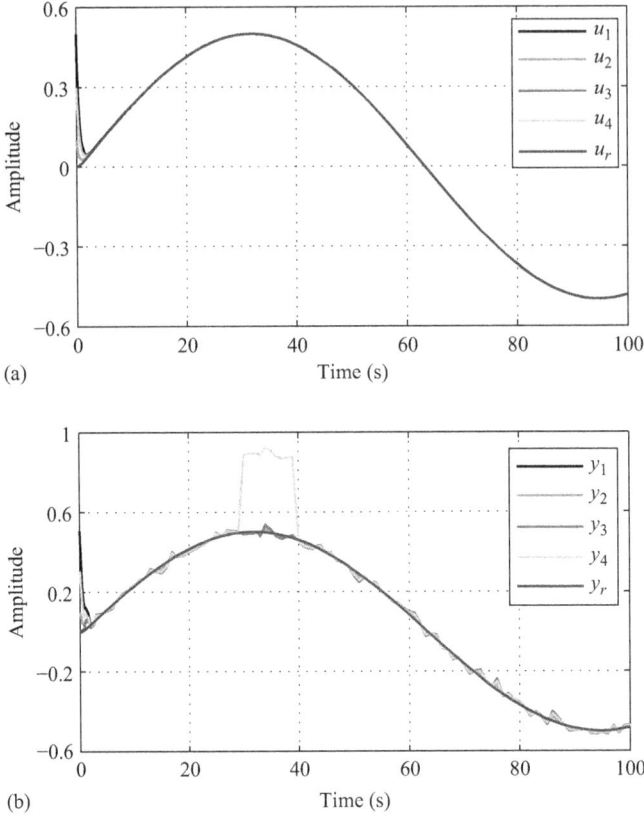

*Figure 7.8* *The states ($u_i(t)$, $u_r(t)$) and outputs ($y_i(t)$, $y_r(t)$) of the agents ($i = 1, 2, 3, 4$) and the reference model*

since $r_2(t)$ and $r_3(t)$ have exceeded their thresholds decisively. Hence, each agent can detect its own faults as well as faults of its nearest neighboring agents. This illustrates that our proposed IFDCC method does indeed satisfy the FD requirements of the network of multiagent AUV systems. Figure 7.8 depicts the states and outputs of agents 1, 2, 3, 4 and the reference model. It can be concluded that the MRC is achieved, which validates the effectiveness of our proposed IFDCC methodology.

## 7.6 Conclusion

In this chapter, we have developed and presented a robust distributed IFDCC problem for a network of multiagent systems using state estimation and relative output information. Using the proposed distributed structure not only each agent's faults but also faults of its nearest neighbors can be detected and isolated. Moreover, the IFDCC problem was solved such that all agents can achieve either SC or MRC. Two case studies corresponding to two teams of AUVs were also presented to demonstrate and illustrate the effectiveness and capabilities of our proposed approach.

## *Chapter 8*
# Perspectives and future directions of research

In this book, our interest has been on design and analysis of integrated fault detection and control (IFDC) strategies for complex linear systems. In order to develop safe and reliable complex systems, fault detection and isolation (FDI) schemes should be employed that are capable of detecting and isolating faults in different parts of these systems. On the other hand, one of the main challenges that one is faced with after detecting and isolating the undesirable anomalies and faults in these systems is to develop fault-tolerant control (FTC) strategies that can ensure and maintain the system stability and performance in presence of external disturbances and faults.

The considered problem in this book, namely, the IFDC design problem, is timely and important due to the fact that there is always a conflict between FDI and FTC design objectives, and an appropriate trade-off must be made between them. The approaches and frameworks that are proposed and presented here are based on different types of robust state observers and filters that lead to less computational complexity and conservativeness as compared to the separate design of FDI and FTC modules. We have formulated and introduced several IFDC problems within the domains of linear time-invariant (LTI), Markovian jump, multiagent, and networked control systems (NCSs) where novel results are obtained.

In Chapters 3–7, various issues in the design of IFDC algorithms for complex linear systems were tackled. These include issues such as robustness with respect to external disturbances, compensations for the over-provisioning of the computational and communication resources, and different IFDC architectures from centralized to distributed, from deterministic to stochastic, and from time-triggered to event-triggered cases and scenarios. Clearly, there is more work that needs to be done to have a complete theory on IFDC design for complex linear systems. In this chapter, first a brief summary of each chapter is provided and a couple of major open problems in each area are identified.

## 8.1 Future research directions

In this section, the major research areas that are addressed in this book are first summarized and some of the open problems in each area are identified.

### 8.1.1    A single dynamic observer-based module for design of integrated fault detection, isolation, and tracking control scheme

In Chapter 3, the problem of integrated fault detection, isolation, and tracking (IFDIT) control design was considered for linear systems subject to both bounded energy and bounded peak disturbances. An $H_\infty/H_-/L_1$ formulation of the problem was developed and presented by using a dynamic observer. In essence, a single dynamic observer module was designed that generates the residuals as well as the control signals.

The objective of the IFDIT module is to ensure that simultaneously the effects of disturbances and control signals on the residual signals are minimized (in order to accomplish the fault detection goal) subject to the constraints that the transfer matrix from the faults to the residuals is equal to a preassigned diagonal transfer matrix (in order to accomplish the fault isolation goal), while the effects of disturbances, reference inputs and faults on the specified control outputs are minimized (in order to accomplish the fault-tolerant and tracking control goals).

Therefore, occurrence of simultaneous faults in the system can also be handled. A set of linear matrix inequality (LMI) feasibility conditions were derived to ensure solvability of the problem. The extended LMI approach was employed for reducing the conservativeness in the problem solution by introducing additional matrix variables to eliminate the couplings of the Lyapunov matrices with the system matrices. Application of the methodology to a linearized longitudinal model of the Subzero II autonomous underwater vehicle (AUV) was also presented to demonstrate and illustrate the effectiveness of the proposed diagnostic and control approach.

The major directions for future research that can be pursued in this area are as follows:

- Development of IFDIT algorithms by taking into account modeling errors and unmodeled dynamics.
- In this chapter, the minimum and maximum values of the evaluation function under the fault free case were taken as threshold values. This is a common and widely used method for choosing the thresholds. However, when the threshold is fixed (static), the detection time (the time between the fault occurrence and the fault declaration) could become high.

  On the other hand, if the threshold is set to a too high value, the fault detection sensitivity will reduce, whereas if the threshold is set to a too low value, false alarm rates will increase. Consequently, dynamic and adaptive thresholds that vary according to changes in the inputs to the system have received much attention in the past few years [29,30,33,34]. This extension is left as a topic of our future research.
- Given that our proposed IFDIT methodology is a passive FTC methodology, the fault detection information is not explicitly used in the controller reconfiguration design. However, the generated diagnostic information could be used subsequently in the framework of design of active FTC systems. This problem is also left as a topic of our future research.

### 8.1.2 Integrated design of fault detection, isolation, and control for continuous-time Markovian jump systems

In Chapter 4, the integrated fault detection, isolation, and control (IFDIC) problem for a continuous-time Markovian jump linear system (MJLS) with uncertain transition probabilities was considered. In essence, the IFDIC problem was formulated as a multiobjective $H_\infty/H_-/L_1$ problem for a continuous-time MJLS. An LMI approach for solving the IFDIC design was introduced, which in addition to mean square stabilizing the closed-loop system, simultaneously guarantees fault detection, isolation, and control objectives. LMI conditions were presented so that no products of Lyapunov matrices and system state-space matrices were involved.

This results in a significant reduction in the conservatism of the IFDIC solution. Moreover, the IFDIC problem was solved so that each element of the residual vector is only sensitive to a specified fault, and therefore, occurrence of simultaneous faults in the system can also be handled and resolved. Application of our methodology to a linearized model of the GE F-404 aircraft engine was also presented to demonstrate and illustrate the capabilities and effectiveness of the proposed approach.

The major directions for future research that can be pursued in this area are as follows:

- Development of the IFDIC algorithms for discrete-time Markovian jump systems.
- Extending the results of this chapter to semi-Markovian jump systems [368], which are more general than Markovian jump systems for modeling practical systems.
- The equality constraint that was used in the proofs of Theorems 4.1 and 4.2, which may lead to some conservativeness in the methodology, could be relaxed by considering other types of filters instead of the classical Luenberger observers in the structure of the IFDIC module.

### 8.1.3 Event-triggered multiobjective control and fault diagnosis: a unified framework

In Chapter 5, a new LMI approach to study the problems of event-triggered multiobjective synthesis of feedback controllers and fault diagnosis filters in a unified framework was developed. Toward this end, a mixed generalized $H_2/H_\infty/H_-/l_1$ formulation of an event-triggered integrated fault detection, isolation, and control (E-IFDIC) problem for LTI systems with reduced communication in the sensor and the actuator networks has been developed and designed. An LMI approach for codesign of the IFDIC problem and event-triggered conditions was also introduced, that in addition to stabilizing the closed-loop system, simultaneously guarantees the fault detection, isolation, and control objectives.

Moreover, it was shown that a number of existing problems in the fields of time and event-trigged control and fault diagnosis can be considered as special cases of the proposed methodology. The effectiveness and capabilities of the proposed design methodology were illustrated by using two industrial case studies and the results were compared with available work in the literature.

The major directions for future research that can be pursued in this area are as follows:

- Applying the recently developed dynamic triggering strategy [369,370], instead of the proposed static triggering mechanism in this chapter to exceed the capabilities of static event-triggering conditions to accomplish further data transmission reductions.
- Extension of the proposed E-IFDIC problem to continuous-time NCSs.

### 8.1.4  Event-triggered fault estimation and accommodation design for linear systems

In Chapter 6, a novel event-triggered active fault-tolerant control (E-AFTC) mechanism for discrete-time linear systems subject to faults and external disturbances was investigated and proposed. To reduce the communication rates, an event-triggered condition was designed to determine whether the newly measured data should be transmitted or not. The proposed approach was achieved by designing the fault estimation and FTC modules and event-triggered conditions collectively via $H_\infty$ optimization technique by using a single-step LMI formulation.

The major directions for future research that can be pursued in this area are as follows:

- In this chapter, the rate of transmission is reduced by considering event detector at the sensor node. However, it is possible to extend these results to a more general framework by developing and constructing two event-triggered strategies at both the sensor and the FTC module nodes to reduce the communication load in the sensor-to-FTC module and the FTC module-to-actuator channels, respectively.
- Development of E-AFTC algorithms for nonlinear dynamical systems. This problem has a potential application to nonlinear NCSs.

### 8.1.5  Integrated fault detection and consensus control design for a network of multiagent systems

In Chapter 7, the problem of robust distributed integrated fault detection and consensus control (IFDCC) design for a network of multiagent systems was considered by using state estimation and relative output measurements. It was shown that by using the proposed distributed structure, not only each agent's faults but also faults of its nearest neighbors can be detected. Moreover, a mixed $H_\infty/H_-$ formulation of the IFDCC problem was presented and solved such that all the agents can achieve either state consensus or model reference consensus. By utilizing a decomposition methodology, it was shown that the computational complexity of the distributed IFDCC design is significantly lower than a centralized methodology. LMIs were used to design the IFDCC filter gains. Two case studies corresponding to two teams of AUVs were also presented to demonstrate and illustrate the effectiveness and capabilities of the proposed approach.

The major directions for future research that can be pursued in this area are as follows:

- Investigation of the effects of time-varying network topologies on the IFDCC scheme for multiagent systems.
- Development and design of the event-triggered simultaneous fault detection and consensus control problem for a network of multiagent systems.
- The proposed filters (7.3) and (7.4) for Requirements 1 and 2, respectively, are indeed "full-order" observers. However, it is conceivable to instead consider "reduced-order" observers so that the problem of distributed IFDCC design can be based on reduced-order observers. This problem may be solved by extending the existing results in the related work in reduced-order observer-based protocol design to consensus of linear multiagent systems as reported in [371].
- The proposed IFDCC methodology is not applicable to multiagent systems with $D_f = 0$. However, it is possible to address this problem by using the existing methodologies in the field of $H_\infty / H_-$ FDI design that consider linear systems with $D_f = 0$.

## 8.1.6   General future directions of research

In this section, we discuss a number of major research directions and open problems for future investigation that can be pursued by extending our proposed results to other complex systems subject to practical constraints that were not considered and addressed in this book.

### 8.1.6.1   IFDC design in presence of constraints in practical applications

Traditionally, fault diagnosis and control algorithms are generally designed based on implicit assumptions of unlimited computational resources, nondelayed sensing and actuation, unlimited bandwidth, and perfect communication environments. However, due to existence of inherent constraints in practical applications and real world, it is necessary to investigate the IFDC design problem in presence of a number of practical factors, such as communication constraints and actuator saturations. In the following, we briefly discuss the possible directions for future research in the area of IFDC design in presence of these practical constraints.

**A. Communication constraints**

Conventional IFDC systems deal with control and fault diagnosis issues in which the data communication links in the sensor-to-IFDC module and IFDIC module-to-actuator channels are considered to be ideal. Indeed, it is assumed that the sensor measurements (IFDC module outputs) are completely and exactly transmitted to the IFDC module (or actuator) due to point-to-point wired connections. However, due to presence of communication networks in today's modern and complex systems such as NCS and multiagent systems, this is not always a correct assumption. Indeed, different communication constraints such as time delays, packet dropouts, and quantization errors are usually occurring in these systems.

***Delay effect:*** It is well known that time delays exist in many applications and can degrade the stability and performance of the IFDC system significantly. The delays could be constant or time varying, uniform, or nonuniform [372]. Therefore, the problems of fault diagnosis and control design with time delays have attracted much attention in recent years.

***Intermittent communications and data packet dropouts:*** In practice, the autonomous agents may only communicate with each other at some disconnected time intervals because of the sensing ranges limitations, effects of obstacles, and failure of physical devices. As far as these practical situations are concerned, the study of intermittent interactions of networked systems has received considerable attention in the recent years [373]. Moreover, data packet dropouts inevitably occur due to network traffic congestion and system components or network links failures. Generally, the characteristics of the data packet dropouts can be categorized into random and deterministic types.

***Quantized communication:*** In NCSs, the information is usually exchanged over digital communication channels. Since digital channels are subject to bandwidth constraints and only finite number of bits of information can be transmitted along each channel; therefore, information transmitted among different parts of the system has to be quantized prior to being sent [374]. Quantization can reduce the quantity of data transmission, whereas it can increase the complexity of the system control. Consequently, the problems of control and fault diagnosis with quantized communication have attracted considerable attention over the past few years [375,376].

In [377], the problem of fault detection design for switched systems subject to quantization effects and packet dropout is studied. With respect to packet dropout, the original switched system has been modeled as a new switched system which switches by both the original switching signal and the case of packet dropout. The FD scheme has been converted into a robust filter design problem. In [378], the problem of stochastic fault detection filter design in finite-frequency domain for a class of NCSs with respect to signal quantization and data packet dropout is addressed by applying the MJLSs framework. Indeed, by considering a logarithmic quantizer and the Markovian packet dropout, the NCS is modeled as a MJLS with quantization error. The sufficient LMI conditions are derived to guarantee the fault detection performance in finite frequency domain for the MJLS.

The problem of IFDC design for switched systems with two quantized signals is studied in [52]. In [379], the FD design problem for networked discrete-time infinite-distributed delay systems with packet dropouts is studied. In the proposed methodology, both sensor-to-controller and controller-to-actuator packet dropouts are described by using two different Bernoulli distributed white sequences, respectively. In [49], the IFDC design problem for a class of uncertain linear systems with state time delay is considered. In [57], the IFDC design problem for networked-based discrete-time switched systems with time-varying delays under an arbitrary switching signal is addressed. In [380], the problem of IFDC design for Itô-type stochastic time-delay systems is studied. A full-order dynamic output feedback controller is designed to achieve the desired control and detection objectives.

**B. Saturation constraint**

It is well known that saturation nonlinearities are ubiquitous in physical and engineering systems, and therefore, most existing complex systems can be subject to input saturation [73,381].

The relationships between the input saturation and the detectability of additive faults in model-based fault diagnosis methods have been studied in [382]. It is shown that the detectability of additive faults decreases as the system becomes nonlinear due to presence of saturation. Moreover, it is demonstrated that this effect is greater for actuator faults as compared to the sensor faults. An integrated fault detection, diagnosis and reconfigurable control design method with explicit consideration of control input constraints is addressed in [383]. LMI conditions are presented to compute the design parameters using integrated design of reconfigurable controller and fault diagnosis module. The set invariance condition given in [384] is used to tackle the actuator saturation.

In [385], the problem of fault detection and accommodation design of saturated actuators for trajectory tracking of underactuated surface vessels in presence of nonlinear uncertainties is studied. A robust fault detection observer is proposed to detect the actuator faults in presence of nonlinear uncertainties and external disturbances. An adaptive fault accommodation scheme using the dynamic surface design is proposed to compensate for the detected faults. The work [386] considers the problem of integrated fault diagnosis and FTC of linear systems subject to actuator faults and control input constraints. An adaptive observer approach is used for accomplishing the joint state-fault estimation. Moreover, a feedback controller is designed to stabilize the closed-loop system in presence of actuator faults as well as actuator constraints.

Based on the above discussion, it follows that the problem of IFDC design for complex systems in presence of communication and saturation constraints is not only theoretically challenging but also practically important. However, the study and extension of novel proposed methodologies in this book for complex systems with communication and saturation constraints is an open problem and is left as a topic of our future research.

## 8.1.6.2   IFDC design for LPV systems

Within the linear parameter varying (LPV) framework, a class of nonlinear/time-varying systems can be represented in terms of parameterized linear dynamics in which the model coefficients depend on a set of measurable variables called scheduling variables [387]. The ability of LPV models to capture nonlinearities in the system dynamics by using linear dynamical relationships that are dependent on time-varying measurable signals makes one to possibly apply linear optimal control techniques to nonlinear systems represented by LPV models [388].

Based on this attractive property of LPV models, it is not surprising that LPV identification and control has attracted considerable attention recently as in

[389–392]. Consequently, LPV models have found a wide variety of applications in real world problems, ranging from aircraft [393], helicopter dynamics [394], wind turbines [395–397], automobiles [398], servo systems [399], hypersonic vehicles [400], web service systems [401], mechatronic systems [402], aerospace applications [403], high purity distillation column [404], robot control [405], motion platforms [406], flight control [407], compressor [408], biomedical applications [409], chemical processes [410], among others.

In the practice of control engineering, there are many applications that require design of a single controller that can guarantee stability and desirable transient response performance for all operating conditions of the plant. A most common approach taken in these situations is that of gain-scheduling. In the context of LPV models, one or more scheduling variables can be used to parameterize the space of operating points of the plant. For each operating point within this space, the gain-scheduling approach will yield a linear system. A similar parameterized controller can then be designed that fulfills the required closed-loop objectives at each operating point and an acceptable behavior while switching between multiple operating points [411]. A survey of different LPV gain-scheduled controller synthesis methods is given in detail in [412–414].

In [415], a parameter-dependent interval observer is designed to solve the FD problem for discrete-time LPV systems subject to bounded disturbances. The fault detection observer design problem for LPV systems in finite frequency domain is addressed in [416]. The work [417] proposed a new approach for design of distributed state estimation and FDI filters for a class of heterogeneous LPV multiagent systems. In [418], the event-triggered constant reference tracking control problem for discrete-time LPV systems is studied. The experimental results for a laboratory tank system are provided that show the effectiveness and capabilities of the proposed methodology. In [56], the IFDC design problem for a switched LPV system with inexact scheduling parameters is studied and simulation results for an aero-engine model is provided to show the effectiveness and capabilities of the proposed methodology. In [41], a two-step procedure is developed to solve the IFDC problem for linear discrete-time systems in presence of polytopic uncertainties.

In this book, a number of existing challenges and limitations in conventional IFDC design problem were addressed by studying the novel IFDC schemes for complex linear systems. However, due to the possibility of applying linear techniques for LPV systems, the proposed results in this book can be applied in studying the IFDC design problem for LPV, networked LPV and LPV multiagent systems.

It is worth noting that these extensions can be somehow challenging and are not straightforward. For example, an IFDC module designed for LPV systems quite often depends on scheduling variables, and this results in certain difficulties in the context of LPV event-triggering IFDC design as compared to LTI systems. In fact, due to this dependency, the scheduling variables must also be sent to the LPV IFDC module, and therefore, unlike LTI systems, the scheduling variables and the system measurements should be considered in the event condition.

### 8.1.6.3 Attack detection and resilient control design for cyber-physical systems

The recent increases in complexity of physical engineering systems that are distributed and computer networked have necessitated development of more sophisticated monitoring and control technologies. Due to major breakthroughs in software engineering technologies, embedded systems are increasingly being utilized in wide range of engineering domains from aerospace systems to automation systems, to smart grids, to chemical processes, to environmental engineering technologies, and broadly speaking to critical infrastructure systems. In particular, distributed control systems provide a promising framework to address issues related to large-scale systems.

For example, supervisory control and data acquisition (SCADA) systems, NCS, and programmable logic controllers (PLCs), are now established areas in for example smart grid in power networks industry and other key industrial infrastructure. The new generation of highly interconnected and networked engineering systems is categorized as cyber-physical systems (CPS), where employing NCS and embedded systems are inevitable, and that has contributed to the recognition of CPS systems as an area of significant national strategic importance [419]. CPS has many applications in different fields, among them transportation networks, power generation and distribution networks, water and gas distribution networks, and advanced communication systems [420].

With the increasing complexity and connectivity of CPS, an important challenge in these systems is that they have many entry points for attacks or intrusions. Attacks are caused by adversaries or hackers with the intention to damage or destroy the CPS [421]. The goal of the adversaries is to steal, alter, or destroy a specific target by hacking a susceptible system and if one or more parts of the system are attacked by an adversary, it is crucial for the CPS to continue operating with minimal degradation in performance. Therefore, different from the traditional sensor, actuator and system faults, the attack behaviors of adversaries are fast becoming a new research field in very recent years.

Some typical examples for attacks in real systems are as follows: the Stuxnet worm threat, multiple recent power blackouts in Brazil, the SQL Slammer worm attack on the Davis–Besse nuclear plant, power and transportation network attacks, and other staged attacks in power generators [422–424]. Perhaps, one of the most popular examples is Maroochy water breach in Queensland, Australia that happened in 2000 [425]. In this incident, a hacker managed to hack into some valves' controllers which led to flooding of the grounds of a hotel, a park, and a river with a million liters of sewage [426]. Consequently, a significant attack on a water system might lead to widespread illness, could affect critical services such as firefighting and health care, and could disrupt other dependent sectors such as energy, transportation, and agriculture [427]. These examples indicate the clear need for strategies and mechanisms to identify and deal with attacks on CPSs.

Various types of attacks in a CPS, depending on their abilities, can be classified into different categories such as denial-of-service (DoS), replay, and deception (or false data injection) attacks [428]. In the DoS, the attacker tries to drop the transmitted data, whereas in the deception attacks, the jammer aims to change the transmitted data. In the reply attacks, the attacker maliciously repeat transmitted data.

By assuming that there exist colluding omniscient attackers that are capable of altering the system dynamics through exogenous inputs, the model of an LTI system in presence of attacks/faults can be presented as follows:

$$\dot{x}(t) = Ax(t) + B_u u(t) + B_d d(t) + B_\vartheta \vartheta^a(t),$$
$$y(t) = Cx(t) + D_d d(t) + D_\vartheta \vartheta^s(t),$$

(8.1)

where $u(t)$ denotes the control input, $d(t)$ denotes the external disturbances and noise, $\vartheta^a(t)$ and $\vartheta^s(t)$ are the unknown fault/attack function. $A$, $B_u$, $B_d$, $C$, and $D_d$ are constant matrices with appropriate dimensions. Note that the attacks $(B_\vartheta \vartheta^a(t), D_\vartheta \vartheta^s(t))$ in (8.1) are designed based on knowledge of the system structure, parameters and states at all times [424].

In the DoS attacks, the goal of the attacker is to prevent the actuator (or controller) from receiving control commands (or sensor measurements). Using the general framework (8.1), a DoS attack on the actuator and sensor can be respectively modeled as [429]:

$$\begin{cases} B_\vartheta = B_u, \\ \vartheta^a(t) = -u(t), \quad \text{and} \\ \vartheta^s(t) = 0, \end{cases} \quad \begin{cases} D_\vartheta = C, \\ \vartheta^a(t) = 0, \\ \vartheta^s(t) = -x(t). \end{cases}$$

(8.2)

In data deception attacks, the goal of attacker is to prevent the actuator (sensor) from receiving an integrity data; therefore, the attacker injects false information to controllers (sensors). A data deception attack on the actuator and sensor can be respectively modeled as [429]:

$$\begin{cases} B_\vartheta = B_u, \\ \vartheta^a(t) = -u(t) + \theta^a, \text{ or } \vartheta^a(t) = -\theta^a, \\ \vartheta^s(t) = 0, \end{cases}$$

(8.3)

$$\text{and} \quad \begin{cases} D_\vartheta = C, \\ \vartheta^a(t) = 0, \\ \vartheta^s(t) = -x(t) + \theta^s, \text{ or } \vartheta^a(t) = -\theta^s, \end{cases}$$

where $\theta^a$ and $\theta^s$ denote false data that the adversary attempts to send to the actuator and the sensor, respectively. Other kinds of attacks such as zero dynamic attacks, bounded energy attacks and also bounded peak attacks can also be modeled using the general model (8.1). It is worth noting that using the proposed framework (8.1) the effects of both failures and attacks against the system can be considered [424].

Attack detection and attack resilient control are two important aspects of the security of CPS. To be more specific, attack detection aims to detect the existence of malicious attacks and further identify their actions. This goal is addressed by designing appropriate estimators or filters [428]. In [430], the problem of detecting multiple cyber-attacks in control systems is investigated. Based on frequency-domain transformation technique and auxiliary detection tools, an algebraic detection approach is proposed to detect the anomalies in the control system. Necessary and sufficient conditions guaranteeing the detectability of the multiple stochastic cyber-attacks are obtained. A comparative study between Kalman filters and dynamic observers for the problem of dynamic state estimation under cyber-attacks is studied in [431]. Based on subspace techniques, the authors of [432] proposed the problem of data-driven attacks on state estimation, assuming that the adversary is capable of monitoring a subset of system measurements without detailed knowledge of the network topology and system parameters.

The problem of detection and isolation of both classical sensor/actuator faults and communication network-induced deception attacks on a water distribution network comprised of cascaded canal pools using a bank of observers is considered in [433]. Each observer is insensitive to one fault/attack mode and sensitive to other modes. Design of observers is achieved by using a delay-dependent LMI method. The problem of detecting cyber-physical attacks on SCADA systems is addressed in [434]. The Variable Threshold Window Limited CUmulative SUM algorithm is utilized to detect the changes in the sequence of residuals generated from either the Kalman filter or the parity space method. The proposed methodology is applied to the problem of cyber-physical attack detection in a simple SCADA water system.

In [435], it is shown that with knowledge of the system model, an intelligent cyber attacker is able to carefully design a data injection sequence, such that the state estimation error increases without triggering the alarm of the monitoring system. To solve this problem, in [436] an iterative optimization algorithm is developed to compute a feasible coding matrix for the original sensor outputs to increase the estimation residues, such that the alarm will be triggered by the detector even under intelligent data injection attacks.

In [437], the problem of designing centralized and distributed attack detection monitors for LTI and large-scale systems is proposed. Moreover, fundamental monitoring limitations are obtained based on system-theoretic and graph-theoretic perspectives. However, the effects of external disturbances and noise which occur in practical applications have not been considered. In [429,430], the problem of attack detection for LTI systems under multiple stochastic cyber-attacks and stochastic noise is considered. Toward this end, a dynamic filter is employed and the parameters of the filter are designed such that the effects of noise on residual signal is attenuated. In [438], a nonlinear $H_\infty$ filter for stochastic cyber-attacks estimation of nonlinear systems is proposed which maximize the sensitivity of the cyber-attacks and simultaneously minimize the effect of the disturbances on the residual signal.

Based on the above review, although there exist certain methodologies in attack detection of linear and even large-scale systems, they suffer from certain limitations and drawbacks. These limitations are summarized as follows:

- In most of the existing methodologies, the effects of external disturbances and noise, communication, and saturation constraints which occur in practical applications have not been considered.
- Most of the existing attack estimators have been simply confined to traditional so-called "static" observers (classical Kalman–Luenberger observer) which are not suitable and effective in presence of intelligent attackers.

- In most of the published work, it is assumed that the state (output) signal of the plant is often measured at sampled points with a constant sampling period, which is called the "time-triggering" sampling and is sometimes less preferable from a resource utilization point of view.

Furthermore, in situations in which attacks occur in system components (e.g., sensors, actuators, and communication links), this may lead to a serious degradation in performance or, even worse, to an overall system failure. In this respect, it is worth noting that standard feedback control systems are not capable of handling the effects of such malicious attacks. Hence, to achieve robust resiliency of complex engineering systems, it is essential to develop resilient control methodologies that can offer robustness to various classes of cyber-attacks that attempt to alter the closed-loop control system behavior using some knowledge of the overall system's dynamics. Consequently, the main goal of attack resilient control is to design control laws which can ensure control system performance despite presence of malicious attacks.

Based on receding-horizon control, a resilient consensus algorithm against replay attacks in operator-vehicle networks is proposed in [439]. In [440], an event-triggered resilient attack controller is proposed to guarantee the stability of multi-input controllable continuous-time linear systems under periodic DoS jamming attacks. The effects of finite energy and bounded attacks on actuator and sensor signals for LTI systems are analyzed in [441,442], respectively. Moreover, the optimal actuator and sensor attacks (worst case attack signals) for both finite and infinite horizon linear quadratic control are obtained.

In [443], a resilient control problem for discrete-time LTI systems under reply attacks and subject to state and input constraints is considered. A variation of the receding-horizon control law is proposed to deal with the replay attacks and to analyze the resulting system performance degradation. A set of sufficient conditions are obtained to guarantee asymptotic and exponential stability.

In [444], the reinforcement learning technique is used to design distributed formation control problem in an operator-vehicle network in presence of attacks. In [445], a distributed cyber-security tracking control problem is addressed for stochastic linear multiagent systems under two types of attacks. Sufficient conditions on robust mean-square exponential consensus tracking are derived through the idea of average dwell time switching between some stable and unstable subsystems that are obtained from graph theory analysis.

A necessary and sufficient condition is obtained in [446], under which the adversary can successfully identify the transfer function of the physical system. Moreover, an algorithm is designed, under which the adversary can compute the physical system model. Finally, a low-rank controller is designed which makes the system unidentifiable to the adversary, while trading off the linear quadratic Gaussian performance. In [447], the input-to-state stabilizing (ISS) control design problem under DoS attacks for NCSs is addressed. The frequency and duration of the DoS attacks under which ISS of the closed-loop system can be preserved is obtained. To achieve ISS, a suitable scheduling of the transmission times is determined.

Based on the above review, although there exist some methodologies in attack resilient control design of CPS, they suffer from a number of limitations and drawbacks. These limitations are summarized as follows:

- It should be noted that most of the proposed attack resilient controllers are designed for simple LTI systems and considering the more complex systems such as large-scale CPS needs more research and consideration.
- It can be seen that most of the existing attack resilient controllers (such as [447]) are based on state-feedback under a restrictive assumption that all the system's states can be measured. However, a full-state measurement is not available in many real physical systems, and therefore, state feedback controllers cannot be used for them, and instead, an output feedback controller should be used.

The results of this book are conductive for conducting research, addressing the timely challenges and developing novel methodologies in cyber security, monitoring, diagnostics, and resilient control recovery of CPS infrastructure to improve and ensure reliability and resilience of these safety critical systems subject to multiple types of faults/malicious attacks. Therefore,

- Based on the results of this book (especially those in Chapter 7), it is possible to design a novel distributed attack monitoring methodology for large-scale CPS in presence of external noise and disturbances. The dynamic filter that was proposed in Chapter 2 can be used toward generating the residual signal to solve the limitations of the classical observers and filters for attack detection. The main challenge is to simultaneously address the unavoidable trade-offs between different goals, namely, the disturbance attenuation and the attack sensitivity.
- Based on the existing results in Chapters 5 and 6, it is possible to design the problem of event-triggered fault/attack monitoring of large-scale CPS.
- The results of this book can be applied to the problem of distributed output-feedback resilient control design of large-scale CPS where not all the states of the system are measurable.
- The results of this book can be applied in integrated design of attack monitoring systems and resilient controllers for complex CPSs.

# References

[1] Meng X, Chen T. Event detection and control co-design of sampled-data systems. International Journal of Control. 2014;87(4):777–786.

[2] Ferrari R. Distributed Fault Detection and Isolation of Large-scale Nonlinear Systems: An Adaptive Approximation Approach. Italy: University of Trieste; 2009.

[3] Heemels WPMH, Donkers MCF. Model-based periodic event-triggered control for linear systems. Automatica. 2013;49(3):698–711.

[4] Zhai D, An L, Dong J, Zhang Q. Simultaneous $H_2/H_\infty$ fault detection and control for networked systems with application to forging equipment. Signal Processing. 2016;125(Supplement C):203–215.

[5] Zhan XS, Zhou ZJ, Wu J, Han T. Optimal modified tracking performance of time-delay systems with packet dropouts constraint. Asian Journal of Control. 2017;19(4):1508–1518.

[6] Wang TB, Wang YL, Wang H, Zhang J. Observer-based $H_\infty$ control for continuous-time networked control systems. Asian Journal of Control. 2016; 18(2):581–594.

[7] Zhai D, An L, Li J, Zhang Q. Fault detection for stochastic parameter-varying Markovian jump systems with application to networked control systems. Applied Mathematical Modelling. 2016;40(3):2368–2383.

[8] Davoodi M, Gallehdari Z, Saboori I, *et al.* An Overview of Cooperative and Consensus Control of Multiagent Systems. In: Wiley Encyclopedia of Electrical and Electronics Engineering. Hoboken, NJ: John Wiley & Sons, Inc.; 2016.

[9] Lapp SA, Powers GA. Computer-aided synthesis of fault-trees. IEEE Transactions on Reliability. 1977;37:2–13.

[10] Magrabi SM, Gibbens PW. Decentralized fault detection and diagnosis in navigation systems for unmanned aerial vehicles. In: IEEE 2000 Position Location and Navigation Symposium; 2000. p. 363–370.

[11] Low XC, Willsky AS, Verghese GL. Optimally robust redundancy relations for failure detection in uncertain systems. Automatica. 1986;22:333–344.

[12] Frank PM. Fault diagnosis in dynamic systems using analytical and knowledge-based redundancy – A survey and some new results. Automatica. 1990;26(3):459–474.

[13] Clark RN. A simplified instrument failure detection scheme. IEEE Transactions on Aerospace and Electronic Systems. 1978;14:558–563.

[14]    Clark RN. Instrument fault detection. IEEE Transactions on Aerospace and Electronic Systems. 1978;14:456–465.

[15]    Hajiyev CM, Caliskan F. Fault detection in flight control systems via innovation sequence of Kalman filter. UKACC International Conference on Control. 1998;2:1528–1533.

[16]    Hajiyev CM, Caliskan F. Fault detection in flight control systems based on the generalized variance of the Kalman filter innovation sequence. Proceedings of the American Control Conference. 1999;1:109–113.

[17]    Caliskan F, Hajiyev CM. EKF based surface fault detection and reconfiguration in aircraft control systems. Proceedings of the American Control Conference. 2000;2:1220–1224.

[18]    Larsona EC, Parker BE, Clark BR. Model-based sensor and actuator fault detection and isolation. Proceedings of the American Control Conference. 2002;5:4215–4219.

[19]    Maybeck DQ, Stevens RD. Reconfigurable flight control via multiple model adaptive control method. IEEE Transactions on Aerospace and Electronic Systems. 1991;27:470–479.

[20]    Zhang Y, Jiang J. Integrated active fault-tolerant control using IMM approach. IEEE Transactions on Aerospace and Electronic Systems. 2001;37: 1221–1235.

[21]    Beard RV. Failure accommodation in linear systems through self-reorganization. PhD dissertation, Massachusetts Inst Tech, Cambridge, MA. 1971.

[22]    Jones HL. Failure detection in linear systems. PhD dissertation, Massachusetts Inst Tech, Cambridge, MA. 1973.

[23]    Massoumnia MA. A geometric approach to the synthesis of failure detection filter. IEEE Transactions on Automatic Control. 1986;31:839–846.

[24]    Isermann R. Process fault detection based on modeling and estimation methods – A survey. Automatica. 1984;20:387–404.

[25]    Chen J, Patton RJ. Robust Model-Based Fault Diagnosis for Dynamic Systems. Boston: Kluwer Academic Publishers; 1999.

[26]    Dai X, Gao Z. From model, signal to knowledge: A data-driven perspective of fault detection and diagnosis. IEEE Transactions on Industrial Informatics. 2013 November;9(4):2226–2238.

[27]    Zhong M, Ding SX, Lam J, Wang H. An LMI approach to design robust fault detection filter for uncertain LTI systems. Automatica. 2003;39: 543–550.

[28]    Wang JL, Yang GH, Liu J. An LMI approach to $H_-$ index and mixed $H_-/H_\infty$ fault detection observer design. Automatica. 2007;43:1656–1665.

[29]    Demetriou MA, Ito K, Smith RC. Adaptive monitoring and accommodation of nonlinear actuator faults in positive real infinite dimensional systems. IEEE Transactions on Automatic Control. 2007;52(12):2332–2338.

[30]    Ding SX. Model Based Fault Diagnosis Techniques – Design Schemes, Algorithms and Tools. Berlin, Heidelberg: Springer Verlag; 2008.

[31]    Meskin N, Khorasani K. Fault Detection and Isolation: Multi-Vehicle Unmanned Systems. New York: Springer; 2011.

[32] Ding SX, Jcinsch J, Frank PM, Ding F.L. A unified approach to the optimization of fault detection systems. International Journal of Adaptive Control and Signal Processing. 2000;14:725–745.

[33] Demetriou MA, Armaou A. Dynamic online nonlinear robust detection and accommodation of incipient component faults for nonlinear dissipative distributed processes. International Journal of Robust and Nonlinear Control. 2012;22(1):3–23.

[34] Ferdowsi H, Jagannathan S. A unified model-based fault diagnosis scheme for non-linear discrete-time systems with additive and multiplicative faults. Transactions of the Institute of Measurement and Control. 2013;35(6):742–752.

[35] Gao Z, Cecati C, Ding SX. A survey of fault diagnosis and fault-tolerant techniques – Part I: Fault diagnosis with model-based and signal-based approaches. IEEE Transactions on Industrial Electronics. 2015 June;62(6): 3757–3767.

[36] Jacobson CA, Nett CN. An integrated approach to controls and diagnostics using the four parameter controller. IEEE Control Systems. 1991;11(6):22–29.

[37] Henry D, Zolghadri A. Design and analysis of robust residual generators for systems under feedback control. Automatica. 2005;41(2):251–264.

[38] Kilsgaard S, Rank ML, Niemann HH, Stoustrup J. Simultaneous design of controller and fault detector. In: Proc. 37th Conf. Decision Control. Kobe, Japan; 1996. p. 628–629.

[39] Hearns G, Grimble MJ, Johnson MA. Integrated fault monitoring and reliable control. In: UKACC International Conference on Control; 1998. p. 1175–1179.

[40] Wang H, Yang GH. Integrated fault detection and control for LPV systems. International Journal of Robust and Nonlinear Control. 2009;19: 341–363.

[41] Wang H, Yang GH. Simultaneous fault detection and control for uncertain linear discrete-time systems. IET Control Theory & Applications. 2009;3:583–594.

[42] Li XJ, Yang GH. Dynamic observer-based robust control and fault detection for linear systems. IET Control Theory & Applications. 2012;6(17):2657–2666.

[43] Davoodi MR, Talebi HA, Momeni HR. A novel simultaneous fault detection and control approach based on dynamic observer. International Journal of Innovative Computing, Information and Control. 2012;8:4915–4930.

[44] Davoodi MR, Golabi A, Talebi HA, Momeni HR. Simultaneous fault detection and control design for switched linear systems based on dynamic observer. Optimal Control Applications and Methods. 2013;34:35–52.

[45] Zhong GX, Yang GH. Robust control and fault detection for continuous-time switched systems subject to a dwell time constraint. International Journal of Robust and Nonlinear Control. 2015;25(18):3799–3817.

[46] Ding SX. Integrated design of feedback controllers and fault detectors. Annual Reviews in Control. 2009;33(2):124–135.

[47]   Jacobson CA, Nett CN. An integrated approach to controls and diagnostics using the four parameter controller. IEEE Control Systems Magazine. 1991;11(6):22–29.

[48]   Davoodi MR, Golabi A, Talebi HA, Momeni HR. Simultaneous fault detection and control design for switched linear systems: A linear matrix inequality approach. Journal of Dynamic Systems, Measurement, and Control. 2012;134(6):1–10.

[49]   Soltani H, Naoui SBHA, Aitouche A, El Harabi R, Abdelkrim MN. Robust simultaneous fault detection and control approach for time-delay systems. IFAC-PapersOnLine. 2015;48(21):1244–1249.

[50]   Zhong GX, Yang GH. Simultaneous fault detection and control for discrete-time switched systems. Circuits, Systems, and Signal Processing. 2015 December;34(12):3811–3831.

[51]   Li J, Yang GH. Simultaneous fault detection and control for switched systems with actuator faults. International Journal of Systems Science. 2016;47(10):2411–2427.

[52]   Li J, Park JH, Ye D. Simultaneous fault detection and control design for switched systems with two quantized signals. ISA Transactions. 2017;66(Supplement C):296–309.

[53]   Du Y, Budman H, Duever TA. Integration of fault diagnosis and control based on a trade-off between fault detectability and closed loop performance. Journal of Process Control. 2016;38(Supplement C):42–53.

[54]   Shokouhi-Nejad H, Ghiasi AR, Badamchizadeh MA. Robust simultaneous fault detection and control for a class of nonlinear stochastic switched delay systems under asynchronous switching. Journal of the Franklin Institute. 2017;354(12):4801–4825.

[55]   Shokouhi-Nejad H, Ghiasi AR, Badamchizadeh MA. Robust simultaneous finite-time control and fault detection for uncertain linear switched systems with time-varying delay. IET Control Theory & Applications. 2017;11(7):1041–1052.

[56]   Zhu K, Zhao J. Simultaneous fault detection and control for switched LPV systems with inexact parameters and its application. International Journal of Systems Science. 2017;48(14):2909–2920.

[57]   Wang S, Wang Y, Pang J, Liu K. Simultaneous fault detection and controller design for networked-based discrete-time switched systems with time-varying delays. In: 29th Chinese Control And Decision Conference (CCDC); 2017. p. 2109–2114.

[58]   Ning Z, Yu J, Wang T. Simultaneous fault detection and control for uncertain discrete-time stochastic systems with limited communication. Journal of the Franklin Institute. 2017;354(17):7794–7811.

[59]   Beard RW, Lawton J, Hadaegh FY. A coordination architecture for spacecraft formation control. IEEE Transactions on Control Systems Technology. 2001;9(6):777–789.

[60]   Sinopoli B, Sharp C, Schenato L, Schaffert S, Sastry SS. Distributed control applications within sensor networks. Proceedings of the IEEE. 2003;91(8):1235–1246.

[61] Yuan J, Zhou ZH, Mu H, Sun YT, Li L. Formation control of autonomous underwater vehicles based on finite-time consensus algorithms. In: Proceedings of Chinese Intelligent Automation Conference. Springer; 2013. p. 1–8.

[62] Davoodi MR, Meskin N, Khorasani K. Simultaneous fault detection and control design for a network of multi-agent systems. In: European Control Conference (ECC); 2014. p. 575–581.

[63] Gallehdari Z, Meskin N, Khorasani K. Cost performance based control reconfiguration in multi-agent systems. In: American Control Conference (ACC); 2014. p. 509–516.

[64] Saif O, Fantoni I, Zavala-Rio A. Flocking of multiple unmanned aerial vehicles by LQR control. In: International Conference on Unmanned Aircraft Systems (ICUAS); 2014. p. 222–228.

[65] Chen YF, Ure NK, Chowdhary G, How JP, Vian J. Planning for large-scale multiagent problems via hierarchical decomposition with applications to UAV health management. In: American Control Conference (ACC); 2014. p. 1279–1285.

[66] Leonard J, Savvaris A, Tsourdos A. Energy management in swarm of unmanned aerial vehicles. Journal of Intelligent & Robotic Systems. 2014;74(1–2):233–250.

[67] Souliman A, Joukhadar A, Alturbeh H, Whidborne JF. Intelligent collision avoidance for multi agent mobile robots. In: Intelligent Systems for Science and Information. Springer; 2014. p. 297–315.

[68] Khong SZ, Tan Y, Manzie C, Nešić D. Multi-agent source seeking via discrete-time extremum seeking control. Automatica. 2014;50(9):2312–2320.

[69] Guo M, Michaelavlanos Z, Dimarogonas DV. Controlling the relative agent motion in multi-agent formation stabilization. IEEE Transaction on Automatic Control. 2014;59(3):820–826.

[70] Lin P, Lu W, Song Y. Distributed nested rotating consensus problem of multi-agent systems. In: The 26th Chinese Control and Decision Conference (2014 CCDC); 2014. p. 1785–1789.

[71] Wen G, Duan Z, Ren W, Chen G. Distributed consensus of multi-agent systems with general linear node dynamics and intermittent communications. International Journal of Robust and Nonlinear Control. 2013;24(16):2438–2457.

[72] Zhang C, Wang Y, Zhao Y. Agent-based distributed simulation technology of satellite formation flying. In: Fourth World Congress on Software Engineering (WCSE); 2013. p. 13–16.

[73] Wang Q, Gao H, Alsaadi F, Hayat T. An overview of consensus problems in constrained multi-agent coordination. Systems Science & Control Engineering. 2014;2(1):275–284.

[74] Cao Y, Yu W, W, Chen G. An overview of recent progress in the study of distributed multi-agent coordination. IEEE Transactions on Industrial Informatics. 2013 February;9(1):427–438.

[75] Murray RM. Recent research in cooperative control of multivehicle systems. Journal of Dynamic Systems, Measurement, and Control. 2007;129(5):571–583.

[76]   Ge X, Yang F, Han QL. Distributed networked control systems: A brief overview. Information Sciences. 2017;380:117–131.

[77]   Tang Y, Qian F, Gao H, Kurths J. Synchronization in complex networks and its application – A survey of recent advances and challenges. Annual Reviews in Control. 2014;38(2):184–198.

[78]   Liu Q, Wang Z, He X, Zhou DH. A survey of event-based strategies on control and estimation. Systems Science & Control Engineering. 2014;2(1): 90–97.

[79]   Qu Z. Cooperative Control of Dynamical Systems: Applications to Autonomous Vehicles. London, U.K.: Springer-Verlag; 2009.

[80]   Mesbahi M, Egerstedt M. Graph Theoretic Methods for Multiagent Networks. Princeton, NJ: Princeton University Press; 2010.

[81]   Ren W, Cao Y. Distributed Coordination of Multi-Agent Networks. Communications and Control Engineering. London, UK: Springer-Verlag; 2011.

[82]   Ryan A, Zennaro M, Howell A, Sengupta R, Hedrick JK. An overview of emerging results in cooperative UAV control. In: Proceedings of the 43rd IEEE Conference on Decision and Control; 2004. p. 602–607.

[83]   Ren W, Beard RW, Atkins EM. A survey of consensus problems in multi-agent coordination. In: Proceedings of the American Control Conference; vol. 3; 2005. p. 1859–1864.

[84]   Feng S, Zhang H. Formation control for wheeled mobile robots based on consensus protocol. In: IEEE International Conference on Information and Automation (ICIA); 2011. p. 696–700.

[85]   Semsar-Kazerooni E, Khorasani K. Team consensus for a network of unmanned vehicles in presence of actuator faults. IEEE Transactions on Control Systems Technology. 2010;18(5):1155–1161.

[86]   Azizi SM, Khorasani K. Cooperative actuator fault accommodation of formation flying vehicles with absolute measurements. In: 49th IEEE Conference on Decision and Control (CDC); 2010. p. 6299–6304.

[87]   Joordens MA, Jamshidi M. Consensus control for a system of underwater swarm robots. IEEE Systems Journal. 2010 March;4(1):65–73.

[88]   Olfati-Saber R. Flocking for multi-agent dynamic systems: Algorithms and theory. IEEE Transactions on Automatic Control. 2006;51(3):410–420.

[89]   Peng J, Sun X, Zhu F, Li X. Multi UAVs cooperative task assignment using multi agent. In: Chinese Control and Decision Conference (CCDC); 2008. p. 4517–4520.

[90]   He B, Wen JT. Cooperative load transport: A formation-control perspective. IEEE Transactions on Robotics. 2010 August;26(4):742–750.

[91]   Xu D, Xia K. Role assignment, non-communicative multi-agent coordination in dynamic environments based on the situation calculus. In: WRI Global Congress on Intelligent Systems (GCIS '09); vol. 1; 2009. p. 89–93.

[92]   de Oliveira IR, Carvalho FS, Camargo JB, Sato LM. Multi-agent tools for air traffic management. In: 11th IEEE International Conference on Computational Science and Engineering Workshops (CSEWORKSHOPS'08); 2008. p. 355–360.

[93] Ren W, Beard RW. Distributed Consensus in Multi-Vehicle Cooperative Control; Theory and Applications. Berlin, Heidelberg: Springer; 2008.

[94] Olfati-Saber R, Fax JA, Murray RM. Consensus and cooperation in networked multi-agent systems. Proceedings of the IEEE. 2007;95(1):215–233.

[95] Jadbabaie A, Lin J, Morse AS. Coordination of groups of mobile autonomous agents using nearest neighbor rules. IEEE Transactions on Automatic Control. 2003 June;48(6):988–1001.

[96] Jaimes B, Jamshidi M. Consensus-based and network control of UAVs. In: 5th International Conference on System of Systems Engineering (SoSE); 2010. p. 1–6.

[97] Lee D, Spong MW. Stable flocking of multiple inertial agents on balanced graphs. IEEE Transactions on Automatic Control. 2007;52(8):1469–1475.

[98] Moreau L. Stability of multiagent systems with time-dependent communication links. IEEE Transactions on Automatic Control. 2005;50(2):169–182.

[99] Shoham Y, Leyton-Brown K. Multiagent Systems: Algorithmic, Game-Theoretic, and Logical Foundations. Cambridge: Cambridge University Press; 2008.

[100] Olfati-Saber R, Murray RM. Distributed cooperative control of multiple vehicle formations using structural potential functions. In: Proc. IFAC World Congress; 2002.

[101] Wei R, Beard RW, Atkins EM. Information consensus in multivehicle cooperative control. IEEE Control Systems. 2007;27(2):71–82.

[102] Smallwood DA, Whitcomb LL. Model-based dynamic positioning of underwater robotic vehicles: theory and experiment. IEEE Journal of Oceanic Engineering. 2004;29(1):169–186.

[103] Olfati-Saber R, Shamma JS. Consensus filters for sensor networks and distributed sensor fusion. In: Proceedings of the 44th IEEE Conference on Decision and Control and European Control Conference. Spain; 2005. p. 6698–6703.

[104] Munz U, Papachristodoulou A, Allgower F. Consensus in multi-agent systems with coupling delays and switching topology. IEEE Transactions on Automatic Control. 2011 December;26(12):2976–2982.

[105] Seuret A, Dimarogonas DV, Johansson KH. Consensus under communication delays. In: 47th IEEE Conference on Decision and Control. Cancun, Mexico; 2008. p. 4922–4927.

[106] Ajorlou A, Momeni A, Aghdam AG. A class of bounded distributed control strategies for connectivity preservation in multi-agent systems. IEEE Transactions on Automatic Control. 2010;55(12):2828–2833.

[107] Ren W, Beard RW. Consensus seeking in multi-agent systems under dynamically changing interaction topologies. IEEE Transactions on Automatic Control. 2005 May;50(5):655–661.

[108] Bošković JD, Li SM, Mehra RK. Formation flight control design in the presence of unknown leader commands. In: Proc. American Control Conference; 2002. p. 2854–2859.

[109] Ren W. Consensus strategies for cooperative control of vehicle formations. IET Control Theory and Applications. 2007;1(2):505–512.

[110]  Ren W. Consensus based formation control strategies for multi-vehicle systems. In: Proceedings of the American Control Conference. USA; 2006. p. 4237–4242.

[111]  Ando H, Oasa Y, Suzuki I, Yamashita M. Distributed memoryless point convergence algorithm for mobile robots with limited visibility. IEEE Transactions on Robotics and Automation. 1999;15(5):818–828.

[112]  Sinha A, Ghose D. Generalization of linear cyclic pursuit with application to rendezvous of multiple autonomous agents. IEEE Transactions on Automatic Control. 2006;51(11):1819–1824.

[113]  Cortes J, Martinez S, Bullo F. Robust rendezvous for mobile autonomous agents via proximity graphs in arbitrary dimensions. IEEE Transactions on Automatic Control. 2009;51(8):1289–1298.

[114]  Reynolds C. Flocks, birds, and schools: A distributed behavioral model. Computer Graphics. 1987;21(4):25–34.

[115]  Vicsek T, Czirók A, Ben-Jacob E, Cohen I, Shochet O. Novel type of phase transition in a system of self-driven particles. Physical Review Letters. 1995;75(6):1226–1229.

[116]  Gazi V, Passino KM. Stability analysis of social foraging swarms. IEEE Transactions on Systems, Man, and Cybernetics—Part B: Cybernetics. 2004;34(1):539–557.

[117]  Gazi V, Passino KM. Stability analysis of swarms. IEEE Transactions on Automatic Control. 2003;48(4):692–697.

[118]  Ren W. Distributed attitude alignment in spacecraft formation flying. International Journal of Adaptive Control and Signal Processing. 2007; 21(2–3):95–113.

[119]  Olfati-Saber R. Distributed Kalman filter with embedded consensus filters. In: Proceedings of the 44th IEEE Conference on Decision and Control and European Control Conference. Spain; 2005. p. 8179–8184.

[120]  Olfati-Saber R. Distributed Kalman filtering for sensor networks. In: Proceedings of the 46th IEEE Conference on Decision and Control. USA; 2007. p. 5492–5498.

[121]  Ren W, Beard RW. Formation feedback control for multiple spacecraft via virtual structures. IEE Proceedings – Control Theory and Applications. 2004;151(3):357–368.

[122]  Olfati-Saber R, Murray RM. Distributed structural stabilization and tracking for formations of dynamic multi-agents. In: Proc. Conference on Decision and Control; 2002. p. 209–215.

[123]  Stipanović DM, Inalhan G, Teo R, Tomlin CJ. Decentralized overlapping control of a formation of unmanned aerial vehicles. Automatica. 2004;40:1285–1296.

[124]  Olfati-Saber R, Murray RM. Flocking with obstacle avoidance: Cooperation with limited communication in mobile networks. In: Proc. Conference on Decision and Control; 2003. p. 2022–2028.

[125]  Olfati-Saber R, Murray RM. Consensus protocols for networks of dynamic agents. In: Proc. American Control Conference; 2003. p. 951–956.

[126] Gazi V. Stability Analysis of Swarms. Columbus, OH: The Ohio State University; 2002.

[127] Olfati-Saber R, Murray RM. Consensus problems in networks of agents with switching topology and time-delays. IEEE Transactions on Automatic Control. 2004;49(9):1520–1533.

[128] Ren W. On consensus algorithms for double-integrator dynamics. In: Proc. Conference on Decision and Control; 2007. p. 2295–2300.

[129] Xiao F, Wang L. Consensus problems for high-dimensional multi-agent systems. IET Control Theory Applications. 2007;1(3):830–837.

[130] Semsar E, Khorasani K. Adaptive formation control of UAVs in the presence of unknown vortex forces and leader commands. In: Proc. American Control Conference; 2006. p. 3563–3568.

[131] Paley D, Leonard NE, Sepulchre R. Collective motion: Bistability and trajectory tracking. In: Proc. Conference on Decision and Control; 2004. p. 1932–1937.

[132] Alighanbari M, How JP. Decentralized task assignment for unmanned aerial vehicles. In: Proc. Conference on Decision and Control and European Control Conference; 2005. p. 5668–5673.

[133] Nett E, Schemmer S. Reliable real-time communication in cooperative mobile applications. IEEE Transactions on Computers. 2003;52:166–180.

[134] Jinyan S, Long W, Junzhi Y. Cooperative control of multiple robotic fish in a disk-pushing task. In: Proc. American Control Conference; 2006. p. 2730–2735.

[135] Tanner HG, Jadbabaie A, Pappas GJ. Flocking in fixed and switching networks. IEEE Transactions on Automatic Control. 2007;52(5):863–868.

[136] Scharf DP, Hadaegh FY, Ploen SR. A survey of spacecraft formation flying guidance and control. Part II: Control. In: Proc. American Control Conference; 2004. p. 2976–2985.

[137] Lawton J, Beard RW, Hadaegh FY. An adaptive control approach to satellite formation flying with relative distance constraints. In: Proc. American Control Conference; vol. 3; 1999. p. 1545–1549.

[138] Ren W, Beard RW. Decentralized scheme for spacecraft formation flying via the virtual structure approach. Journal of Guidance, Control and Dynamics. 2004;127(1):73–82.

[139] Fierro R, Das AK. A modular architecture for formation control. In: Proc. IEEE Workshop on Robot Motion and Control; 2002. p. 285–290.

[140] Beard RW, Lawton J, Hadaegh FY. A feedback architecture for formation control. In: Proc. American Control Conference; vol. 6; 2000. p. 4087–4091.

[141] Hadaegh FY, Ghavimi AR, Singh G, Quadrelli M. A centralized optimal controller for formation flying spacecraft. In: Proc. Int. Conf. Intel. Tech.; 2000.

[142] Scharf DP, Acikmese AB, Ploen SR, Hadaegh FY. A direct solution for fuel-optimal reactive collision avoidance of collaborating spacecraft. In: Proc. American Control Conference; vol. 2; 2006. p. 5201–5206.

[143] Hadaegh FY, Scharf DP, Ploen SR. Initialization of distributed spacecraft for precision formation flying. In: Proc. Conference on Control Applications; vol. 2; 2003. p. 1463–1468.

[144] Mesbahi M, Hadaegh FY. Reconfigurable control for the formation flying of multiple spacecraft. In: Proc. International Multi-Conference on Circuits, Systems, and Control; 1999.

[145] Smith RS, Hadaegh FY. A Distributed parallel estimation architecture for cooperative vehicle formation control. In: Proc. American Control Conference; 2006. p. 4219–4224.

[146] Smith RS, Hadaegh FY. Distributed parallel estimation architecture for cooperative vehicle formation control. In: Proc. IFAC World Congress; 2006.

[147] Smith RS, Hadaegh FY. Distributed estimation, communication and control for deep space formations. IET Control Theory Applications. 2007;1(2):445–451.

[148] Fax JA, Murray RM. Information flow and cooperative control of vehicle formations. IEEE Transactions on Automatic Control. 2004;49(9):1465–1476.

[149] Arcak M. Passivity as a design tool for group coordination. IEEE Transactions on Automatic Control. 2007;52(8):1380–1390.

[150] Decastro GA, Paganini F. Convex synthesis of controllers for consensus. In: Proc. American Control Conference; 2004. p. 4933–4938.

[151] Raffard RL, Tomlin CJ, Boyd SP. Distributed optimization for cooperative agents: Application to formation flight. In: Proc. Conference on Decision and Control; 2004. p. 2453–2459.

[152] Inalhan G, Stipanović DM, Tomlin CJ. Decentralized optimization, with application to multiple aircraft coordination. In: Proc. Conference on Decision and Control; 2002. p. 1147–1155.

[153] Speyer JL, Zika N, Franco E. Determination of the value of information in large vehicle arrays. In: Proc. American Control Conference; 2006. p. 1–7.

[154] Speyer JL, Seok I, Michelin A. Decentralized control based on the value of information in large vehicle arrays. In: Proc. American Control Conference; 2008. p. 5047–5054.

[155] Ren W. Distributed attitude consensus among multiple networked spacecraft. In: Proc. American Control Conference; 2006. p. 1760–1765.

[156] Rahmani A, Mesbahi M. On the controlled agreement problem. In: Proc. American Control Conference; 2006. p. 1376–1381.

[157] Hussein II, Bloch AM. Dynamic coverage optimal control for interferometric imaging spacecraft formations (part II): The nonlinear case. In: Proc. American Control Conference; 2005. p. 2391–2396.

[158] Xiao L, Boyd S. Fast linear iterations for distributed averaging. Systems and Control Letters. 2004;53(1):65–78.

[159] Panyakeow P, Mesbahi M. Decentralized deconfliction algorithms for unicycle UAVs. In: Proceedings of the American Control Conference. USA; 2010. p. 794–799.

[160] Rezaee H, Abdollahi F, Talebi HA. $\mathcal{H}_\infty$ based motion synchronization in formation flight with delayed communications. IEEE Transactions on Industrial Electronics. 2014;61(11):6175–6182.

[161] Rezaee H, Abdollahi F. A decentralized cooperative control scheme with obstacle avoidance for a team of mobile robots. IEEE Transactions on Industrial Electronics. 2014;61(1):347–354.

[162] Barnes LE, Fields MA, Valavanis KP. Swarm formation control utilizing elliptical surfaces and limiting functions. IEEE Transactions on Systems, Man, and Cybernetics—Part B: Cybernetics. 2009;39(6):1434–1445.

[163] Cui R, Ge SS, How BV, Choo YS. Leader-follower formation control of underactuated autonomous underwater vehicles. Ocean Engineering. 2010;37(17–18):1491–1502.

[164] Do KD. Practical formation control of multiple underactuated ships with limited sensing ranges. Robotics and Autonomous Systems. 2011; 59(6):457–471.

[165] Singh SN, Pachter M. Adaptive feedback linearizing nonlinear close formation control of UAVs. In: Proc. American Control Conference; 2000. p. 854–858.

[166] Leonard NE, Fiorelli E. Virtual leaders, artificial potentials and coordinated control of groups. In: Proc. Conference on Decision and Control; 2001. p. 2968–2973.

[167] Wang PKC. Navigation strategies for multiple autonomous mobile robots moving in formation. In: Intelligent Robots and Systems '89, Proceedings of the Autonomous Mobile Robots and Its Applications. 1991;8: 177–195.

[168] Baleh T, Arkin RC. Behavior-based formation control for multi-robot teams. IEEE Transactions on Robotics and Automation. 1998;14:926–939.

[169] Mesbahi M, Hadaegh FY. Formation flying of multiple spacecraft via graphs, matrix inequalities, and switching. AIAA Journal of Guidance, Control, and Dynamics. 2001;24(2):369–377.

[170] Wu CW. Agreement and consensus problems in groups of autonomous agents with linear dynamics. In: Proc. IEEE International Symposium on Circuits and Systems (ISCAS); 2005. p. 292–295.

[171] Tanner HG, Pappas GJ, Kumar V. Leader-to-formation stability. IEEE Transactions on Robotics and Automation. 2004;20(3):443–455.

[172] Tanner HG, Kumar V, Pappas GJ. The effect of feedback and feedforward on formation ISS. In: Proc. IEEE International Conference on Robotics and Automation; 2002. p. 3448–3453.

[173] Lin Z, Ding W, Yan G, Yu C, Giua A. Leader-follower formation via complex Laplacian. Automatica. 2013;49(6):1900–1906.

[174] Mesbahi M, Hadaegh FY. Formation flying control of multiple spacecraft via graphs, matrix inequalities, and switching. Journal of Guidance, Control, and Dynamics. 2001;24(2):369–377.

[175] Fidan B, Gazi V, Zhai S, Cen N, Karataş E. Single-view distance-estimation-based formation control of robotic swarms. IEEE Transactions on Industrial Electronics. 2013 December;60(12):5781–5791.

[176] Rezaee H, Abdollahi F, Menhaj MB. Model-free fuzzy leader-follower formation control of fixed wing UAVs. In: Proceedings of the 13th Iranian Conference on Fuzzy Systems. Iran; 2013. p. 1–5.

[177]    Tan K, Lewis MA. Virtual structures for high-precision cooperative mobile robotic control. In: Proceedings of the IEEE/RSJ International Conference on Intelligent Robots and Systems. Japan; 1996. p. 132–139.

[178]    Do KD. Bounded assignment formation control of second-order dynamic agents. IEEE/ASME Transactions on Mechatronics. 2014 April;19(2): 477–489.

[179]    Rezaee H, Abdollahi F. Synchronized cross coupled sliding mode controllers or cooperative UAVs with communication delays. In: Proceedings of the 51st IEEE Conference on Decision and Control. USA; 2012. p. 3116–3121.

[180]    Behroozi F, Gagnon R. A computer-assisted study of pursuit in a plane. American Mathematical Monthly. 1975;82:804–812.

[181]    Behroozi F, Gagnon R. Cyclic pursuit in a plane. Journal of Mathematical Physics. 1979;20(11):2212–2216.

[182]    Marshall JA, Broucke ME, Francis BA. Formations of vehicles in cyclic pursuit. IEEE Transactions on Automatic Control. 2004;9(11):1963–1974.

[183]    Chen Z, Zhang H. A remark on collective circular motion of heterogeneous multi-agents. Automatica. 2013 May;49(5):1236–1241.

[184]    Rezaee H, Abdollahi F. Pursuit formation of double-integrator dynamics using consensus control approach. IEEE Transactions on Industrial Electronics. 2015;62(7):4249–4256.

[185]    Lin Z, Broucke ME, Francis BA. Local control strategies for groups of mobile autonomous agents. IEEE Transactions on Automatic Control. 2004 April;49(4):622–629.

[186]    Marshall JA, Tsai D. Periodic formations of multi vehicle systems. IET Control Theory and Applications. 2010;5(2):389–396.

[187]    Sepulchre R, Paley DA, Leonard NE. Stabilization of planar collective motion with limited communication. IEEE Transactions on Automatic Control. 2008 April;53(3):706–719.

[188]    Morbidi F, Giannitrapani A, Prattichizzo D. Maintaining connectivity among multiple agents in cyclic pursuit: A geometric approach. In: Proceedings of the 49th IEEE Conference on Decision and Control. USA; 2010. p. 7461–7466.

[189]    Rezaee H, Abdollahi F. Pursuit formation control scheme for double-integrator multi-agent systems. In: Proceedings of the 19th IFAC World Congress. South Africa; 2014. p. 10054–10059.

[190]    Yamaguchi H. Adaptive formation control for distributed autonomous mobile robot groups. In: Proc. International Conference on Robotics and Automation; 1997. p. 2300–2305.

[191]    Olfati-Saber R, Murray RM. Graph rigidity and distributed formation stabilization of multi-vehicle systems. In: Proc. Conference on Decision and Control; 2002. p. 2965–2971.

[192]    Koo TJ, Shahruz SM. Formation of a group of unmanned aerial vehicles (UAVs). In: Proc. American Control Conference; 2001. p. 69–74.

[193]    Smith RS, Hadaegh FY. Closed-loop dynamics of cooperative vehicle formations with parallel estimators and communication. IEEE Transactions on Automatic Control. 2007;52(8):1404–1414.

[194] Smith RS, Hadaegh FY. Parallel estimators and communication in spacecraft formations. In: Proc. IFAC World Congress; 2005.

[195] Samar S, Boyd S, Gorinevsky D. Distributed estimation via dual decomposition. In: Proc. European Control Conference; 2007. p. 1511–1519.

[196] Moshtagh N, Jadbabaie A, Daniilidis K. Distributed geodesic control laws for flocking of nonholonomic agents. In: Proc. Conference on Decision and Control; 2005. p. 2835–2840.

[197] Tanner HG, Jadbabaie A, Pappas GJ. Stable flocking of mobile agents, Part I: Fixed topology. In: Proc. Conference on Decision and Control; 2003. p. 2010–2015.

[198] Tanner HG, Jadbabaie A, Pappas GJ. Stable flocking of mobile agents, Part II: Dynamic topology. In: Proc. Conference on Decision and Control; 2003. p. 2016–2021.

[199] Lee D, Spong MW. Stable flocking of multiple inertial agents on balanced graphs. In: Proc. American Control Conference; 2006. p. 2136–2141.

[200] Lee D, Spong MW. Stable flocking of inertial agents on balanced communication graphs. IEEE Transactions on Automatic Control. 2007;52(8):1469–1475.

[201] Marschak J. Elements of a theory of teams. Management Science. 1955;1:127–137.

[202] Ho YC, Chu KC. Team decision theory and information structures in optimal control problems—Part I. IEEE Transactions on Automatic Control. 1972;17(1):15–22.

[203] Ho YC, Chu KC. Team decision theory and information structures in optimal control problems—Part II. IEEE Transactions on Automatic Control. 1972;17(1):22–28.

[204] Marschak J, Radner R. Economic Theory of Teams. New Haven, CT: Yale University Press; 1972.

[205] Radner R. Team decision problems. Annals of Mathematical Statistics. 1962;33(3):857–881.

[206] Yoshikawa T. Decomposition of dynamic team decision problems. IEEE Transactions on Automatic Control. 1978;23:627–632.

[207] Ren W, Beard RW. Consensus seeking in multiagent systems under dynamically changing interaction topologies. IEEE Transactions on Automatic Control. 2005;50(5):655–661.

[208] Fagnani F, Zampieri S. Asymmetric randomized gossip algorithms for consensus. In: Proc. IFAC World Congress; 2008. p. 9051–9056.

[209] Fang L, Antsaklis PJ, Tzimas A. Asynchronous consensus protocols: Preliminary results, simulations, and open questions. In: Proc. Conference on Decision and Control-European Control Conference (CDC-ECC); 2005. p. 2194–2199.

[210] Hirche S, Hara S. Stabilizing interconnection characterization for multi-agent systems with dissipative properties. In: Proc. IFAC World Congress; 2008. p. 1571–1577.

[211]    Rahmani A, Mesbahi M. Pulling the strings on agreement: Anchoring, controllability, and graph automorphisms. In: Proc. American Control Conference; 2007. p. 2738–2743.

[212]    Rahmani A, Mesbahi M. A graph-theoretic outlook on the controllability of the agreement dynamics. In: Proc. European Control Conference; 2007.

[213]    Liu B, Xie G, Chu T, Wang L. Controllability of interconnected systems via switching networks with a leader. In: Proc. IEEE Conference on Systems, Man, and Cybernetics; 2006. p. 3912–3916.

[214]    Angeli D, Bliman PA. Tight estimates for non-stationary consensus with fixed underlying spanning tree. In: Proc. IFAC World Congress; 2008. p. 9021–9026.

[215]    Cao M, Spielman DA, Morse AS. A lower bound on convergence of a distributed network consensus algorithm. In: Proc. Conference on Decision and Control; 2005. p. 2356–2361.

[216]    Olfati-Saber R. Ultrafast consensus in small-world networks. In: Proc. American Control Conference; 2005. p. 2371–2378.

[217]    Moreau L. Stability of continuous time distributed consensus algorithms. In: Proc. Conference on Decision and Control; 2004. p. 3998–4003.

[218]    Lee D, Spong MW. Agreement with non-uniform information delays. In: Proc. American Control Conference; 2006. p. 756–761.

[219]    Münz U, Papachristodoulou A, Allgöwer F. Nonlinear multi-agent system consensus with time-varying delays. In: Proc. IFAC World Congress; 2008. p. 1522–1527.

[220]    Tsitsiklis JN. Problems in Decentralized Decision Making and Computation. Boston, MA: Massachusetts Institute of Technology; 1984.

[221]    Tsitsiklis JN, Bertsekas DP, Athans M. Distributed asynchronous deterministic and stochastic gradient optimization algorithms. IEEE Transactions on Automatic Control. 1986;31(9):803–812.

[222]    Blondel V, Hendrickx JM, Olshevsky A, Tsitsiklis JN. Convergence in multi-agent coordination, consensus, and flocking. In: Proc. Conference on Decision and Control-European Control Conference (CDC-ECC); 2005. p. 2996–3000.

[223]    Ren W. Second-order consensus algorithm with extensions to switching topologies and reference models. In: Proc. American Control Conference; 2007. p. 1431–1436.

[224]    Xie G, Wang L. Consensus control for a class of networks of dynamic agents: Switching topology. In: Proc. American Control Conference; 2006. p. 1382–1387.

[225]    Shi H, Wang L, Chu T. Coordinated control of multiple interactive dynamical agents with asymmetric coupling pattern and switching topology. In: Proc. IEEE/RSJ International Conference on Intelligent Robots and Systems; 2006. p. 3209–3214.

[226]    Wang L, Xiao F. A new approach to consensus problems for discrete-time multiagent systems with time-delays. In: Proc. American Control Conference; 2006. p. 2118–2123.

[227] Bauso D, Giarre L, Pesenti R. Mechanism design for optimal consensus problems. In: Proc. Conference on Decision and Control; 2006. p. 3381–3386.

[228] Kingston DB, Beard RW. Discrete-time average-consensus under switching network topologies. In: Proc. American Control Conference; 2006. p. 3551–3556.

[229] Gazi V. Stability of an asynchronous swarm with time-dependent communication links. IEEE Transactions on Systems, Man, and Cybernetics Part B (Cybernetics). 2008;38(1):267–274.

[230] Sun YG, Wang L, Xie G. Average consensus in directed networks of dynamic agents with time-varying communication delays. In: Proc. Conference on Decision and Control; 2006. p. 3393–3398.

[231] Ren W, Beard RW. Consensus of information under dynamically changing interaction topologies. In: Proc. American Control Conference; 2004. p. 4939–4944.

[232] Wang JZ, Mareels I, Tan Y. Robustness of distributed multi-agent consensus. In: Proc. IFAC World Congress; 2008. p. 1510–1515.

[233] Ma CQ, Li T, Zhang JF. Leader-following consensus control for multi-agent systems under measurement noises. In: Proc. IFAC World Congress; 2008. p. 1528–1533.

[234] Li T. Asymptotically unbiased average consensus under measurement noises and fixed topologies. In: Proc. IFAC World Congress; 2008. p. 2867–2873.

[235] Bauso D, Giarre L, Pesenti R. Robust control in uncertain multi-inventory systems and consensus problems. In: Proc. IFAC World Congress; 2008. p. 9027–9032.

[236] Horn RA, Johnson CR. Matrix Analysis. Cambridge: Cambridge University Press; 2012.

[237] Saboori I. Cooperative and Consensus-Based Control for a Team of Multi-Agent Systems. Montreal: Concordia University; 2016.

[238] Gahinet P, Apkarian P. A linear matrix inequality approach to $H_\infty$ control. International Journal of Robust and Nonlinear Control. 1994;4:421–448.

[239] Zhou K, Doyle JC. Essentials of Robust Control. Upper Saddle River, New Jersey: Prentice Hall; 1998.

[240] Wang Y, Zhang H. $H_\infty$ control for uncertain Markovian jump systems with mode-dependent mixed delays. Progress in Natural Science. 2008;18(3):309–314.

[241] Lin P, Jia Y, Du J, Yu F. Distributed leadless coordination for networks of second-order agents with time-delay on switching topology. In: American Control Conference; 2008. p. 1564–1569.

[242] Liu J, Wang JL, Yang GH. An LMI approach to minimum sensitivity analysis with application to fault detection. Automatica. 2005;41(11):1995–2004.

[243] Wang H, Yang GH. A finite frequency domain approach to fault detection observer design for linear continuous-time systems. Asian Journal of Control. 2008;10(5):559–568.

[244] Davoodi MR, Meskin N, Khorasani K. Simultaneous fault detection and control design for an autonomous unmanned underwater vehicle. In: 7th IEEE GCC Conference and Exhibition; 2013. p. 529–534.

[245]    Gao H, Lam J, Xie L, Wang C. New approach to mixed $H_2/H_\infty$ filtering for polytopic discrete-time systems. IEEE Transactions on Signal Processing. 2005 August;53(8):3183–3192.

[246]    Dai X, Gao Z, Breikin T, Wang H. Zero assignment for robust $H_2/H_\infty$ fault detection filter design. IEEE Transactions on Signal Processing. 2009;57:1363–1372.

[247]    Park JK, Shin DR, Chung TM. Dynamic observers for linear time-invariant systems. Automatica. 2002;38(6):1083–1087.

[248]    Pertew AM, Marquez HJ, Zhao Q. Design of unknown input observers for Lipschitz nonlinear systems. In: Proceedings of the American Control Conference; 2005. p. 4198–4203.

[249]    Lien CH. An efficient method to design robust observer-based control of uncertain linear systems. Applied Mathematics and Computation. 2004; 158(1):29–44.

[250]    Khosrowjerdi MJ, Nikoukhah R, Safari-Shad N. Fault detection in a mixed $H_2/H_\infty$ setting. IEEE Transactions on Automatic Control. 2005; 50(7):1063–1068.

[251]    Casavola A, Famularo D, Franze G. A robust deconvolution scheme for fault detection and isolation of uncertain linear systems: An LMI approach. Automatica. 2005;41:1463–1472.

[252]    Lien CH. Robust observer-based control of systems with state perturbations via LMI approach. IEEE Transactions on Automatic Control. 2004;49(8):1365–1370.

[253]    Hyakudome T. Design of autonomous underwater vehicle. International Journal of Advanced Robotic Systems. 2011;8(1):131–139.

[254]    Budiyono A. Advances in unmanned underwater vehicles technologies: Modeling, control and guidance perspective. Indian Journal of Marine Sciences. 2009;38(3):282–295.

[255]    Ernits J, Dearden R, Pebody M. Automatic fault detection and execution monitoring for AUV missions. In: IEEE/OES Autonomous Underwater Vehicles (AUV); 2010. p. 1–10.

[256]    Alessandri A, Hawkinson T, Healey AJ, Veruggio G. Robust model-based fault diagnosis for unmanned underwater vehicles using sliding mode-observers. In: 11th International Symposium on Unmanned Untethered Submersible Technology; 1999. p. 22–25.

[257]    Alessandri A, Caccia M, Veruggio G. Fault detection of actuator faults in unmanned underwater vehicles. Control Engineering Practice. 1999; 7(3):357–368.

[258]    Kim JH, Beale GO. Fault detection and classification in underwater vehicles using the $T^2$ statistic. In: Mediterranean Control and Automation Conference; 2001.

[259]    Antonelli G, Caccavale F, Sansone C, Villani L. Fault diagnosis for AUVs using support vector machines. In: Proc. IEEE International Conference on Robotics 8 Automation; 2004. p. 4486–4491.

[260]    Liang X, Li W, Su L, Yin H, Zhao J. Thruster fault diagnosis of autonomous underwater vehicles based on least disturbance wavelet neural network.

In: Proceedings of 2010 Second International Conference on Computer Modeling and Simulation; 2010. p. 78–82.

[261] Feng Z, Allen R. Reduced order $H_\infty$ control of an autonomous underwater vehicle. Control Engineering Practice. 2004;12:1511–1520.

[262] Fossen TI. Marine Control Systems. Guidance, Navigation and Control of Ships, Rigs and Underwater Vehicles. Trondheim, Norway: Marine Cybernetics AS; 2002.

[263] Healey AJ, Lienard D. Multivariable sliding mode control for autonomous diving and steering of unmanned underwater vehicles. IEEE Journal of Oceanic Engineering. 1993;18(3):327–339.

[264] Davoodi MR, Meskin N, Khorasani K. Simultaneous fault detection, isolation and control tracking design using a single observer-based module. In: Proc. American Control Conference. Portland, Oregon, USA; 2014. p. 3047–3052.

[265] Davoodi M, Meskin N, Khorasani K. A single dynamic observer-based module for design of simultaneous fault detection, isolation and tracking control scheme. International Journal of Control. 2018;91(3):508–523.

[266] Zhang X, Polycarpou MM, Parisini T. A robust detection and isolation scheme for abrupt and incipient faults in nonlinear systems. IEEE Transactions on Automatic Control. 2002;47:576–593.

[267] Chen W, Saif M. Fault detection and isolation based on novel unknown input observer design. In: Proceedings of the American Control Conference. Minnesota, USA; 2006. p. 5129–5134.

[268] Gagliardi G, Casavola A, Famularo D. A bank of observers based LPV fault detection and isolation method for spark injection engines. In: Proceedings of the IEEE International Conference on Control Applications. Yokohama, Japan; 2010. p. 561–566.

[269] Meskin N, Khorasani K. Actuator fault detection and isolation for a network of unmanned vehicles. IEEE Transactions on Automatic Control. 2009;54(4):835–840.

[270] Shames I, Teixeira AH, Sandberg H, Johansson KH. Distributed fault detection for interconnected second-order systems. Automatica. 2011;47:2757–2764.

[271] Scherer C, Gahinet P, Chilali M. Multiobjective output-feedback control via LMI optimization. IEEE Transactions on Automatic Control. 1997;42(7):896–911.

[272] Commault C, Dion JM, Sename O, Motyeian R. Observer-based fault detection and isolation for structured systems. IEEE Transactions on Automatic Control. 2002;47(12):2074–2079.

[273] Davoodi MR, Khorasani K, Talebi HA, Momeni HR. Distributed fault detection and isolation filter design for a network of heterogeneous multi-agent systems. IEEE Transactions on Control Systems Technology. 2014;22(3):1061–1069.

[274] Trinh DH, Chafouk H. Fault detection and isolation using Kalman filter bank for a wind turbine generator. In: 19th Mediterranean Conference on Control Automation (MED); 2011. p. 144–149.

[275]   Oliveira MCD, Bernussou J, Geromel JC. A new discrete-time robust stability condition. Systems & Control Letters. 1999;37(4):261–265.

[276]   Oliveira MCD, Geromel JC, Bernussou J. Extended $H_2$ and $H_\infty$ norm characterizations and controller parameterizations for discrete-time systems. International Journal of Control. 2002;75(9):666–679.

[277]   Pipeleers G, Demeulenaere B, Swevers J, Vandenberghe L. Extended LMI characterizations for stability and performance of linear systems. Systems & Control Letters. 2009;58:510–518.

[278]   Mazars E, Jaimoukha IM, Li Z. Computation of a reference model for robust fault detection and isolation residual generation. Journal of Control Science and Engineering. 2008;2008:1–12.

[279]   Gao H, Chen T. Network-based $H_\infty$ output tracking control. IEEE Transactions on Automatic Control. 2008;53(3):655–667.

[280]   Chung WH, Speyer JL. A game theoretic fault detection filter. IEEE Transactions on Automatic Control. 1998;43(2):143–161.

[281]   Long Y, Yang GH. Fault detection for a class of nonhomogeneous Markov jump systems based on delta operator approach. Proceedings of the Institution of Mechanical Engineers, Part I: Journal of Systems and Control Engineering. 2013;227(1):129–141.

[282]   Guo J, Huang X, Cui Y. Design and analysis of robust fault detection filter using LMI tools. Computers & Mathematics with Applications. 2009;57:1743–1747.

[283]   Boyd S, El Ghaoui L, Feron E, Balakrishnan V. Linear Matrix Inequalities in System and Control Theory. Philadelphia, PA: Studies in Applied Mathematics. SIAM; 1994.

[284]   Feng Z, Allen R. $H_\infty$ autopilot design for an autonomous underwater vehicle. In: Proceedings of the IEEE International Conference on Control Applications. Glasgow; 2002. p. 350–354.

[285]   Naik MS, Singh SN. State-dependent Riccati equation-based robust dive plane control of AUV with control constraints. Ocean Engineering. 2007;34(11–12):1711–1723.

[286]   Davoodi MR, Meskin N, Khorasani K. Integrated fault detection, isolation and control design for continuous-time Markovian jump systems with uncertain transition probabilities. In: 53rd IEEE Conference on Decision and Control. Los Angeles, USA; 2014. p. 5743–5749.

[287]   Davoodi M, Meskin N, Khorasani K. Integrated design of fault detection, isolation, and control for continuous-time Markovian jump systems. International Journal of Adaptive Control and Signal Processing. 2017. Available from: http://dx.doi.org/10.1002/acs.2806.

[288]   Zhang H, Wang J, Shi Y. Robust sliding-mode control for Markovian jump systems subject to intermittent observations and partially known transition probabilities. Systems & Control Letters. 2013;62(12):1114–1124.

[289]   Boukas EK. Stochastic Switching Systems: Analysis and Design. Boston: Birkhauser; 2006.

[290]   Zhong M, Ye H, Shi P, Wang G. Fault detection for Markovian jump systems. IEE Proceedings – Control Theory and Applications. 2005;152(4):397–402.

[291] Zhang L, Boukas EK, Baron L. Fault detection for discrete-time Markov jump linear systems with partially known transition probabilities. In: 47th IEEE Conference on Decision and Control; 2008. p. 1054–1059.

[292] Saijai J, Abdo A, Damlakhi W, Ding SX. Fault detection scheme for discrete-time Markov jump linear systems with mode-independent residual. In: IEEE International Conference on Control Applications; 2012. p. 1674–1679.

[293] Long Y, Yang GH. Fault detection for a class of nonlinear stochastic systems with Markovian switching and mixed time-delays. International Journal of Systems Science. 2014;45(3):215–231.

[294] Dong H, Wang Z, Gao H. Fault detection for Markovian jump systems with sensor saturations and randomly varying nonlinearities. IEEE Transactions on Circuits and Systems I: Regular Papers. 2012 October;59(10):2354–2362.

[295] Meskin N, Khorasani K. A geometric approach to fault detection and isolation of continuous-time Markovian jump linear systems. IEEE Transactions on Automatic Control. 2010;55(6):1343–1357.

[296] Meskin N, Khorasani K. Fault detection and isolation of discrete-time Markovian jump linear systems with application to a network of multi-agent systems having imperfect communication channels. Automatica. 2009;45(9):2032–2040.

[297] Li YH, Jiang BY, Li F, Zhou B. Robust $L_1/H_\infty$ control of linear Markovian jump systems. In: International Conference on Machine Learning and Cybernetics (ICMLC); vol. 2; 2010. p. 975–980.

[298] Haddad MM, Chellaboina VS. Mixed-norm $H_2/L_1$ controller synthesis via fixed-order dynamic compensation: A Riccati equation approach. In: IEEE Conference on Decision and Control; 1997. p. 452–457.

[299] Curry TD, Collins EG. Robust fault detection and isolation using robust $l_1$ estimation. Journal of Guidance, Control, and Dynamics. 2005; 28(6):1131–1139.

[300] Xiong J, Lam J, Gao H, Ho DWC. On robust stabilization of Markovian jump systems with uncertain switching probabilities. Automatica. 2005;41(5):897–903.

[301] Cao YY, Lam J. Robust $H_\infty$ control of uncertain Markovian jump systems with time-delay. IEEE Transactions on Automatic Control. 2000;45(1):77–83.

[302] He S, Liu F. Fuzzy model-based fault detection for Markov jump systems. International Journal of Robust and Nonlinear Control. 2009;19(11):1248–1266.

[303] Yu J, Liu M, Yang W, Shi P, Tong S. Robust fault detection for Markovian jump systems with unreliable communication links. International Journal of Systems Science. 2013;44(11):2015–2026.

[304] Luo M, Liu G, Zhong S. Robust fault detection of Markovian jump systems with different system modes. Applied Mathematical Modelling. 2013;37(7):5001–5012.

[305] Ma L, Wang Z, Bo Y, Guo Z. Robust $H_\infty$ sliding mode control for nonlinear stochastic systems with multiple data packet losses. International Journal of Robust and Nonlinear Control. 2012;22:473–491.

[306]    Wang Z, Shen B, Shu H, Wei G. Quantized $H_\infty$ control for nonlinear stochastic time-delay systems with missing measurements. IEEE Transactions on Automatic Control. 2012;57(6):1431–1444.

[307]    Wang Z, Liu Y, Liu X. $H_\infty$ filtering for uncertain stochastic time-delay systems with sector-bounded nonlinearities. Automatica. 2008;44(5):1268–1277.

[308]    Haddad WM, Chellaboina VS. Mixed-norm $H_2/L_1$ controller synthesis via fixed-order dynamic compensation: A Riccati equation approach. International Journal of Control. 1998;71(1):35–59.

[309]    Collins Jr EG, Song T. Robust $l_1$ estimation using the Popov-Tsypkin multiplier with application to robust fault detection. International Journal of Control. 2001;74(3):303–313.

[310]    Davoodi MR, Meskin N, Khorasani K. Event-triggered fault detection, isolation and control design of linear systems. In: IEEE International Conference on Event-Based Control, Communication, and Signal Processing. Poland; 2015.

[311]    Davoodi M, Meskin N, Khorasani K. Event-triggered multiobjective control and fault diagnosis: A unified framework. IEEE Transactions on Industrial Informatics. 2017 February;13(1):298–311.

[312]    Liu Q, Wang Z, He X, Zhou DH. Event-based distributed filtering with stochastic measurement fading. IEEE Transactions on Industrial Informatics. 2015 December;11(6):1643–1652.

[313]    Chen X, Hao F. Observer-based event-triggered control for certain and uncertain linear systems. IMA Journal of Mathematical Control and Information. 2013;30(4):527–542.

[314]    Donkers MCF, Heemels WPMH. Output-based event-triggered control with guaranteed $L_\infty$-gain and improved and decentralized event-triggering. IEEE Transactions on Automatic Control. 2012;57(6):1362–1376.

[315]    Panagi P, Polycarpou MM. A coordinated communication scheme for distributed fault tolerant control. IEEE Transactions on Industrial Informatics. 2013 February;9(1):386–393.

[316]    Tahir M, Mazumder SK. Self-triggered communication enabled control of distributed generation in microgrids. IEEE Transactions on Industrial Informatics. 2015 April;11(2):441–449.

[317]    Peng C, Han QL. A novel event-triggered transmission scheme and control co-design for sampled-data control systems. IEEE Transactions on Automatic Control. 2013;58(10):2620–2626.

[318]    Anta A, Tabuada P. To sample or not to sample: Self-triggered control for nonlinear systems. IEEE Transactions on Automatic Control. 2010;55(9):2030–2042.

[319]    Meng X, Chen T. Event triggered robust filter design for discrete-time systems. IET Control Theory Applications. 2014;8(2):104–113.

[320]    Wang X, Lemmon MD. Event-triggering in distributed networked control systems. IEEE Transactions on Automatic Control. 2011;56(3):586–601.

[321]    Abdelrahim M, Postoyan R, Daafouz J, Nešić D. Co-design of output feedback laws and event-triggering conditions for linear systems. In: IEEE Conference on Decision and Control; 2014. p. 3560–3565.

[322] Wu L, Yao X, Zheng WX. Generalized $H_2$ fault detection for two-dimensional Markovian jump systems. Automatica. 2012;48(8):1741–1750.

[323] Zheng Z, Wu L, Xiong Y. Fault detection of switched systems with repeated scalar non-linearities. IET Control Theory Applications. 2014;8(9): 755–764.

[324] Li S, Sauter D, Xu B. Co-design of event-triggered $H_\infty$ control for discrete-time linear parameter-varying systems with network-induced delays. Journal of the Franklin Institute. 2015;352(5):1867–1892.

[325] Li S, Sauter D, Xu B. Fault isolation filter for networked control system with event-triggered sampling scheme. Sensors. 2011;11:557–572.

[326] Liu J, Yue D. Event-based fault detection for networked systems with communication delay and nonlinear perturbation. Journal of the Franklin Institute. 2013;350(9):2791–2807.

[327] Hajshirmohamadi S, Davoodi MR, Meskin N. Event-triggered fault detection for discrete-time linear systems. In: IEEE Conference on Control Applications (CCA); 2015. p. 990–995.

[328] Hajshirmohamadi S, Davoodi MR, Meskin N, Sheikholeslam F. Event-triggered fault detection and isolation for discrete-time linear systems. IET Control Theory and Applications. 2016;10(5):526–533.

[329] Davoodi MR, Meskin N, Khorasani K. Simultaneous fault detection, isolation and control tracking design using a single observer-based module. In: American Control Conference; 2014. p. 3047–3052.

[330] Peng C, Yang TC. Event-triggered communication and $H_\infty$ control co-design for networked control systems. Automatica. 2013;49(5):1326–1332.

[331] Caccavale F, Iamarino M, Pierri F, Tufano V. Control and Monitoring of Chemical Batch Reactors. London: Springer-Verlag; 2011.

[332] Walsh GC, Ye H. Scheduling of networked control systems. IEEE Control Systems Magazine. 2001;21(1):57–65.

[333] Feng Z, Allen R. Reduced order $H_\infty$ control of an autonomous underwater vehicle. Control Engineering Practice. 2004;12(12):1511–1520.

[334] Davoodi M, Meskin N, Khorasani K. Event-triggered fault estimation and accommodation design for linear systems. In: Second International Conference on Event-based Control, Communication, and Signal Processing (EBCCSP); 2016. p. 1–6.

[335] Miskowicz M. Event-Based Control and Signal Processing. Boca Raton, FL: CRC Press; 2015.

[336] Heemels WPMH, Donkers MCF, Teel AR. Periodic event-triggered control for linear systems. IEEE Transactions on Automatic Control. 2013;58(4):847–861.

[337] Zhang H, Feng G, Yan H, Chen Q. Observer-based output feedback event-triggered control for consensus of multi-agent systems. IEEE Transactions on Industrial Electronics. 2014 September;61(9):4885–4894.

[338] Hajshirmohamadi S, Davoodi MR, Meskin N, Sheikholeslam F. Event-triggered fault detection for discrete-time linear multi-agent systems. In: American Control Conference; 2016. p. 5945–5950.

[339]    Almeida J, Silvestre C, Pascoal AM. Self-triggered output feedback control of linear plants in the presence of unknown disturbances. IEEE Transactions on Automatic Control. 2014 November;59(11):3040–3045.

[340]    Jiang B, Staroswiecki M, Cocquempot V. Fault accommodation for nonlinear dynamic systems. IEEE Transactions on Automatic Control. 2006 September;51(9):1578–1583.

[341]    Davoodi MR, Meskin N, Khorasani K. Event-triggered fault detection, isolation and control design of linear systems. In: International Conference on Event-based Control, Communication, and Signal Processing (EBCCSP); 2015. p. 1–6.

[342]    Davoodi M, Meskin N, Khorasani K. Event-triggered multi-objective control and fault diagnosis: A unified framework. IEEE Transactions on Industrial Informatics. 2017 February;13(1):298–311.

[343]    Khosrowjerdi MJ, Nikoukhah R, Safari-Shad N. A mixed $H_2/H_\infty$ approach to simultaneous fault detection and control. Automatica. 2004;40(2): 261–267.

[344]    Bouarar T, Marx B, Maquin D, Ragot J. Fault-tolerant control design for uncertain Takagi–Sugeno systems by trajectory tracking: A descriptor approach. IET Control Theory & Applications. 2013 September;7(14):1793–1805.

[345]    Edwards C, Tan CP. Sensor fault tolerant control using sliding mode observers. Control Engineering Practice. 2006;14(8):897–908.

[346]    Odgaard PF, Stoustrup J. Fault tolerant control of wind turbines using unknown input observers. In: 8th IFAC Symposium on Fault Detection, Supervision and Safety of Technical Processes; 2012. p. 313–318.

[347]    Shi P, Liu M, Zhang L. Fault-tolerant sliding-mode-observer synthesis of Markovian jump systems using quantized measurements. IEEE Transactions on Industrial Electronics. 2015;62(9):5910–5918.

[348]    Dong Q, Zhong M, Ding SX. Active fault tolerant control for a class of linear time-delay systems in finite frequency domain. International Journal of Systems Science. 2012;43(3):543–551.

[349]    Chang JL. Applying discrete-time proportional integral observers for state and disturbance estimations. IEEE Transactions on Automatic Control. 2006 May;51(5):814–818.

[350]    Zhang K, Jiang B, Shi P. Observer-based integrated robust fault estimation and accommodation design for discrete-time systems. International Journal of Control. 2010;83(6):1167–1181.

[351]    Tabatabaeipour SM, Bak T. Robust observer-based fault estimation and accommodation of discrete-time piecewise linear systems. Journal of the Franklin Institute. 2014;351(1):277–295.

[352]    Jiang B, Zhang K, Shi P. Integrated fault estimation and accommodation design for discrete-time Takagi–Sugeno fuzzy systems with actuator faults. IEEE Transactions on Fuzzy Systems. 2011 April;19(2):291–304.

[353]    Corless M, Tu J. State and input estimation for a class of uncertain systems. Automatica. 1998;34(6):757–764.

[354]    Davoodi MR, Meskin N, Khorasani K. Simultaneous fault detection and control design for a network of multi-agent systems. In: European Control Conference; 2014. p. 575–581.

[355] Davoodi M, Meskin N, Khorasani K. Simultaneous fault detection and consensus control design for a network of multi-agent systems. Automatica. 2016;66:185–194.

[356] Zhang X. Decentralized fault detection for a class of large-scale nonlinear uncertain systems. In: American Control Conference; 2010. p. 5650–5655.

[357] Meskin N, Khorasani K, Rabbath CA. A hybrid fault detection and isolation strategy for a network of unmanned vehicles in presence of large environmental disturbances. IEEE Transactions on Control Systems Technology. 2010;18(6):1422–1429.

[358] Davoodi MR, Khorasani K, Talebi HA, Momeni HR. A robust semi-decentralized fault detection strategy for multi-agent systems: An application to a network of micro-air vehicles. International Journal of Intelligent Unmanned Systems. 2013;1(1):21–35.

[359] Li Z, Duan Z, Huang L. $H_\infty$ control of networked multi-agent systems. Journal of Systems Science and Complexity. 2009;22(1):35–48.

[360] Li Z, Duan Z, Chen G, Huang L. Consensus of multiagent systems and synchronization of complex networks: A unified viewpoint. IEEE Transactions on Circuits and Systems I: Regular Papers. 2010;57(1):213–224.

[361] Alavi SMM, Saif M. A QFT-based decentralized design approach for integrated fault detection and control. IEEE Transactions on Control Systems Technology. 2012;20(5):1366–1375.

[362] Massioni P, Verhaegen M. Distributed control for identical dynamically coupled systems: A decomposition approach. IEEE Transactions on Automatic Control. 2009;54(1):124–135.

[363] Zhang H, Lewis FL, Das A. Optimal design for synchronization of cooperative systems: State feedback, observer and output feedback. IEEE Transactions on Automatic Control. 2011 August;56(8):1948–1952.

[364] Li Z, Mazars E, Zhang Z, Jaimoukha IM. State–space solution to the $H_-/H_\infty$ fault-detection problem. International Journal of Robust and Nonlinear Control. 2012;22:282–299.

[365] Liu Y, Jia Y, Du J, Yuan S. Dynamic output feedback control for consensus of multi-agent systems: An $H_\infty$ approach. In: American Control Conference; 2009. p. 4470–4475.

[366] Zhang P, Ding SX. An integrated trade-off design of observer based fault detection systems. Automatica. 2008;44:1886–1894.

[367] Varrier S. Robust Control of Autonomous Underwater Vehicles. France: Gipsa-Lab, INRIA; 2010.

[368] Li F, Wu L, Shi P. Stochastic stability of semi-Markovian jump systems with mode-dependent delays. International Journal of Robust and Nonlinear Control. 2014;24(18):3317–3330.

[369] Girard A. Dynamic triggering mechanisms for event-triggered control. IEEE Transactions on Automatic Control. 2015;60(7):1992–1997.

[370] Dolk VS, Borgers DP, Heemels WPMH. Output-based and decentralized dynamic event-triggered control with guaranteed $L_p$-gain performance and Zeno-freeness. IEEE Transactions on Automatic Control. 2017;62(1):34–49.

[371]   Zhou B, Xu C, Duan G. Distributed and truncated reduced-order observer based output feedback consensus of multi-agent systems. IEEE Transactions on Automatic Control. 2014;59(8):2264–2270.

[372]   Mu B, Li H, Ding J, Shi Y. Consensus in second-order multiple flying vehicles with random delays governed by a Markov chain. Journal of the Franklin Institute. 2015;352(9):3628–3644.

[373]   Huang N, Duan Z, Zhao Y. Consensus of multi-agent systems via delayed and intermittent communications. IET Control Theory Applications. 2015; 9(1):62–73.

[374]   Guo XG, Wang JL, Liao F, Suresh S, Narasimalu S. Quantized insensitive consensus of Lipschitz nonlinear multi-agent systems using the incidence matrix. Journal of the Franklin Institute. 2015;352(11):4845–4863.

[375]   Fu J, Wang J. Adaptive coordinated tracking of multi-agent systems with quantized information. Systems & Control Letters. 2014;74:115–125.

[376]   Li D, Liu Q, Wang X, Yin Z. Quantized consensus over directed networks with switching topologies. Systems & Control Letters. 2014;65:13–22.

[377]   Li J, Park JH, Ye D. Fault detection filter design for switched systems with quantisation effects and packet dropout. IET Control Theory Applications. 2017;11(2):182–193.

[378]   Long Y, Yang GH. Fault detection for networked control systems subject to quantisation and packet dropout. International Journal of Systems Science. 2013;44(6):1150–1159.

[379]   Wan X, Fang H, Fu S. Observer-based fault detection for networked discrete-time infinite-distributed delay systems with packet dropouts. Applied Mathematical Modelling. 2012;36(1):270–278.

[380]   Meng XJ, Yang GH. Simultaneous fault detection and control for stochastic time-delay systems. International Journal of Systems Science. 2014;45(5):1058–1069.

[381]   Morales-Caporal R, Pacas M. Suppression of saturation effects in a sensorless predictive controlled synchronous reluctance machine based on voltage space phasor injections. IEEE Transactions on Industrial Electronics. 2011 July;58(7):2809–2817.

[382]   Blazquez LF, de Miguel LJ. Saturation effects detecting additive faults. In: 2001 European Control Conference (ECC); 2001. p. 2340–2345.

[383]   Fan J, Zhang Y, Zheng Z. Integrated robust fault detection, diagnosis and reconfigurable control system with actuator saturation. In: Annual Conference of the Prognostics and Health Management (PHM) Society; 2011. p. 1–11.

[384]   Hu T, Lin Z, Chen BM. An analysis and design method for linear systems subject to actuator saturation and disturbance. Automatica. 2002;38(2):351–359.

[385]   Park BS, Yoo SJ. Fault detection and accommodation of saturated actuators for underactuated surface vessels in the presence of nonlinear uncertainties. Nonlinear Dynamics. 2016 July;85(2):1067–1077.

[386]   Fan J, Zhang Y, Zheng Z. Adaptive observer-based integrated fault diagnosis and fault-tolerant control systems against actuator faults and saturation. Journal of Dynamic Systems, Measurement, and Control. 2013;135(4):1–13.

[387]  Meisami-Azad M, Mohammadpour J, Grigoriadis KM, Harold MP, Franchek MA. LPV gain-scheduled control of SCR aftertreatment systems. International Journal of Control. 2012;85(1):114–133.

[388]  Millerioux G, Rosier L, Bloch G, Daafouz J. Bounded state reconstruction error for LPV systems with estimated parameters. IEEE Transactions on Automatic Control. 2004;49(8):1385–1389.

[389]  Shamma JS, Athans M. Guaranteed properties of gain scheduled control for linear parameter-varying plants. Automatica. 1991;27(3):559–564.

[390]  Zhou K, Doyle JC, Glover K. Robust and Optimal Control. vol. 40. New Jersey: Prentice Hall; 1996.

[391]  Apkarian P, Adams RJ. Advanced gain-scheduling techniques for uncertain systems. IEEE Transactions on Control Systems Technology. 1998; 6(1):21–32.

[392]  Borges J, Verdult V, Verhaegen M, Botto MA. Separable least squares for projected gradient identification of composite local linear state-space models. In: Proc. of the 16th Intl. Symp. Math. Theory of Networks and Systems; 2004.

[393]  Marcos A, Balas GJ. Development of linear-parameter-varying models for aircraft. Journal of Guidance, Control, and Dynamics. 2004;27(2):218–228.

[394]  Verdult V, Lovera M, Verhaegen M. Identification of linear parameter-varying state-space models with application to helicopter rotor dynamics. International Journal of Control. 2004;77(13):1149–1159.

[395]  Bianchi FD, Mantz RJ, Christiansen CF. Gain scheduling control of variable-speed wind energy conversion systems using quasi-LPV models. Control Engineering Practice. 2005;13(2):247–255.

[396]  Meisami-Azad M, Grigoriadis KM. Anti-windup linear parameter-varying control of pitch actuators in wind turbines. Wind Energy. 2015;18(2):187–200.

[397]  van Wingerden JW. Control of Wind Turbines with 'Smart' Rotors: Proof of Concept & LPV Subspace Identification. Delft: Delft University of Technology; 2008.

[398]  Baslamisli S, Köse IE, Anlaş G. Gain-scheduled integrated active steering and differential control for vehicle handling improvement. Vehicle System Dynamics. 2009;47(1):99–119.

[399]  Wijnheijmer F, Naus G, Post W, Steinbuch M, Teerhuis P. Modelling and LPV control of an electro-hydraulic servo system. In: Proc. of the 2006 IEEE Conf. on Computer Aided Control System Design; 2006. p. 3116–3121.

[400]  Fiorentini L, Serrani A, Bolender MA, Doman DB. Nonlinear robust adaptive control of flexible air-breathing hypersonic vehicles. Journal of Guidance, Control, and Dynamics. 2009;32(2):402–417.

[401]  Tanelli M, Ardagna D, Lovera M. LPV model identification for power management of web service systems. In: Proc. of the 2008 IEEE Intl. Conf. on Control Applications; 2008. p. 1171–1176.

[402]  Paijmans B, Symens W, Brussel HV, Swevers J. Identification of interpolating affine LPV models for mechatronic systems with one varying parameter. European Journal of Control. 2008;14(1):16–29.

[403]   Barker JM, Balas GJ. Comparing linear parameter-varying gain-scheduled control techniques for active flutter suppression. Journal of Guidance, Control, and Dynamics. 2000;23(5):948–955.

[404]   Bachnas AA, Tóth R, Ludlage JHA, Mesbah A. A review on data-driven linear parameter-varying modeling approaches: A high-purity distillation column case study. Journal of Process Control. 2014;24(4):272–285.

[405]   Nishimura H, Funaki K. Motion control of three-link brachiation robot by using final-state control with error learning. IEEE/ASME Transactions on Mechatronics. 1998;3(2):120–128.

[406]   Wassink MG, van de Wal M, Scherer C, Bosgra O. LPV control for a wafer stage: Beyond the theoretical solution. Control Engineering Practice. 2005;13(2):231–245.

[407]   Pellanda PC, Apkarian P, Tuan HD. Missile autopilot design via a multi-channel LFT/LPV control method. International Journal of Robust and Nonlinear Control. 2002;12(1):1–20.

[408]   Giarré L, Bauso D, Falugi P, Bamieh B. LPV model identification for gain scheduling control: An application to rotating stall and surge control problem. Control Engineering Practice. 2006;14(4):351–361.

[409]   Takahashi K, Massaquoi SG. Neuroengineering model of human limb control gain-scheduled feedback control approach. In: Proc. of the 46th IEEE Conf. on Decision and Control; 2007. p. 5826–5832.

[410]   Gao J, Budman HM. Design of robust gain-scheduled PI controllers for nonlinear processes. Journal of Process Control. 2005;15(7):807–817.

[411]   Sename O, Gaspar P, Bokor J. Robust Control and Linear Parameter Varying Approaches: Application to Vehicle Dynamics. vol. 437. Berlin, Heidelberg: Springer; 2013.

[412]   Leith DJ, Leithead WE. Survey of gain-scheduling analysis and design. International Journal of Control. 2000;73(11):1001–1025.

[413]   Mohammadpour J, Scherer CW. Control of Linear Parameter Varying Systems with Applications. Berlin, Heidelberg: Springer Science & Business Media; 2012.

[414]   Rugh WJ, Shamma JS. Research on gain scheduling. Automatica. 2000;36(10):1401–1425.

[415]   Zhang ZH, Yang GH. Fault detection for discrete-time LPV systems using interval observers. International Journal of Systems Science. 2017;48(14):2921–2935.

[416]   Chen J, Cao YY, Zhang W. A fault detection observer design for LPV systems in finite frequency domain. International Journal of Control. 2015; 88(3):571–584.

[417]   Chadli M, Davoodi M, Meskin N. Distributed state estimation, fault detection and isolation filter design for heterogeneous multi-agent linear parameter-varying systems. IET Control Theory Applications. 2017;11(2): 254–262.

[418]   Golabi A, Meskin N, Toth R, Mohammadpour J, Donkers T, Davoodi M. Event-triggered constant reference tracking control for discrete-time LPV

systems with application to a laboratory tank system. IET Control Theory Applications. 2017;11(16):2680–2687.

[419] Hu F, Lu Y, Vasilakos AV, *et al.* Robust cyber-physical systems: Concept, models, and implementation. Future Generation Computer Systems. 2016;56:449–475. Available from: http://www.sciencedirect.com/science/article/pii/S0167739X15002071.

[420] Kim KD, Kumar PR. Cyber-physical systems: A perspective at the centennial. Proceedings of the IEEE. 2012 May;100(Special Centennial Issue):1287–1308.

[421] Ding D, Wang Z, Dong H, Liu Y, Ahmad B. Performance analysis with network-enhanced complexities: On fading measurements, event-triggered mechanisms, and cyber attacks. Abstract and Applied Analysis. 2014;2014:1–10.

[422] Amin S, Litrico X, Sastry S, Bayen AM. Cyber security of water SCADA systems – Part I: Analysis and experimentation of stealthy deception attacks. IEEE Transactions on Control Systems Technology. 2013 September;21(5):1963–1970.

[423] Teixeira A, Sandberg H, Johansson KH. Networked control systems under cyber attacks with applications to power networks. In: American Control Conference; 2010. p. 3690–3696.

[424] Pasqualetti F, Dorfler F, Bullo F. Cyber-physical security via geometric control: Distributed monitoring and malicious attacks. In: 51st Annual Conference on Decision and Control; 2012. p. 3418–3425.

[425] Fawzi H, Tabuada P, Diggavi S. Secure estimation and control for cyber-physical systems under adversarial attacks. IEEE Transactions on Automatic Control. 2014 June;59(6):1454–1467.

[426] Cardenas AA, Amin S, Sastry S. Research challenges for the security of control systems. In: Proc. 3rd Conf. Hot Topics Security; 2008. p. 1–6.

[427] Van Leuven L. Water/Wastewater Infrastructure Security: Threats and Vulnerabilities. In: Clark RM, Hakim S, Ostfeld A, editors. Handbook of Water and Wastewater Systems Protection. vol. 2 of Protecting Critical Infrastructure. New York: Springer; 2011. p. 27–46.

[428] Zhu M, Martinez S. Distributed Optimization-Based Control of Multi-Agent Networks in Complex Environments. Berlin, Heidelberg: Springer; 2015.

[429] Li Y, Voos H, Rosich A, Darouach M. A Stochastic Cyber-Attack Detection Scheme for Stochastic Control Systems Based on Frequency-Domain Transformation Technique. In: Au M, Carminati B, Kuo CCJ, editors. Network and System Security. vol. 8792 of Lecture Notes in Computer Science. Berlin, Heidelberg: Springer International Publishing; 2014. p. 209–222.

[430] Li Y, Voos H, Darouach M, Hua C. An algebraic detection approach for control systems under multiple stochastic cyber-attacks. IEEE/CAA Journal of Automatica Sinica. 2015 July;2(3):258–266.

[431] Taha AF, Qi J, Wang J, Panchal JH. Dynamic State Estimation Under Cyber Attacks: A Comparative Study of Kalman Filters and Observers. arXiv:150807252. 2015.

[432]    Kim J, Tong L, Thomas RJ. Subspace methods for data attack on state estimation: A data driven approach. IEEE Transactions on Signal Processing. 2015 March;63(5):1102–1114.

[433]    Amin S, Litrico X, Sastry SS, Bayen AM. Cyber security of water SCADA systems—Part II: Attack detection using enhanced hydrodynamic models. IEEE Transactions on Control Systems Technology. 2013 September;21(5):1679–1693.

[434]    Do VL, Fillatre L, Nikiforov I. Two sub-optimal algorithms for detecting cyber/physical attacks on SCADA systems. In: Proceedings of the Tenth International Conference on System Identification and Control Problems (SICPRO '15). Moscow; 2015. p. 1–10.

[435]    Kwon C, Liu W, Hwang I. Security analysis for cyber-physical systems against stealthy deception attacks. In: American Control Conference; 2013. p. 3344–3349.

[436]    Miao F, Zhu Q, Pajic M, Pappas GJ. Coding sensor outputs for injection attacks detection. In: 53rd Annual Conference on Decision and Control; 2014. p. 5776–5781.

[437]    Pasqualetti F, Dorfler F, Bullo F. Attack detection and identification in cyber-physical systems. IEEE Transactions on Automatic Control. 2013 November;58(11):2715–2729.

[438]    Li Y, Voos H, Pan L, Darouach M, Hua C. Stochastic cyber-attacks estimation for nonlinear control systems based on robust $H_\infty$ filtering technique. In: 27th Chinese Control and Decision Conference; 2015. p. 5590–5595.

[439]    Zhu M, Martinez S. On resilient consensus against replay attacks in operator-vehicle networks. In: American Control Conference; 2012. p. 3553–3558.

[440]    Foroush HS, Martinez S. On multi-input controllable linear systems under unknown periodic DoS jamming attacks. In: Proc. of the Conference on Control and Its Applications; 2013. p. 222–229.

[441]    Djouadi SM, Melin AM, Ferragut EM, Laska JA, Dong J, Drira A. Finite energy and bounded actuator attacks on cyber-physical systems. In: European Control Conference; 2015. p. 3659–3664.

[442]    Djouadi SM, Melin AM, Ferragut EM, Laska JA, Dong J. Finite energy and bounded attacks on control system sensor signals. In: American Control Conference; 2014. p. 1716–1722.

[443]    Zhu M, Martinez S. On the performance analysis of resilient networked control systems under replay attacks. arXiv:13072790. 2013;p. 1–20.

[444]    Zhu M, Martinez S. Attack-resilient distributed formation control via online adaptation. In: 50th IEEE Conference on Decision and Control and European Control Conference; 2011. p. 6624–6629.

[445]    Feng Z, Hu G. Distributed tracking control for multi-agent systems under two types of attacks. In: 19th IFAC World Congress. South Africa; 2014. p. 5790–5795.

[446]    Yuan Y, Mo Y. Security in cyber-physical systems: Controller design against known-plaintext attack. In: Control Decision Conference. Japan; 2015.

[447]    De Persis C, Tesi P. Input-to-state stabilizing control under denial-of-service. IEEE Transactions on Automatic Control. 2015;60(11):2930–2944.

# Index